The New Beetle

Mom,

I would love to buy you a new Beetle — this book is going to have to suffice!

Merry Christmas 2001.

Love,
Boss.

xoxo

ISBN 1-85868-681-4

Editor: Tim Dedopulos
Senior Art Editor: Zoë Maggs
Designer: Brian Flynn
Picture Research: Lorna Ainger
Production: Alexia Turner

Printed & Bound in Dubai

ACKNOWLEDGEMENTS

Without the help of Volkswagen, in Germany, America and Britain, this book would not have been possible. Special thanks go to Tony Fouladpour at *VW of America* Public Relations, who was a great help with Concept 1 and Concept Cabriolet, Rüdiger Folten of the Design and Strategy department at *Volkswagen AG Germany*, who gave me a fantastic insight into the design of the New Beetle, and Alison Kempster and Paul Bucket of the *Volkswagen UK* press office, for all their efforts. Thanks also to Dean Kirsten, Technical Editor of *Dune Buggies and Hot VWs Magazine*, Mike Pye for loaning me his Apple Mac, Nigel Fryatt, *VolksWorld*'s Publisher, for granting me permission to write this book, Keith Seume, total VW-head and prior Editor of *VolksWorld,* for his help over the years, and David West, of parts specialist *German and Swedish* in London, England, for handing over the keys to his New Beetle for a four day road test which saw just under 1000 miles added.

PICTURE CREDITS

The publishers would like to thank the following sources for their kind permission to reproduce the pictures in this book:

Blue Nelson 40, 41, 42; Bo Bertilsson 61, 62, 65; BR-Foto/M. Berrang 66; *Car Magazine*/Richard Newton 6t, 10, 57, 68, 71, 72t, 73, 74/Glenn Paulina 6b, 8, 9, 37b, 47, 52, 54, 79; Mike Key 63, 69, 72b, 76; Dean Kirsten 20, 21, 24, 35b, 36, 37t; Jim Maxwell 60, 64, 67; National Motor Museum, Beaulieu 7, 14, 16, 18; Terry Shuler/VW Historian 43, 44, 45, 46; Volkswagen Archive 12, 13, 15, 17, 19, 20b, 22, 23, 26-34, 35t, 38, 39, 70, 75, 77, 78; VW America 4, 48-51, 53, 55, 56, 58; VW Design, Germany 3;

Every effort has been made to acknowledge correctly and contact the source and/copyright holder of each picture, and Carlton Books Limited apologises for any unintentional errors or omissions which will be corrected in future editions of this book.

The New Beetle

Ivan McCutcheon

CARLTON

Contents

Introduction

The world-renowned VW Beetle was designed by Ferdinand Porsche during the 1930s. It rose from the ashes after World War II in a truly amazing fashion, against formidable odds. The car persevered during its early years of production to finally become established as the world's best-selling car to date.

The car's success was due to three main factors – good world wide sales, excellent service back up, and the craft with which it was advertised by Doyle Dane Bernbach. The campaigns were honest and never over-promoted the car. The virtues of build quality, durability, reliability, economic ownership, and superb after-sales back up were presented in simple terms and the car sold in great numbers.

The dash features a flower vase

Soon the VW wasn't the oddest looking car on the block – it blended in with all the other Beetles. With a production figure which currently stands at over 22,000,000 it is no surprise the Beetle is the most easily recognisable automobile in the world.

Most of us have had some kind of contact with a Beetle, whether that be having owned one or more, travelled in one, known someone who owns one or, in quite a few cases, having been born in one! There are innumerable Beetle tales in countless countries and a myriad of memories, both good and bad, about these quirky cars.

The original Beetle was masterful in any weather conditions

Today the Volkswagen Beetle is one of the few classic vehicles you are still likely to see in use in significant numbers on a daily basis. This car has the largest enthusiast following of any model, and the most single brand monthly publications devoted to it. In addition, it still supports the largest after-market parts and maintenance industry bar none. By 1966 in the United States alone Volkswagen and its distributors and dealers employed over 29,000 staff. If this figure were to be added to the numbers of people in the unofficial support work force which have made a living selling parts and maintaining the vehicle, it could easily be doubled.

On the open road, the New Beetle turns heads

With such a fantastic record it was no surprise that the public and press went wild about the VW concept car for the 1994 Detroit Motor Show. The New Beetle has gone on to give the company more good press than any other model it has made. Since being launched on January 5th 1998, the New Beetle has been on more front covers and television screens than any new car in recent times. This has created mayhem for the dealers and must be quite upsetting to other manufacturers – after all, Volkswagen has gone and done it again!

What remains to be seen is whether the public interest will burn out or keep going, as the new VW Beetle has not yet been officially launched in many markets. It is uncertain whether this very trendy car will be 'in' for a short time or whether it will sell for decades like its older brother did. Previous VW advertising claimed the Beetle fad had died out as so many had been sold that it wasn't a unique car any longer – but even this didn't stop sales.

We will look into what has happened with the VW Beetle so far, from the Concept 1 vehicle at Detroit through to the launch of the New Beetle. We will also explore the factory where it is built and take a look at the way the car has been such a successful exercise for VW. The future will take its own course, and no doubt there will be many more interesting chapters in the life of this great little car. My only regret is I can not cover acts which have not yet taken place.

I have never seen myself as a soothsayer, but I feel this fantastic car is the Millennium Bug we would be only to happy to own! If you are one of those who have already taken delivery of a New Beetle I hope you enjoy every moment you have together. After all, you are driving not only one of the most talked-about new cars of the decade, but a piece of automotive history in the making!

VW has come up with a unique car again!

The VW 30 prototype from the front and rear

1

Heritage

The in-depth history of the Beetle has been covered in many publications, but there will be a number of new enthusiastic owners who have become interested in the car's history and heritage solely through the New Beetle. Although this book focusses on the New Beetle, we will take a brief look at the Beetle's history in the same way VW looked to its roots for the basic theme of this sensational car.

The Beetle was originally designed by Dr. Ferdinand Porsche. The first prototype car which bore any resemblance to the Beetle was the Porsche Type 32, built in 1933 for NSU. This car had a very similar shape, an air-cooled four cylinder boxer engine (horizontally opposed pairs of cylinders) mounted at the rear and displacing 1470cc, linked to a swing-axle transmission. It also featured the Porsche-designed front and rear torsion bar suspension systems, which would go on to be used on the Beetle. This prototype would never actually see production as a proper vehicle, but it proved to be a valuable exercise for what lay ahead.

At the beginning of 1933, Ferdinand Porsche met with the new Chancellor of Germany, Adolf Hitler, to discuss the Auto-Union racing car he was working on at the time. The meeting was to request that the funding of the car be supplied by the government, and everything went according to plan.

When Hitler decided he wanted to mobilise the German population at all levels, and not just those wealthy enough afford the luxury of a motor car, he needed someone to design the vehicle. His thoughts turned to Dr. Porsche. Ironically enough, Porsche had already been trying to get a Peoples car into production, having had the same idea.

The VW 38 prototype from 1938 was basically the same as the car that went into production

Porsche was asked to put his ideas down on paper and submit them for review. The main recommendations of the paper, which was presented in January 1934, were that the car should be built to a very high standard, but also be low priced. The car should not be under-powered, and should not be a small car but a four-seater hardtop with the best possible handling and suspension, able to reach a maximum cruising speed of around 100Km/h and have a climbing ability of around 30 per cent. Porsche also stressed the vehicle should be simple to maintain and cheap to run.

In May 1934 Porsche was asked to meet with Hitler in Berlin. At this meeting, Dr. Porsche discovered that Hitler liked what he had read. He wanted Porsche to produce a Peoples car which had the ingredients Porsche had suggested – for the price of a motorcycle!

Though he felt it was impossible to meet these requirements, Porsche made designs for a prototype vehicle. This became known as the Porsche Type 60, and he submitted it for Hitler's approval. The

This 1950 Beetle shows how little was changed from the VW 38. Note that the front turn signals are later additions

design was passed and handed over to the R.D.A, a governing body for the vehicle manufacturing industry. Porsche was contracted to develop the People's car in June 1934.

Porsche was given a grant to produce three working prototypes from the Type 60 design, but by today's standards the amount he was given was farcical. However, Porsche took the contract and the grant. The three prototypes were given Verusch (experimental) type designations V1 and V2.

By October 1935 the prototypes were being built in secret and, although a little behind schedule, they were ready for testing. There were a few mechanical problems which appeared during rigorous daily testing in excess of 400 miles, but generally the car proved to be satisfactory.

Hitler then passed control of the Peoples car project to the DAF, the German workers front. He charged the German workers an extra 1.5 per cent tax and set up a fund for the project which he called the Society for the development of Volkswagens or, to use the German abbreviation, 'Gesuvor'.

Although this enforced finance deal was a bit underhand, it provided funds to produce further prototypes, known as the VW 30, in early 1937. These prototypes were actually produced by Daimler Benz
and looked similar to the car which we know as the Beetle. The prototypes were tested by 120 SS servicemen over more than a million and a half miles – a feat unheard of at the time. With such rigorous
testing, any problems would certainly have been found and fixed.

1938 saw the Volkswagen Beetle in the final stages of development with the production of 44 VW
38 prototypes. In February, Adolf Hitler ordered the head of the Gesuvor, Dr Bodo Lafferentz, to locate
an area in which to build a production factory for the Peoples Car. The ideal site required enough surrounding land for workers' accommodation. The area he chose was Fallersleben, and the town is now
known as Wolfsburg, close to the Mittelland Canal and the Autobahn between Hanover and Berlin.

Strength Through Joy

The factory site was under construction by February 1938 and on May 26th,
Adolf Hitler attended a lavish ceremony where he officially unveiled the new
prototype cars and also laid the foundation stone of the factory. At this point
the factory and town were not known as Wolfsburg, but the Kraft-durch-Freude
Stadt, or Strength Through Joy Town. If you wonder exactly what that is supposed to mean, you'll be thinking along the same lines as Dr Porsche was when
he heard Hitler announce the name for the vehicle. It was not going to be the
People's car, or Volkswagen, but it would be called the KdF-wagen, or Strength
Through Joy car! Despite the fact that the name was really rather silly, sounding
odd even in the original German, from May 26th the car was officially known
as the KdF-wagen!

The Kdf-wagens were then driven around Germany and shown off to the
public. A KdF savers' scheme was set-up by Dr Robert Ley, head of DAF. The idea

1953 saw the oval rear window

was you simply bought stamps and stuck them onto cards. When you had saved enough stamps, you
got your car. It didn't work quite that smoothly. In reality, you expected to get your car and then started waiting – hundreds of thousands of Germans entered the scheme and not one got a KdF-wagen.

During 1939, construction of the factory was going very well indeed, and by April the American-
made machinery was arriving and being fitted. However, the outbreak of World War II stopped work
on the factory. It had yet to produce a car. By the end of the year, the factory had made a loss equivalent to two million dollars! Throughout the war years, from 1940 to 1945, the unfinished factory pro-

duced military vehicles based mainly on KdF-wagen mechanicals, also designed by Porsche. In total 50,788 Kübelwagens and 14,276 amphibious Schwimmwagens were made in 1940. By the end of the war just 630 Kdf-wagens were produced. Many of these were strictly for high ranking Nazi party members, or for use in the field.

Above: the Oval window Main pic: from August 1957, the 'Big window'

The British Years

Production of military vehicles made the factory a target for the allied bombers. By 1944, two thirds of it had been destroyed in daylight bombing raids. However, as the factory's facade was over a mile long, production continued despite the poor conditions.

At the end of the war, the allies took control of all German industry and the Volkswagen factory was put under the control of the British army. Under the command of Colonel Charles Radcliffe, and Colonel Michael McEvoy of the Royal Electrical and Mechanical Engineers, poverty stricken locals were given the task of clearing up the damaged factory and repairing allied vehicles and rebuilding engines. The Beetle production machinery was found, taken out of storage, and two Beetles were constructed by hand so that the car could be evaluated.

Although the Beetle persevered, if received a mixed reception. Apart from those at the factory, including REME Major Ivan Hirst, no-one seemed to think very much of the car. A leading British motor manufacturer, Humber, evaluated the Volkswagen and thought it was a waste of time saying, 'It does not meet the fundamental technical requirements of a motor car. As regards to performance and design, it is quite unattractive to the average motor car buyer. It is too ugly and too noisy. To build the car commercially would be a completely uneconomic enterprise.' According to Humber, the future of the Beetle looked bleak. The officers of REME had faith, however. During 1945 the Volkswagen town was renamed Wolfsburg, and the factory was called the Wolfsburg Motor Works.

In early 1946 Wing Commader Dick Berryman, the production engineer, was given orders by Radcliffe to get the Volkswagen factory working again. By March 1946 the factory had produced 1000 units and by the end of the year, nearly 8000 had been built and given the designation Type 11. These vehicles were destined for occupying forces, known as The Control Commission for Germany. The factory had also been producing Kübelwagens and various variants from spare parts, again for the C.C.G.

An uncertain future?

In 1947, Major Hirst was given permission to sell Volkswagens to British servicemen. This was followed by perhaps the most significant step towards the future of VW – the managing officers at the factory decided to exhibit the Volkswagen at the Hannover Trade Fair. For the first time, the German public could actually buy the car, even though few had enough money. In addition, this important move successfully secured a contract with the Pon brothers from the Netherlands, who were to be the first VW dealers and importers.

At this stage the future of the factory was still not certain, and by the end of 1947 the British were looking for a German to take control of Wolfsburg. Major Hirst eventually managed to track down the right man, Heinz Nordhoff, and as the future would prove, appointing this man as Managing Director secured the future success of Volkswagen.

By May 1948, 25,000 Volkswagens had been produced. Volkswagen also exported vehicles – 23% of its production was shipped to Luxembourg, Belgium, Switzerland, Denmark and Sweden, bringing in valuable foreign currency. In Germany Volkswagen accounted for over half the cars built, and in 1948, a total of 19,244 cars were built in total that year. Nordhoff had created service and sales organisations wherever the VW was sold, and this became the foundation of the world wide success of the VW. In 1949 the C.C.G. control of Volkswagen was relinquished and the Federal Government was given control of the operation.

Nordhoff arranged for Ben Pon to take the first VW to the U.S.A, to be displayed at a German Industrial Exhibition in New York. However, the car was not greeted with much enthusiasm and Pon ended up selling it to pay for his hotel bill before returning to Europe. This was bad news for Nordhoff, as he wanted to break into the largest car market in the world and obtain US Dollars to fund further purchases of machinery.

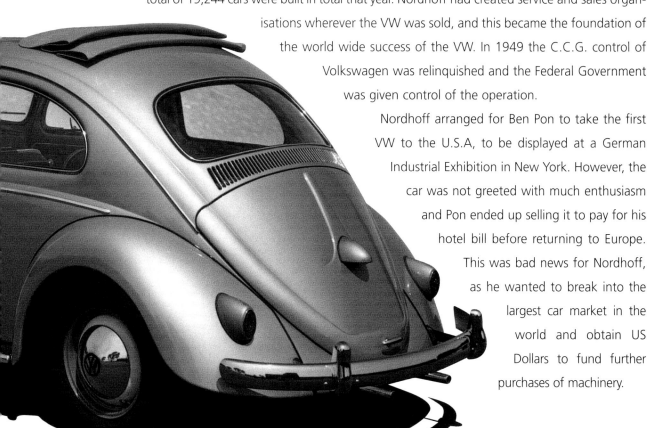

This situation changed in the middle of the fifties, as the American market looked beyond massive highly-chromed vehicles with powerful engines. The Volkswagen Beetle came across as a class-free and unconventional car which had no non-essential extras. This contrasted with the American vehicles, which typically had extensive electrical specifications, and included power steering, powered windows, power seats and sometimes even powered convertible roofs. With the Beetle, less was more, which appealed to individualists and those not able to afford the monsters Detroit had thrived on.

Sales take off

The bonus of owning a VW was that it looked the same every year, unlike the American cars which the manufacturers changed every year in a competition for customers. So if you bought a Beetle and kept it a few years most people wouldn't even know what year the car had been made. Parts were readily available, through the huge dealer network which had been formed. Service bills were cheap, and the Beetle was a very reliable car. It was no surprise that the word started to get around!

Official sale started in Britain in 1953. Surrey-based motor trader John Colborne Baber had been bringing used Beetles into Britain since 1948 and dressing them up with chrome trim and vinyl interiors. Many of these Beetles were converted to right-hand drive. Official imports didn't start until 1953, when the ban on car imports, set after World War II, was released.

In 1955 the Millionth Beetle rolled of the Wolfsburg production line, and by 1959 Volkswagen was producing over 575,000 Beetles a year. The waiting list in the U.S.A. was up to six months. In 1960 Volkswagen reached another landmark – 500,000 Beetles had been sold in the U.S.A.

This was an amazing achievement when you consider the fact the company hadn't seriously advertised the car! When VW did, sales soared even further, mainly thanks to the ideas of Doyle Dayne Bernbach. The advertising was honest, unpretentious and fun. DDB's technique made prospective customers feel they were being addressed sensibly and on their own level, rather than bragging that life would be more glamorous if they owned a Beetle. Consequently, 1962 saw the 5 millionth VW produced and annual production up to a million. In 1963 VW opened the headquarters of Volkswagen of America in Englewood Cliffs, New Jersey.

In 1964, Volkswagen formed VW de Mexico S.A. de C.V in Peubla, Mexico and started building VWs to German standards at low prices for local consumption. This was one of many factories worldwide which produced Beetles either from scratch or from kits sent from Germany. Peubla went on to play a very large part in the story of the New Beetle. By 1965, 10 million Volkswagens had been pro-

The 50th Year Beetle from 1985 was built in Mexico

duced. In February 1972, VW took over the record for highest production of a vehicle from the Ford Model T, with the 15,007034th Beetle. The cars continued to roll off the production line.

After some 30 years, the Wolfsburg factory ceased Beetle production in 1974 to make way for the new VW Golf. However the Beetle was still built in several places – including Emden in Germany, Belgium, Peubla in Mexico, South Africa, Brazil, Australia, The Philippines, Uruguay, Nigeria, Malaysia, Peru, Venezuela and Indonesia! On the 19th of January 1978, the last Beetle to be built in Europe rolled of the line at Emden, although Mexican imports were still brought in to Europe. On the 15th of May 1981, the 20 millionth Beetle was built in Mexico. Beetles were shipped into Germany until August 1985, by which time demand had dropped, and imports became uneconomical. Production didn't cease though, and in June 1992 numbers were up to 21 million. Today the air-cooled Beetle is still in production at Peubla, and over 22 Million have been produced.

The Beetle has the strongest following of any car. There are over a thousand enthusiast clubs world wide, and hundreds of VW-only car meetings are held each year. There are countless unofficial after-market parts manufacturers, stores, and service and restoration companies. No model has been more written about, customised, raced, tuned or restored to perfection. It is without a doubt the most pop-ular classic car ever produced, and the amazing thing is that is still in production today.

2

Concept 1

Concept 1 and a 1946 Beetle, by Dean Kirsten

When the doors of the North American International Motor Show in Detroit, Michigan opened to the motoring press in January of 1994, Volkswagen of America provided a real shock to the motor world in the shape of something very close to being familiar. It was the design study known as Concept 1, the vehicle presented to the world by Ulrich Seiffert, head of Research and Development at Volkswagen.

However, the story starts long before the 1994 show. To use the words of Rüdiger Folten, Head of Strategy and Design for Volkswagen AG, Wolfsburg, 'When we first started with Concept 1 it was only an idea. We initially wanted to produce a zero emission vehicle. VW sales were falling down in the U.S.A and so we wanted to do something which would make people remember we were still there.

Design rendition of Concept 1

We also wanted the shape to resemble something the American people remembered and were fond of, but it was also to be designed with modern style.' Concept 1 was scratch built and although many publications at the time thought it used the Polo platform, Rüdiger Folten assured that 'Concept 1 was without relation, it was not Polo or anything to do with Polo.'

Concept 1 was designed under top secret conditions at the VWOA Design Centre facility in Simi Valley, California, part of what Volkswagen AG refers to as its

North American Region. They were responsible for the project from start to finish.

Simi Valley was opened by Volkswagen and Audi in January 1991. The aim of the design centre was to improve the position of VW and Audi in the North American market. Volkswagen wanted to develop closer links with the market and to look into the trends of what is considered to be one of the hotbeds of automotive design.

In March 1992, work initially started at Simi Valley on a vehicle concept which relied on alternative power systems – the goal being zero emission. By September, VW had carried out extensive market research on the American automotive world and found that one thing kept coming up. When Volkswagen was mentioned, the instant reaction was the Beetle. It was a legendary car which remained strong in the memories of the American population. The Beetle was the synonym for VW right across North America.

Volkswagen realised it could act on the good feeling that the Beetle gave people, and in turn per-haps bring the VW name back into the limelight in the USA. The idea was given thought and the out-come was to combine the zero-emission study with a car which would revive the Beetle legend. This was not to be a simple retrospective design, even though the first impression would shout out Beetle! The curved roofline which slopes steeply at the back of the car, and the rounded bonnet in contour with the semicircular wings both indicate a strong link with the original Beetle, but at the same time

Similar lines with 48 years between them

Side view shows where the 'three circles' idea used on the New Beetle logo originated.

look new and refreshing.

By March 1993 the design had gone from artwork to three dimensional 1:4 scaled models, and Hartmut Warkuss, project initiator had met with VW Chief Executive Dr Ferdinand Piëch to present the study. The outcome of the meeting was very positive. Having seen the Concept 1 design study and obviously realising the potential high impact it could have on the automotive world, Dr. Piëch ordered for the study to be brought up to a full scale car, which would have to be ready to be debuted by VWOA at Detroit in January 1994.

The design team leader responsible for Concept 1, J.C.Mays – who has since moved to The Ford Motor Company – said 'We wanted to combine the past and the future.' The theory was to create a vehicle which everyone could immediately connect with, but one which also featured up to date technology. This was the foundation for the use of hybrid and electric engine ideas.

The design study brief was to give a choice of three power trains. First was the four cylinder 1.9 Litre/66kW Turbo Diesel with 5 speed Ecomatic transmission. For those unfamiliar with the Ecomatic it fundamentally consists of a computer controlled automatic shutdown system. If the computer decides the vehicle does not need the engine power at that moment, it simply switches the engine off. As the vehicle would pull up to a stop light the engine would cut out, and then as soon as the driver puts a foot on the accelerator pedal the engine would instantly fire-up again. Not only did this ingenious idea

prove to be more fuel efficient, but it also cut emissions by 36 per cent. The Ecomatic initially feels strange to drive, but after an hour or so most people adjust. It certainly poses no problems for a confident driver. In Europe, VW offered the Ecomatic in the Golf, and although it was a worthwhile idea, the public in general didn't take to it. Within a year the Ecomatic was dropped. In Concept 1, the 1.9 TDI was capable of 180km/h (which equates to 111mp/h) and was able to go from 0-62mp/h in 12.8 seconds.

The TDI engine with Ecomatic system

Secondly, the fully electric drive train which Concept 1 would theoretically offer was an AC Induction electric-powered two speed automatic which created 37kw of

VW Concept 1 design renditions

power. This package, which relied on the AEG Sodium/Nickel Chloride high temperature battery, could propel Concept 1 to a top speed of 77.67 mp/h and give an urban range of just under 95 miles. Tests for constant speed range at 31 mp/h produced a range of 155 miles.

The third power train was a hybrid which consisted of a three cylinder 1400cc TDI which produced 50kW linked to the E-Motor, an AC Induction 18kw electric Nickel/Metalhydride powered unit. The diesel motor had a top speed of 102.5mp/h, while when switched over to electric power the maximum speed was 65.24mp/h.

By July 1993 the full scale clay mock-up of the Concept 1 was completed. It was a much smaller car than the production model New Beetle, the dimensions of which we will look at later. Initial comparisons highlight immediate differences between the original Beetle and Concept 1. The length of Concept 1 was 3824mm or 12.54 feet, which is in fact smaller in length than original rear-engined Beetle, which measured 13.35 feet. The width was 5.36 feet, wider than the original Beetle by 31 inches. An interesting point is that the original VW Beetle and Concept 1 are virtually the same height.

Preparation work continued to get the yellow show car ready in time for the show. The design

The interior of the Concept 1 show car

incorporated elements which harked back to the Beetle, but also many which were futuristic statements. To use VW's own words, 'The rounded forms have, wherever possible, been brought as close as possible to two basic geometrical shapes – the circle and the sphere. Smooth surfaces remain smooth, with no swage lines or ledges, deriving their subtle charm solely from their generous curves. The panel configuration is also logical and geometrically clear. There is a pleasing tension between the simplicity of the smooth surfaces and the few straight lines on the one hand, and the rounded shapes on the other – the lines of the side windows are a good example.'

When in silhouette or side view, Concept 1 would reveal something which was very different to the original Beetle. The windscreen was not upright! On Concept 1 the windscreen gave a smooth transition from the front of the vehicle to the roof. This would give the driver and passengers a far more spacious feel and panoramic view. Anyone who has driven the original VW Beetle will tell you they felt very close to the upright windscreen.

The interior of the car was to feature black leather upholstery, muting the seats and highlighting the symmetrical layout of the very basic dashboard. The central air vent outlet/radio pod and large single display gauge were dark grey plastic, whilst co-ordinated yellow sections to the left and right made the dashboard blend in with the doors. The passenger's side of the dashboard featured a large grab handle, with a flower vase to its left. This was a simple trick to tie in with the original Beetle, as glass or china dash mounted flower vases were very popular during the fifties and early sixties. However, although very popular, they were not a standard Beetle feature, and were only offered as accessories. Be this as it may, this simple idea gained a huge amount of press coverage for VW, and made it into production on the New Beetle!

Running from the bulkhead under the dash between the front seats and through to the rear seat, Concept 1 had a colour coded yellow tunnel cover which incorporated two beverage cup holders in front of the transmission shifter, along with a further cup holder in the rear. Both of the inner door sills were also colour coded. Both interior door panels were basic, featuring simple door pulls, winder handles and door lock pulls. The inner door tops were colour coded as well, and this theme carried on through to the underside of the rear quarter windows.

On the outside

Moving out of the cockpit, the concept car was equipped with very large 18 inch six-spoked wheels with 155/75 Goodyear tires. The width of the tires helped give the impression that they were tall and skinny, and therefore similar to the original Beetle. Exterior details included two Beetle styled horn-grilles. A direct link was present with the aluminium grilles found on the front wings/fenders of the original VW Beetle. One of these had the horn behind it, but the other was just blanked off – there for reasons of symmetry. These did not make it into production.

The bonnet/hood of Concept 1 was shorter and less bulbous than later production models, giving the vehicle a much more curved shape in profile than the production car. The headlights incorporated the front turn signals. The front also featured what would appear to be a removable lower front panel or apron, which was designed to be changed as per the registration or licence plate requirements of different markets. The apron also featured small high power mini spot lights.

All four wings or fenders were designed to be interchangeable from corner to corner. For example, one front fender could be swapped for the opposite rear fender and vice-versa.This was a fantastic idea which would have halved dealer panel stocks and orders, but unfortunately it proved impractical in series production. The rear bumper was smooth and followed the curve of the wing/fender, assisting with the three curve profile of the vehicle. Again this would be lost in production. As with the front bumper, the rear featured a separate apron, but rather than having lights, the rear apron had an exhaust tail pipe cut out and, on the opposite side, a towing hook. Moving to the boot lid, Concept 1 was equipped with a third brake light, VW roundel and boot lock.

At this stage we should consider the feelings of the VW enthusiasts (or should that be zealots?), as part of the design of Concept 1 was to move away from the rear mounted air-cooled engine linked to a swing axle or independent rear suspension transmission. The new car had a front mounted engine and transmission and was to be front wheel drive. This was a large departure which made many feel that although the car might look like a Beetle, it would never be a real Beetle. The Beetle design had not really been changed since 1977 except for updates and improvements. However, if the model had remained the mainstay of Volkswagen sales in Europe and America, it could well have progressed to the front engine/front wheel drive layout. As it was dropped from the European and American markets by the Golf and its derivatives, and only produced for South American markets, we will never know what might have happened to the Beetle and its layout.

The basic flight deck of Concept 1

What we do know is that the story was about to become even more exciting, as Concept 1 was ready to go public at the Detroit Auto Show. VW admitted the response exceeded all expectations. The car created enormous interest. Of course part of the reason was the obvious association with its predecessor. The attention to detail and quality in the design was also obvious.

At the launch, Ulrich Seiffert, VW's Executive Director of Technology and Development, told the press, 'The intention is to emphasise the typical qualities of a VW; its honest, reliable, timeless and youthful design. We will never bring the Beetle back, but we would like to go back to our roots with an honest, reliable car. It will influence the design of our smaller product range.'

At this stage there was no statement made about the possibility of putting Concept 1 into production, mainly because VW hadn't made any plans to do so! However, the reaction from the public, press and VW dealers all across North America was extremely positive. Everybody seemed to want VW to build this car! Volkswagen clearly had to start thinking very seriously about taking steps to put the vehicle into production. When the press asked Steiffert if VW would put the Concept 1 into production, his first, rather muted reaction was that it was 'possible'. The most important question at VW was, as Rüdiger Folten puts it 'will we be able to take this shape, which is well known, and make it work?'

The next step was to unveil a further study on the Concept 1 theme, the Concept 1 Cabriolet. The success at Detroit was continued at the Geneva Motor Show. If going back to its roots worked so well for VW with the sedan, there was little doubt over whether or not a Cabriolet version would create a

The Concept 1 cabriolet – the logical progression

stir. The prototype was heralded as a glimpse into the future of Volkswagen, a car which was in a class of its own, simple, unique and modern, looking forward optimistically but still showing its heritage.

It was at this stage that Dr Piëch promised the motoring world that Concept 1 'need not remain just a vision.' A press release from Wolfsburg followed shortly after and stated: 'Volkswagen AG Approves Concept 1. Wolfsburg, Germany – Volkswagen's popular Concept 1 automobile, a design study car that created a public and media sensation upon its unveiling at the 1994 North American International Auto Show in Detroit, Michigan, will be produced and on the market before the year 2000, Volkswagen confirmed. The Volkswagen AG Board of Management gave its approval in November 1994 for the development of Concept 1. The decision came in response to appeals from enthusiasts all over the world who embraced the design concept. Volkswagen said the United States will be one of the primary markets for the new car.'

VW later said the economic production and marketable pricing would only be possible on the basis of what Volkswagen call a platform. In addition, it was decided to base the production vehicle on what Rüdiger Folten referred to as the 'future Golf, which later became "Golf 4."' For this to become reality, Concept 1 would need to go through many changes. But as Folten humorously said, 'When Dr. Piëch says we will produce this car, 7,000 people go to work.'

But the Concept 1 wasn't finished with yet. The Cabriolet version was shown in the U.S. at the New York Auto Show. At that time, VW told us, 'the Concept 1 Cabriolet embodies the spirit of Volkswagen with the flair of open air motoring. From the original people's car to people's cars, Volkswagen has continued its motoring evolution. Dependable, reliable, well engineered. Today, Concept 1 and Concept 1 Cabriolet capture the future ideas of what a Volkswagen could be. The growth of the fourth largest auto manufacturer in the world continues with a keen look at future driving needs – needs that blend the heritage of the company with advanced technology.'

The Volkswagen design team in the United States set out to design a car that brought together the past and future. It was to be a grand design, a car that people could relate to and put their trust in, whilst it still offered them the impact of advanced technology. The Concept 1 and Concept 1 Cabriolet were the results of that effort. In profile the car has

Concept 1 Cabriolet in side profile. Fabulous!

VW, you simply have to build a New Beetle Cabriolet!

a clean, simple shape. It is formed from three cylinders – two where the wheels are positioned, and one forming the passenger cabin. There are no aggressive lines.

The car was designed with the motor up front, dual airbags, side impact beams and ABS brakes, and it seated four adults. The 18inch tires were 155/75. All interior lighting on the instrument panel was electroluminescent, which goes a long way to help lessen eye strain. Driver information came from a single round gauge that encompassed the speedometer, engine temperature indicator, fuel gauge, and headlight switch. The Concept 1 Cabriolet also included an AM/FM stereo and a compact disc player with six speakers. Furthermore, the seats and side panels featured impressively stylish woven black and white cloth inlays.

There can be no doubt in the matter – it was the stunning design of Concept 1 which provided the impetus that drove VW to put the New Beetle into construction.

3

Design & Manufacture

The story now moves away from Simi Valley to the design studios of Wolfsburg in Germany, where the task of turning the motor show concept into a tangible production vehicle settled. At a glance, Concept 1 may appear very similar to the production models of the New Beetle, but when you actually compare the two, little remained unchanged. The 'future Golf' platform was virtually a foot longer than the overall length of Concept 1. As Rüdiger Folten, puts it 'The car needed to be bigger – larger, higher, longer – so we had to make it to grow a little.'

Design rendition of the New Beetle

The Wolfsburg team set about re-designing, starting with renditions and then moving onto computer designs. Then a platform was set up as a base for a full size clay model which was worked from a industrial plasticine mass to the required shape by hand using scrapers, modelling tools and templates. Once the design was finalised, further models, firstly without windows and later with them, were produced by casting a substance known as Epowood, which is actually a form of plastic.

Next the Epowood presentation model was measured with a data probe to form the basis for the first prototypes. The data was then programmed into computers and digitised, to be used for the man-

Early design drawings retained Concept 1 mirrors

ufacture of bodywork tooling. Typically, this data also assists with further studies into detail, such as crash simulations and other intricate tests.

The next step was to move onto the interior of the car. The technical design team had to create the internal elements of the body, including load-bearing structures, reinforcements, pillars, sills and inner wings, along with components such as the dashboard, door and lid hinges, interior door panels and bulkhead. To give you an idea of the amount of work involved, the body of a Volkswagen consists of around 300 stamped or pressed steel panels, all of which must pass stringent testing for quality and safety, along with being economically viable to produce!

Interior design was equally important. The inside had to be appealing. After all, this is the part of the car which the driver and passengers see the most of. Visually, the vehicle was ergonomically designed to be 'perfect.' Apart from being designed with up to the minute safety features, the functions had to be 'self explanatory.' In short, this was a very hard task for those concerned. In the same way that the exterior evolved, the interior started with the Concept 1 idea, from which further artists renditions were drawn, and then a full scale wood and plasticine model of the interior – which VW refers to as the 'seat box' – was built. The seat box is where small details and functions are designed.

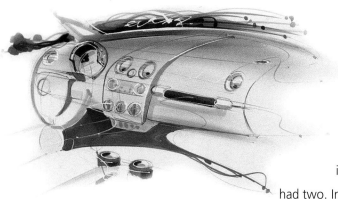

Dashboard design renditions

This particular model obviously needed aesthetic flair, but it also had to be completely functional. Therefore, whilst the dashboard is similar to that of the Concept 1, it has been highly modified. The steering wheel now has three spokes, unlike Concept 1 which had two. In addition, the door panels of the New Beetle were designed to incorporate netted door pockets, moulded armrests and door pulls, and were given the added safety feature of a red warning light which illuminates when the door is open. When attending to the layout of the engine and drive train the Wolfsburg Design team also had to take into account the things we take for granted, such as checking the oil and windscreen washer levels.

The Colour and Trim Department were the next team to work on the project. Working from a virtually infinite choice, they had to work out which colours this car would available in, taking into account current trends and attempting to offer a colour which should meet every customer's needs. Eight colours were chosen for the production vehicle. The interior materials, colours and finishes were all carefully chosen to be harmonious with not only the exterior finishes, but also each other. The Design and Trim Department is like a Paris fashion house, as it must not only meet with the customer's needs, but also come up with new and exciting trim ideas and trends. The first full scale exterior and interior models were presented to the Volkswagen Board of Management in November 1994 for approval.

A new beginning

The next stage should actually be seen as 'a new chapter in the New Beetle story,' says Folten. A sheet metal concept car was built with the aim of unveiling the model in October 1995 at the Tokyo Motor Show, the first showing of Concept 1 or its derivatives in Asia. Although Rüdiger Folten stresses 'there were no secrets about the new Beetle,' not much had been said since March 1994. This made the press sceptical that VW would be able to keep the attraction of Concept 1 during the transition from show car to production vehicle. VW dispelled that rumour decisively at Tokyo with the black prototype car!

Volkswagen had to make comprehensive changes at the front end of the vehicle, mainly for

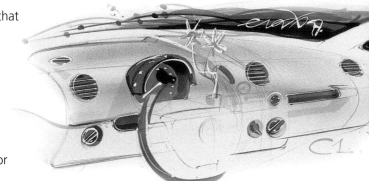

safety reasons, but managed to keep it as close as possible to the lines of Concept 1. At this stage the car was known simply as Concept, and had a full tinted glass roof, which worked very well with the black paint work. What the media did not know when they came away from Tokyo was that the black Concept car was, as Rüdiger Folten states, 'very close to the final production model.' The Japanese public gave the Concept a truly amazing reception, with an astonishing 20,000 advance orders placed at the show. VW were on to a winner. The positive stream of enthusiasm from VW dealers, the media and prospective buyers had paid off.

Dr Piëch told the world it would produce the Concept car before the millennium. He also announced that Volkswagen was to produce the vehicle at its North American production facility at Puebla, Mexico. Further more, Puebla would produce the vehicle not only to serve the needs of the North American market, but for markets world-wide. On the 29th of October 1995, VW issued a press release which was entitled 'Volkswagen announces production site for Concept 1.' It contained some

The VW Concept, built for Tokyo, was close to the later production model

Little was changed at this stage

of Dr Piëch's written remarks released at the Tokyo Motor Show press conference. He said, 'You will undoubtedly be aware that we are currently in the process, and I say process advisedly because it is an ongoing thing, of developing the Concept 1. Naturally we are doing this with the most advanced technologies. However, instead of keeping the car under wraps up to the time of its launch we want our customers to participate in the product development of the Concept 1. We invite you to take an active role. This exemplifies the customer orientation of the Volkswagen Group.

'When we first presented Concept 1 at the 1994 Detroit Motor Show, we were overwhelmed by the positive response. Since then we have talked to customers and dealers all over the world, and taken careful note of the recommendations of the media and motoring press. Why? Because we want to put a car on the markets of the world that perfectly corresponds to the requirements, the desires and dreams of our customers. Today we invite you to acquaint yourselves with the current status of Concept 1. Look at the car, and when doing so, please bear in mind that the memories of yesteryear can also form the basis for the dreams of tomorrow. Concept 1 embodies a vision that redefines the memories of yesteryear.

'As a proof of our confidence in the Concept 1, we are today announcing the scheduled production location for this car: the Concept 1 will be built for all markets at our Puebla factory in Mexico. This will strengthen the consistent endeavour of the Volkswagen Group to deploy global presence and to create enthusiasm among customers with our products.'

The New Beetle appears

The Mexican factory certainly made the most sense, as the main target market for the new model was just over the border, rather than across an ocean! The tax situation would also be better for customers based within the United States as the car would not be an import from Europe.

Five months after the production news, VW was back at the Geneva Motor Show and the car was officially named the 'New Beetle.' Volkswagen had honed the Concept design and

The start of production in Mexico

as Rüdiger Folten told us, 'this was, apart from detail changes, the final stage of development.' An interesting feature of the Geneva New Beetle prototype was the drive train, which was a TDI engine linked to four wheel drive. The vehicle again met with a positive reception from all sides.

At the end of the production line

The interior of the Geneva New Beetle had a grey-yellow combination of fabric and leather, which gave an impression of lightness. The fuel tank capacity was 55 litres, which combined with the 90bhp TDI engine to give a highly respectable driving range before further fuel had to be added. Drag coefficient was also mentioned for the first time at Geneva, and despite the unmistakably unique shape, the drag coefficient was well below the critically sought-after 40 mark.

Now it was just a simple case of producing the car. That wasn't entirely straight-forward, as Volkswagen had to build an entirely new production line alongside the original Beetle production line at Puebla. Build quality to European standards had to be assured, and the factory needed to be able to cope with a rise in production. The Volkswagen Group plans to have released somewhere in the range of 100,000 – 160,000 units by the end of the year 2000. Each car leaving the factory wears a small decal in each door window which states 'Proudly Produced With High Quality Volkswagen de Mexico.' The car wears its production heritage proudly, in full view.

New Beetles at the Puebla factory gate

Launch, Hype & Sales

'Reborn January 5,1998' was the single line statement from the cover of the New Beetle brochure at the launch of the model at the Detroit Auto Show. The VW press release said of the New Beetle, 'The engine's in the front, but its heart's in the same place.' Seven New Beetles were on display at Detroit, and the atmosphere was said to have been 'electric.' The hot news was that the car was set to arrive in the spring of 1998 'at Volkswagen dealers across America.' The New Beetle proves that something very good in a past life can come back as something even better, something totally new.

**The engine's in the front,
but its heart's in the same place.**

One of the many witty VW postcards

While it rekindles the magic of its legendary namesake, the New Beetle is not an update of the original but a completely new and modern car – a Beetle driving forward into the 21st century, optimistically inviting us to follow. As a contemporary creation, the New Beetle is more than a ray of sunny originality in an all-too-serious car market. It is also fully functional, with plenty of creature comforts and the very latest advancements in small-car safety.

It is significantly larger than the original, both inside and out, and shares no parts with its predecessor. It is pow-

This cut-away shows the heart of the Golf 4...

ered by a front mounted 115 horsepower four cylinder engine or, optionally, a highly advanced Turbo direct-injection diesel that delivers an EPA rating of 48 miles per gallon on the highway and a driving range of nearly 700 miles.

The New Beetle uses front wheel drive and is equipped with a wealth of standard items, including CFC-free air conditioning, a pollen and odour filter, a six speaker stereo with CD-control capability, beverage holders, an anti-theft alarm, halogen projector-beam headlights, four wheel disc brakes and central locking system with remote. It can be ordered with a further wealth of extras, including electronic ABS brakes, alloy wheels, cruise control, leather seating, integrated fog lamps, heatable front seats, and one touch up and down power windows.

As a priority in its development, the New Beetle's advanced occupant safety system includes energy absorbing crush zones, pre-tensioning front seat belts, front and rear headrests, daytime running lights, dual airbags and front seat-mounted side air bags.

This New Beetle was autographed at the Los Angeles launch

Based on Volkswagen's new Golf chassis – Europe's number-one selling car – the New Beetle is built to the highest standards. This is clearly expressed in its solid, one piece appearance, which results from narrow panel gap tolerances in production. By using state-of-the-art production techniques like laser welding, Volkswagen has given the New Beetle unmatched torsional and body rigidity. Its fully galvanised body allows for a 12-year warranty against corrosion or rust perforation.

Clive Warrilow, President and CEO, Volkswagen of America Inc, said, 'Some may have predicted a retro car, but as you can see, the New Beetle is a completely modern design, almost futuristic. It is designed to appeal to people who fondly recall the past, as well as young people who have no connection at all to the original. Where the original provided basic transportation, the New Beetle is an upmarket, lifestyle vehicle. It's highly emotional, a car that makes the experience of driving fun again.'

Volkswagen of America Inc. handed out 1000 special Press Packs at Detroit. These were held in black cloth bags with the New Beetle logo on the front. Inside, along with a press release, photographs and slide, there was a Compact Disc entitled 'Dream the same Dream,' which contained the music used at the launch, and a plastic bud vase, just like the one fitted to the dashboard of the New Beetle. This came with its own logo-embossed metal stand.

On the same day as the New Beetle was launched at Cobo Hall, Detroit, Volkswagen launched the car in Los Angeles California at the L.A. Auto Show. Part of the launch involved a card game that all visitors could enter. Winners were given the opportunity to sign their name on the paintwork of the black Beetle on display with a special white pen, and the signed car was later sent to a VW museum. There was also a chance to race on a four-lane VW Golf slot car track. Heat winners were photographed sitting in a red New Beetle on display, and the photograph was then mounted in a Beetle-shaped frame. The whole launch was meant to be fun for all.

The New Beetle lured all types of people to the stand and everyone was excited. People came from different generations and from a diverse range of occupations, but all were enthusiastic about the New Beetle. One particular long term VW enthusiast called Blue Nelson from Marina Del Rey, California,

On show to the public for the first time!

January 9th, 1998 – the Vancouver New Beetle launch

heard that the Canadian New Beetle launch was to be held in Vancouver, British Columbia four days later, on the 9th of January. He immediately dropped everything and booked a flight to Vancouver. His association with the New Beetle didn't stop at being one of the only enthusiasts to visit two launches, but you will find out more later.

The Volkswagen Canada Inc. stand consisted of a revolving display with a metallic blue New Beetle, but as with the L.A. stand it was hard to get up close to the car due to the sheer number of people surrounding it. A second car was on the floor and was open for the public to get inside and examine – the only problem was that the queue was very long! The same enthusiastic buzz which had been witnessed at Detroit was present at both L.A. and Vancouver. Over-worked sales representatives were on hand to take orders and deposits. After the four years of media hype since Detroit 1994, the launch of the New Beetle was about to become the best publicity the company could have wished for. The press went absolutely ballistic!

The New Beetle was in the news in every conceivable format. The car was covered in numerous motoring magazines, in local and national tabloid newspapers and on the regional, national and international television. In Britain, the launch of New Beetle was given a piece on the major Independent Television News programme *News at Ten*. This proves its magnitude, as it is difficult to recall another Volkswagen making it onto this programme. The New Beetle must also go down in history as one of the first cars – if not the very first – to have its launch covered across the Internet.

New Beetles would continue to be in the news throughout the year but media coverage was only the tip of the iceberg. Public interest, which had always been high, was about to step up a gear, and it is doubtful whether VW was ready for the New Beetle-mania which was about to break out. The response to this car became overwhelming and in some cases very emotional. America wanted to buy the car, straight away. The race was on to be the first to own one, the first in the country, the first in the area, the first in the town, the first in the street. Of course some astute people had already got their names down, while others had to find a VW dealer who hadn't already got a huge waiting list. Most dealers had a minimum of 50 or 60 orders, and many had far more. Some were into the hundreds and figured that the waiting list could be as long as a year and a half!

The first retail releases

The New Beetle was basically designed for the American market, and so America was the first to receive the car. VW said it would start to see the first batch of around 4,000 cars delivered to the 599 dealers in the North American network towards the end of March. The rest of the world would not see the car on sale for at least another six months. Many of those in the long line of customers wishing to buy the New Beetle were willing to pay over and above the listed 'sticker' price to have the privilege of being one of the first to be seen in this much talked about VW.

The base 'suggested retail price' from Volkswagen for a standard 2.0 litre was listed at $15,200. The diesel was listed at $16,475, but on top of both prices a destination charge of $500 was to be added. Then there were taxes, dealer charges, registration charges and any optional extras to be added.

All the cars in the first batch to be delivered were equipped with the optional $410 Sports Package. This consisted of projector beam fog lights and alloy wheels, along with the Convenience Packs, which gave power windows and cruise control. So the average 'sticker' price of the first batch of cars was around $17,000. However, as soon as the first batch

Clive Warrilow, President of VWOA, at Detroit

43

The New Beetle press pack is already collectable!

were driven out of the dealerships, New Beetle owners were being stopped in the street and offered an instant profit if they would sell the car. In some cases, this inducement was as much as $10,000!

America had gone mad about the New Beetle. If a VW dealer had an unsold car in the showroom, prospective buyers were bidding against each other for the cars. In many cases the sales staff either sat back and watched the prices rise and rise until one of the bidders gave up, or became referees and broke up the ensuing squabble! One rumour was that two people were arguing over the price of a car for so long that a third person came in and bought the car from another salesman. At other dealerships the customers were getting so heated that they had to be escorted off the premises. Some people wanted a Beetle so much that they followed delivery transporters to dealerships.

Demand causes problems

As with anything which is in demand, there are always a few shrewd business people out there who can smell a profit. So it wasn't long before New Beetles were being sold through non-franchised dealers at marked-up prices. The Internet came into play, and advertisements for 'turn key' cars for $20,000 plus started to appear. Of course, by buying one these cars you get an auto quickly, but there is every likelihood you may find you have lost the 100,000 mile/10 year warranty.

While many of the VW dealers made hay while the sun was shining, commanding even more than $20,000 for New Beetles, Volkswagen of America were powerless to do anything about the mania. Of course, when production really gets going it might be possible to walk into a VW dealership and buy a car in the desired colour for the 'sticker price,' but this day could be a while away yet.

Meanwhile, those who were amongst the first to own the New Beetle – such as Jeff Grewing from Sacramento, Randy Carlson from Brea, California, Bill Kessler from Detroit, Pamela Brown from Cleveland, James Howard from Kansas City, Brian Burrows and Luke Theochari from London, England and Romano Schmidt from Germany – can soak up the envious and admiring looks of every other road user while they quite literally stop traffic in their New Beetles.

When it came to advertising the New Beetle, Volkswagen took a very 'tongue in cheek' route. As with the original campaigns of Doyle Dayne and Bernbach, Volkswagen of America went for simple statements to get the point across. One such advert read 'Nought to 60? Yes!' Another interesting approach was simply to have the text line 'What colour do you dream in?' or 'One for each day of the week' alongside a photograph of a New Beetle in every available colour. Other nice lines were 'Less flower more power,' 'A work of art with side impact airbags... and a bud vase,' and 'The engine's in the front, but its heart's in the same place.'

In a way, VW doesn't need to do too much in the way of promotion. The media has more or less taken care of that for them. In fact, VW itself took care of the advertising when it designed the car, as every New Beetle on the street is advertising. VW has created something unique and instantly recognisable, unlike most other new models, which look very similar to the rest of the pack. Have you recently found yourself wondering what a new car is and then got a surprise when you got close enough to see the badge on the back? There is not much chance of that happening with the New Beetle – just like the original.

This VW postcard tells it like it is!

A positive form of ID.

The New Beetle was introduced to Europe's press at Volkswagen's headquarters at the Wolfsburg factory between the 2nd and 6th of November 1998, and only a fortunate few were invited to this unique launch. VW flew journalists to Hanover airport, and then laid on a fleet of VW buses to ferry them to Wolfsburg. After a German buffet lunch in the management restaurant, we were able to take either a 1.9TDi or 2.0Litre Euro model for a 45 minute test drive.

The Euro Beetle differs in many ways from the U.S. model, mostly to meet the different regulations. For instance, the headlights have side lights, and the running lights/turn signals are simply turn signals. Further, each Euro Beetle is also fitted with side-mounted turn signal indicators set in the quater panel on each side, under the rear view mirrors. The rear bumper has been re-designed to accept the longer, thinner licence plates used in many European countries and is missing the running lights at each corner. One of the back-up/reversing lights has been changed to a fog light. One particularly appealing feature is the new badging – both the front and rear badges have become chrome bases with white enamel VW logoes on a dark blue enamel background.

Following on from the US '99 model, VW also announced that a glass tilting/sliding sunroof is now an optional extra. Further options included cruise control and a VW accessory roof rack. Two new body colours were also announced, Lemon Yellow and Cameo Blue, both solid pastel colours. The European models were distributed from November 1998 onwards.

Perhaps the most noteworthy addition announced for the European model was ESP, or the Electronic Stability Program. This is a system which 'reliably suppresses unwanted under- or oversteer when close to the handling limits.' However, the most exciting news came as a quote in the VW Press Pack – 'An additional power unit is also planned for introduction in the course of the coming year (1999): a 110kW, 150bhp V5 with a displacement of 2.3 Litres is to be the new top engine option.'

After an evening presentation by the VW Board, VW laid on the most memorable launch party ever. This fantastic sixties-themed stage show included songs from the musical "Hair," a Beatles tribute band and a German lady emulating Janis Joplin. The launch was an experience I will never forget.

5

On the Road

Cars are not produced simply to be talked about, they exist to be driven and to get you

to your destination as safely as possible. However, we all like something with a bit of style

and charisma, and on the road the New Beetle gives the driver an experience very few cars

can provide – the sensation of being on the receiving end of virtually undivided attention from

most of the people around you, both pedestrians and other road users.

Phenomenal head-turner!

There are not many cars which will turn heads in the way the New Beetle does. In fact, it creates such a stir that it can be unnerving at times. Have you ever driven through a built up area such as a town and actually had people pointing at you, waving at you, clapping their hands and shouting at you? Well if you have, then the chances are that you are either a famous celebrity, or you are driving a New Beetle. This car really does cause a commotion. You will genuinely see people in the street pointing at the car, and if you have the windows down you can hear them saying 'Look, that's the New Beetle.' Even with the windows shut, you can often read their lips. Personally speaking, I have never witnessed this phenomenon behind the wheel of any other car. If you don't like a lot of attention, or suffer from paranoia, then the New Beetle is certainly not the car for you – at least, not for a few years yet!

Out on the open road, the New Beetle is a pleasure to drive

In traffic, other drivers ask questions, or honk their horns, or simply smile and give you the thumbs up. When you stop for fuel, the attendants ask how it drives and whether you like it. If you stop any-where, you will be greeted like an old friend by people who want to look at this new car. Complete strangers will come up to you and tell you how they used to have a Beetle, and how much they loved it. Others will actually try to get you to take them for a ride in the car.

The weirdest thing is that we are talking about people from all walks of life and age groups. It does-n't seem to matter if it is a student, a business person, a retired couple, a young parent – whoever they are, they seem to know exactly what this car is. Now this could be put down to the high profile media coverage, but it is doubtful every admiring glance is due to good press. From the minute you first set eyes on the New Beetle, you know what it is – it's the New Beetle, it can't be anything else. The Volkswagen master plan has paid off stunningly!

Metallic paint finishes make the New Beetle look even more chic

However, it is when people get a close look at the New Beetle that things become really interesting. Many find the car is 'far bigger' than they expected. Most feel it's pretty or cute; in my experience, only a few people thought it was ugly! On sitting inside the first thing that most people comment on is the dashboard. As they lean forward and touch it, the comments are generally 'this looks really big!', or 'Wow! This dash is huge!' Next it seems to be the headroom, touching the headliner and saying 'This is high up' and 'You could wear a top hat in here.'

One area which does seem to create negative feelings is the plastic roof liner which runs down the

'C' post in each of the rear corners: 'I don't like this, this feels weird on the side of my head.' The most prominent negative factor however was the capping on the door and rear window, which was an unpopular shade of red. The turbo-charged 1.8T has optional GLX trim to replace the

Build quality is, in general, high

standard New Beetle's GLS, including six-spoke alloy wheels, a great speed-activated rear spoiler, leather seating, and much more besides.

The predominant feeling towards the New Beetle is definitely positive. In virtually every case, people didn't just look at the vehicle, the felt that they had to touch it. It was as though they were meeting up with an old friend.

First impressions

When given the keys to the New Beetle, you find out its good and bad points very quickly. Firstly, when walking around the car, the lines, tight gaps and smooth exterior features, such as the way the headlights and taillights are flush fitted, are appealing to the eye. The slightly different hue or tone of the colour of the nearly-matched door mirrors is the only let-down from the outside. Getting into the car proves to be the second disappointment, for those used to central locking. The remote electronic key – which can sound the horn in the form of a panic system, set the alarm and lock both doors – has one drawback; it doesn't open the passenger door! To open the passenger door you need to either 'unlock' the driver's door twice with the normal key or use the remote to open the driver's door and then reach in and press the electric central locking switch. Both options offer little comfort to a passenger standing in the rain! The rear hatch can be opened via a very nice little switch which boasts a New Beetle logo, located on the driver's door, or from the outside of the car by rotating the VW roundel to expose the keyhold for the boot lock and using the ignition key to unlock it.

Once inside the New Beetle and behind the steering wheel for the first time you notice something – the dashboard! It is huge. The driver can not reach the windscreen. In fact it is doubtful that an orang-utan could! The dash looks very interesting though. Initially one could be tricked into thinking they were about to set off on a mission in a spacecraft of some kind, rather than a drive in a car. The materials used in the dashboard, especially those in the pronounced centre section – which houses the

Note how different this model looks without fog lights

stereo, two ventilation outlets and various controls – do not look or feel very good. This is quite sad, as the dash is the part of the car the driver sees the most. Personally speaking, I think that VW could have gone for something which looked a little more sporty and expensive, such as carbon fibre, and made the dash into a truly impressive feature.

The feel of the steering wheel and the passenger grab handle are satisfying, even thought they are made from the same composite material. Behind the steering wheel and in front of the passenger are two areas at the front of the dash which have a rough finish plastic and look great. Beyond these you find more tacky smooth plastic, which for some reason is flat. The opinion of many people is that this could easily have been a great area for trinket storage, although you'd have to stretch a long way to retrieve anything. Instead it is a sun trap, and on a fine sunny day it creates quite bad reflection on the windscreen, which can be trying. There is a spot for eyeglasses on the '99 rear view mirror, though.

You can not help but like the single instrument binnacle which incorporates the speedometer, fuel gauge, tachometer and mileage/trip read out. It is functional without being hard to read in any way. The binnacle also provides warning lights which let you know whether the lights are on, the indicators are flashing, the trunk is open and so on. For '99, the clock is well-placed below the rear view mirror.

The much talked-about flower vase and the area it sits do not, sadly, meet expectations. Again,

they look a bit cheap. The stereo, however, is great, both in design and performance – easy to get the hang of and not tricky to use. Ergonomically speaking, it is a success. All three ventilation controls are simple to operate, as are the air-conditioning and re-circulation switches. The only hope is that air-conditioning is not dropped from the standard specification on certain markets, as demisting the windscreen quickly and safely is made very much easier by good air-conditioning.

Moving to the doors, and without dwelling on the unfortunate door cappings, the controls are again very straightforward. The electronic mirror controls are situated in the capping, in front of the door locking switch and the door handle, which is chromed. Below this you find the door panel itself which blends in nicely with the lines of the dashboard. The door panel on the driver's side houses the fuel filler release and trunk release switches, electric window switches or window winder handle. Below these you will find the netted storage bins, which are large enough for a cellular phone, your favourite cassette and a packet of cigarettes, although the standard is to have no ashtray. In the '99 model, a handy joystick allows you to adjust the wing mirrors.

Settling in

The column mounted switches which control the indicators and wipers are simple to use, as are the main light switch and dash dimmer switch. Many of these components were found to have been made by Volkswagen's German suppliers, and the seat belts were made in Austria.

Both front seats offer the driver and passenger a great range of positions. By the side of each seat

is a lever which will bring the height of the seat up or down. The backrest can also be adjusted, as can the reach of the seat. With a little work the ideal driving position can be found, especially as the steering column offers reach and height adjustments. The headrests work too, particularly the '99 ring versions. You can actually rest your head on them when driving.

Rear seat space is adequate, even for a six footer, although occasional adult use for short journeys is probably the best bet. Children will have no problems in the back of the New Beetle. The rear seat can also folded be to provide extra luggage space, although for design reasons the width of entry to the trunk is limited. The rear seats also feature reading lights in the '99 model, which is a great touch for any book-loving passenger.

Between the seats, the central console features three cup holders. These are fixed and cannot be removed for cleaning. The material used is again fairly cheap looking. The sturdy hand-brake lever has a good grip and features a large chrome button. The lever recesses into the console, rather than standing proud.

In motion

Driving the New Beetle is a super experience even without the "Wow!" factor of being seen in it. However, you can't start the car unless the clutch pedal is depressed! In 2.0 litre, 115bhp form it isn't particularly fast, but the 1.8 litre turbocharged version yields an exciting 150bhp. The car is very quiet, in terms of both mechanical and wind noise. The suspension is comfortable without being too comfortable and the handling is good when tight corners come into play.

However, tight corners can be a problem as a blind spot is created by the 'A' pillars and the

Stunning car, stunning skyline!

exterior mirrors, which are mounted quite high up on the car. This obstruction can not be overcome, and is something you have to come to terms with. The design of the car also requires the driver to become accustomed to the large rear 'C' pillars for the same reason. Whilst diagonal views may be slightly impaired, the front, rear and side views are very good. New Beetle drivers may find it tricky to get used to the fact the windscreen is further away and what lies ahead of it is an unknown quantity – the furthest forward component in view being the windscreen wipers. It can be hard to work out how wide the car is and where the front bumper is. Parking is therefore best accomplished with a certain degree of careful consideration.

The car is very comfortable, and even after a 300 mile journey there is little, if any feeling of fatigue. A few have said the seats are too firm, but many others have found this not to be the case. One of the first questions about the New Beetle was whether or not it would be skittish and wander over the road in cross winds, like the original. The answer to both questions is no. The car drives with the same easy competence as the Golf it shares its running gear with. Even at speeds over 100mph (160kmh) – reached purely for research purposes on private ground of course – on long sweeping roads with a good side wind, the New Beetle felt completely safe. It must be said that the tail end of the car did feel lighter at high speed, but this was certainly not unnerving in anyway. Fortunately, the brakes are very good, with discs all on all four wheels – all ABS in the '99 model. Most importantly, the stopping power is certainly as good as the average motorist will ever need, even in an emergency stopping situation.

For many owners the New Beetle becomes more exciting to drive at night, and this is due to the fantastic dash illumination. The first time the dash is seen at night the comment is usually 'Wow!' The speedo binnacle lights up in a sort of neon blue, along with the readout on the cassette. The speedo

The car to be seen in around town

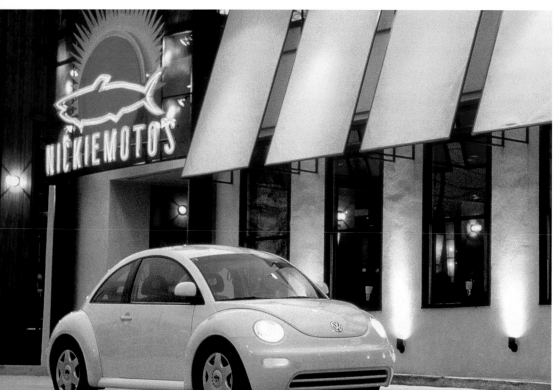

It will soon be time to switch the lights on – good news for the driver!

and gauge needle glow red, as do all the cassette, ventilation and other controls. This not only looks truly superb but is also very easy to drive with. There is no eye strain or fatigue associated with this feature, just a whole load of style and charisma.

On top of the basic price of the New Beetle, there are a number of alternatives which will offer higher performance, greater luggage capacity and extra internal space, but try to select those which don't disappear into irrelevancy once you're out on the open road. The bottom line is that it is the car that makes the difference, not the accessories that go with it. If you want to stand out from the crowd, you want a New Beetle. It's as simple as that.

6

Options

Virtually as soon as the New Beetle was in the hands of the public, the race was on to modify the cars. This phenomenon has been part of the original VW Beetle scene for decades, but there are two very separate areas of modification to consider. The first centres on performance, and the second caters for those who prefer classic vintage VWs.

From the start, after-market companies made every kind of accessory you could think of to spice up the basic specification of the old Beetle. One such item was the bud vase, which is part of the standard specification of the new version. During the fifties German companies such as Rosenthall offered porcelain vases for the original Beetle which were very popular. Today, many vintage Beetle enthusiasts will pay dearly for an original Rosenthall vase, rather than settle for a cheaper reproduction. Other items offered for Beetles in days gone by included slatted wooden roof racks, which again are in big demand and available as reproductions. A company based in Utah, called 'Parts', already offers a slatted wooden roof rack for the New Beetle. Many more original Beetle-style options are on their way.

Randy Carlson's New Beetle on CEC alloys

The performance after-market for the New Beetle is also big business, and from the way things are going, it's certainly going to grow. Randy Carlson has to get the credit for being first out of the starting blocks. He picked up his metallic blue New Beetle and started to modify it straight

McKenna's flamed Beetle – the first custom paint job

away. The car was soon seen wearing a set of CEC 18 x 8.25 inch alloy wheels with 235/40 18 Pirelli P7000 tires, sitting on 1.25 inch lower upgraded springs by Eibach. The suspension was also beefed up with upgraded sway-bars. The standard exhaust tail pipe was changed for a Remus unit. Since then a number of other little tricks have been added to the car, and there are no signs of him stopping yet. Randy has since started the New Beetle club, which is based in California and is already very popular. If you are fortunate enough to own one, you can see his details – along with other interesting contacts – in the data section at the end of this book.

A Volkswagen dealership in Huntington Beach, California, called McKennas VW, was next to join in the fun with a Hot Rod styled New Beetle. McKennas took a black car to a local custom paint specialist where it was given a fifties-styled flamed paint job from the front fenders to the rear of the doors. The result was amazing to say the least, especially when the company added a set of chromed

This flamed New Beetle matches Randy Zelany's other car, an original Beetle!

17 inch Alba multi-spoke alloy wheels. In a flash, McKennas turned a head-turner into a neck-twister!

The flame paint job theme has seemed to catch on across the country, with another owner of a black New Beetle having his car painted with orange flames, to match his 1961 Beetle. Have you got a white New Beetle? Well how about candy purple flames with anodised purple five-spoke alloy wheels? Too late, that has already been done! What about using the old Herbie theme on your white New Beetle, with a red and blue stripe right over the car and the famous number 53 on each door and across the bonnet? Too late, that's been done too! Brian Burrows, a VW event organiser, enthusiast and owner of a World Record holding drag racing Beetle, was the first to sneak a New Beetle out of Canada and into Britain. Having the first New Beetle in the country wasn't enough – he also wanted to be the first to race a New Beetle in competition. The car arrived in Britain on Wednesday 25th April. By Thursday, the car was being drilled to fit a fire extinguisher and battery cut-off switch. Then Brian had it sign-written and striped. Having planned the whole thing out way in advance, he had already secured S 53 as his racing number in the Street Class Championship of the British Volkswagen Drag Racing Club, so the Herbie number is official. The car was then driven to the Avon Park raceway where, on the Saturday morning, it became the first New Beetle in competition. The car covered the quarter mile in 18.37 seconds, which was nothing amazing. However, consider the owner's manual for a minute: 'During the first few operating hours, the engine's internal friction is higher than later when all the moving parts have been broken in. How well this break in process is done depends to a con-siderable extent on the way the vehicle is driven during the first 600 miles. As a rule of thumb: Do not use full throttle. Do not drive faster than three quarters of top speed. Avoid high engine speeds.' Brian clearly followed these instructions to the letter...

Brian Burrows' Herbie was the first to be raced

The VW Adventure

Another New Beetle first was The VW Adventure, a journey which followed the famous Route 66 from California to New York. This was the brainchild of Blue Nelson, who, as mentioned earlier, went to the launches in L.A. and Vancouver. He got together with fellow vintage VW enthusi-ast Eric Meyer, owner of a shoe company called Simple, and the adventure began. The journey was covered by many local newspapers along the way, and updates were posted on a VW Adventure Web Site. Blue Nelson shipped the silver car – which has been a test bed for many prototype accessories – over to Britain in August 1998 for the second part of the VW Adventure. Blue now plans to drive the car around Europe, and hopefully the rest of the world.

By the summer of 1998, performance modifications started to appear on New Beetles at VW events across America. Companies such as Autobahn Designs, or ABD, brought out rear spoilers which mount below the rear window along with small additional spoilers to fit to each corner of the front bumper. ABD also looked at engine tweaks and enhancements. Instead of the standard air box, ABD offered a high-flow foam element-type filter. The smooth plastic engine cover with VW logo and 2.0 emblem was re-moulded by ABD in carbon fibre, which gave the stock engine a racy look. ADB also

Kristen Short from Las Vegas and her supercharged New Beetle

offer a range of higher lift camshafts and will continue to develop products for the New Beetle.

The most recent New Beetle products which have hit the market have include Wolf car covers, heat protective dash mats, 'bras' which fit either across the whole front of the car or just over the leading edge of the bonnet/hood, rear bumper spoilers, and front hoods with mean looking power bulges. It looks like the after-market options for the car will be huge business. It is predicted that VW itself will bring out some exciting accessories.

Of course, there is always VW's 1.8T-engined model, which VWOA have scheduled for launch at Detroit 1999. The 1.8 Litre twenty valve turbocharged unit is the same award-winning base engine used in the Audi A4, Passat and Golf 4. At 5,700rpm the engine produces 150bhp, adding 45bhp over the stock 2.0 Litre petrol engine, which produces 115bhp at 5,500rpm. The new engine will push the maximum top speed up to 126mph. Although at the time of writing the name has not been definitely settled, the 1.8T engined car may well end up being referred to as the New Beetle GTi.

Rumours of four wheel drive systems linked to either VR5 or VR6 engines have also been circulating. For some New Beetle owners waiting for performance models is not an option. This is where a company in Hermosa Beach, California, aptly named 'Dr. Boltz', comes into play. The company specialise in street and racing upgrades for VW, BMW and Porsche cars, but saw the market for New Beetle upgrades. Dr. Boltz started with the car it called Project Number 1, which was a turbocharged 2.0 Litre model producing 200bhp. The installation was designed to be a very neat fit, and as you can guess, it made the New Beetle perform very well indeed. The turbocharger 'kit' is now on the market and will cost around $2000 plus fitting charges. You can not simply add another 85bhp to your New Beetle without thinking about doing a little extra work to the suspension, though. A set of upgraded springs and anti-roll bars are recommended, along with a set of 17 or 18 inch wheels with wider tyres.

The next project car to come from Dr. Boltz was a VR6 powered New Beetle, and Bo Bertillson, a regular contributor to VolksWorld Magazine who is based in California, was one of the first to take a

Dr. Boltz' 'first' VR6-powered New Beetle – note the rear bumper extension

test ride in the car. Once fitted, the VR6 power plant looks although it was installed at the factory, it fits in so well. The engine has been 'chipped' and has high flow intake and exhaust systems, and is said to push up the power output from 178bhp to around 200bhp. Whilst this is around the same output as the Dr. Boltz turbo kit, there is one extremely significant improvement over the turbo – the VR6 has far more power in the lower rev range.

The price of a full Dr. Boltz VR6 conversion with full suspension, body styling and those lovely looking 17 inch Momo wheels is approximately $60,000. This price is of course 'turn key,' which means that includes the cost of the original car itself. Rumour has it that the Wolfsburg design team have obtained themselves a nice, white Dr. Boltz VR6 New Beetle for examination and testing. Whether or not this is fact or fiction, it would be nice to think that VW would release a New Beetle VR5 or 6. We all know how it works with VW now – all you have to do is make enough noise about something and convince enough people, and they'll do it...

Ross Palmer's team was the first to circuit-race a New Beetle

Kristen Short's highly modified New Beetle – 1 FAT BUG!

The New Beetle isn't only causing a stir when comes to modifications. It is also becoming a cult vehicle, and with the already-mentioned New Beetle Club picking up steam very quickly, an enthusiast following is assured. VW event organisers are already adding New Beetle classes for owners who wish to show their customised cars to other enthusiasts, and some of the results of the customisation process look absolutely stunning.

On the 12th February 1998 VW said, 'In terms of the number sold, compared with its predecessor, the New Beetle will never be more than a niche vehicle – but in a niche all of its own.' Despite this somewhat cautious, even pessimistic assessment, the future certainly looks bright for the most sensational new car in modern motoring history.

Vehicle Data

Concept 1

Vehicle Dimensions

Length	3824mm
Width	1636mm
Height	1500mm
Wheel base	2525mm
Weight	907kg
Track F&R	1488mm
Front Over Hang	664mm
Rear Over Hang	636mm
Front Ground Clearance	178mm
Rear Ground Clearance	241mm

Engine/Motor Specifications

TDI Version:

Design	Direct-Injection Diesel Four Cylinder
Displacement	1900cc
Maximum Power	66kw
Maximum Torque	202Nm
Transmission	5-Speed Ecomatic
Maximum Speed	180Km/h
Fuel Consumption	5.1 L/100km
Acceleration	12.8 seconds 0-100 km/h

Electric version:

AC Induction

Maximum Power 37kW

Maximum Torque 130Nm

Transmission . . . . 2-Speed Automatic

Battery Type . . . . Na/NiCl

Battery Weight 260kg

Stored Energy 22kWh

Rated Capacity 90Ah

Load Voltage 248Volts

Maximum Speed 125km/h

Urban Range 250km

Hybrid Version TDI-motor/Electric-motor:

Design . . . . Direct Injection diesel three cylinder/ AC Induction

TDI Displacement 1400cc

Maximum Power 50kW

Maximum Torque 140Nm

Transmission . . . . 5-Speed Semi-Automatic

AC Induction Max Power 18kW

Battery Type . . . . Ni/MeH

Battery Weight 180kg

Stored Energy 10kWh

Rated Capacity 55Ah

Off-Load Voltage 180Volts

Max Speed TDI 165km/h

Max Speed AC 105km/h

New Beetle

Body, Chassis and Suspension

Type	Unitised construction, bolt on fenders
Front Suspension	Independent McPherson struts, coil springs, telescopic shock absorbers, stabiliser bar
Rear Suspension	Independent torsion beam axle, coil springs, telescopic shock absorbers, stabiliser bar
Service Brakes	Power assisted, dual diagonal circuits, 280mm vented front discs and 239mm solid rear discs
Anti-Lock Braking	Optional, all four wheels
Parking Brake	Mechanical, effective on rear wheels
Wheels	6 1/2 J x 16, steel, with full covers, 5 bolts
Tires	205/55R 16H all season
Drag Coefficient	0.38

Dimensions

Wheel base, in	98.9
Track Front, in	59.6
Rear, in	58.7
Overall Length, in	161.1
Overall Width, in	67.9
Overall Height, in	59.5
Ground clearance, in	4.2

	5-Speed manual	4-Speed Automatic
Transmission	5-Speed manual	4-Speed Automatic
Curb Weight, lb	2712	2,778
Payload, lb	992	936

Fuel Consumption

Transmission	5-Speed manual		4-Speed Automatic	
Fuel	Gasoline	Diesel	Gasoline	Diesel
City, mpg *	23	41	22	34
Highway, mpg*	29	48	27	44

* May vary.

Engine

	Gasoline	Diesel
Type	2.0L, 4 Cyl, in-line	1.9L, 4-Cyl, in-line
Bore, in	3.25	3.13
Stroke, in	3.65	3.76
Displacement	1984cc	1896cc
Compression Ratio	10.0:1	19.5:1
Horsepower (SAE) @ rpm	115@5200	90@4000
Max. torque, lbs-ft @ rpm	122@2600	149@1900
Fuel Required	Regular Unleaded	Diesel

Engine Design

Arrangement	Front Mounted, Transverse
Cylinder Block	Cast Iron
Crank Shaft	Cast Iron, 5 main bearings
Cylinder Head	Aluminumalloy, cross flow
Valves train	Single overhead cam shaft, spur belt driven, 2valves per cylinder, maintenance free hydraulic lifters
Cooling	Water cooled, water pump, cross flow radiator, thermostatically controlled electric 2-speed radiator fan
Lubrication	Rotary gear pump, chain drive, oil cooler

Fuel/Air Supply. . . . . Sequential multi-port fuel injection (Motronic)

Emissions System . . . . OBD II, 3-way catalytcic converter with 2 oxygen sensors (upstream and downstream), enhanced evaporation system, Onboard Refuelling Vapour recovery (ORVR). TLEV (transitional low emissions vehicle) concept for California, Massachusetts and New York only

Transmission

5-Speed Manual	Gasoline	Diesel
1st	3.78:1	3.50:1
2nd	2.12:1	1.94:1
3rd	1.36:1	1.23:1
4th	1.03:1	0.84:1
5th	0.84:1	0.68:1
Reverse	3.60:1	3.60:1
Final Drive	4.24:1	3.89:1

4-Speed Automatic	Gasoline	Diesel
1st	2.74:1	2.71:1
2nd	1.55:1	1.44:1
3rd	1.00:1	1.44:1
4th	0.68:1	0.74:1
Reverse	2.11:1	2.88:1
Final Drive	4.88:1	3.63:1

Electrical System

Alternator, Volts/Amps. 14/90

Battery, Volts/Amps 12/60

Ignition . . . . Digital electronic, distributorless coil block, with knock sensor

Firing Order 1-3-4-2

Capacities

Engine Oil (with Filter), qt 4.8

Fuel Tank, gal 14.5

Cooling System. qt 6.7

Steering

Type. . . . . Rack and Pinion, power assisted

Turns (lock to lock) 3.2

Turning Circle (curb to curb), ft. 32.8

Ratio 17.8:1

Interior

Seating Capacity . . . . Four

Head room front. . . . . 41.3 inches

Head room rear . . . . 34.6 inches

Shoulder Room. . . . . 52.8 inches

Leg Room front . . . . 39.4 inches

Leg room rear. . . . . 33.0 inches

Exterior Features

Antenna	Roof mounted, amplified
Corrosion protection	26-step paint/corrosion process
Glass	Tinted
Horn	Dual tone horns
Lights, Front/Rear	Daytime running lights (DRL) upon start-up of vehicle headlights are engaged with reduced power, IP lighting, parking lights and taillights remain off. To engage all lights with full power the light switch must be turned to the 'on' position
	Halogen projector beam headlamps with clear polycarbonate lens
	Optional: Halogen projector lens front foglights
Mirrors	Body colour, power adjustable and heatable
Paint work	Solid or Metallic: Standard
Roof	Optional: Power glass sunroof, tilt and slide, tinted glass, with sunshade and power lock operated convenience closing feature
Alloy Wheels	Optional: 6 1/2 j x 16 inch alloy wheels, six spoke with VW logo hub cap and Anti theft wheel locks
Wipers/Washers	2 speed windshield wipers with variable intermittent wipe feature
	Optional: Heatable windshield washer nozzles (only in combination with heatable front seats)

Interior Comfort and Convenience Features

Air Conditioning . . . . CFC free with variable displacement A/C
compressor

Alarm/Anti Theft . . . . Anti theft alarm system for doors, hood,
trunk lid, radio and starter interrupt,
with warning LED in drivers' door top
sill and with activation 'beep'

Ashtray . . . . Dealer fitted option only

Assist handles . . . . Large assist handle in instrument panel
above glovebox, 2 rear assist straps on
'b' pillar

Clock . . . . Digital clock with blue display mounted
in central forward headliner

Cruise Control . . . . Optional: effective at speeds above 22mph

Defroster . . . . Electric heated rear window

Doors/Side panels . . . . Moulded door trim in leatherette with
upper sill moulding in exterior body

colour, black on vehicles with white exterior
Integrated arm rests in front door panels

Integrated arm rests in rear door/side panels

Floormats . . . . Front and rear carpet style, colour co-ordinated

Fuel filler . . . . Remote fuel filler release located on driver's
door inner, flap connected to fuel filler
neck to protect body

Instrument cluster . . . . Speedometer, tachometer, odometer, trip
odometer, fuel gauge, gear indicator (Automatic
transmission only), warning lights

Red illumination for controls (switches and
buttons), blue illumination for displays
(speedometer cluster and radio display)

headlights-on warning tone (upon opening of
driver's door when ignition key is removed)

Keys . . . . Valet Key

Lighting . . . . Combination interior and reading lamp located
in bottom sill of rear view mirror with time delay

Lighting continued	Glove box light
	luggage compartment light
Locks	Central power locking system (doors) with key operated closing feature for optional sunroof (if available), opening & closing feature for power windows (if so equipped) and selective unlocking at driver or passenger door
	Door mounted lock/unlock switches for central locking system
	Radio-frequency remote locking system with lock, unlock, trunk release and panic buttons on transmitter
Interior Mirrors	Driver and front passenger visor vanity mirror illuminated with cover
Power outlets	2 power outlets (SAE size) in centre console
Radio/Audio	AM/FM cassette stereo sound system with control capability for (optional) CD changer, theft-deterrent warning light and coding system. 6 speakers: 2 tweeters in 'A' pillar, 2 mid-woofers in front doors, 2 mid-woofers in rear quarter panel
	CD-changer preparation (cable from radio to luggage compartment)
Restraint System	Driver and front passenger airbag supplemental restraint system
	Front 3- point safety belts
	Height adjustment for front safety belts
	Rear outboard 3-point safety belts
	Emergency tensioning retractors for front safety belts
	Child seat tether anchorage system (rear seat)
	ALR, ELR (automatic locking retractor, emergency locking retractor) for front passenger and rear outboard safety belts, to secure child seat in place under normal driving conditions. Driver and front passenger side airbag supplemental restraint system
Seating, Front	Front seats, fully reclining, with height adjustment and 'ring style' headrests, easy entry system

integrated into front seat mechanism. Allows seat to move forward easily for access to rear seats

Optional: Heatable front seats (only with heated washer nozzles)

Seating, Rear. . . . . One-piece folding rear seat

rear seat outboard ring style headrests

Steering wheel . . . . 3-spoke padded steering wheel

Optional: Leather wrapped steering wheel (only in combination with leather interior)

height adjustable and telescopic steering column

Steering wheel deformable upon impact

Collapsible steering column

Storage, interior . . . . Front door storage nets

Front passenger seatback magazine/storage pockets

Glovebox, lockable with interior shelf

Centre console with 3 front beverage holders and rear concealed beverage holder, colour co-ordinated

Trim (Details). . . . . Optional: Leather wrapped steering wheel

Optional: leather shift knob and boot (manual trans. only) and hand brake cover

Trunk/Cargo area . . . . Luggage compartment carpeting on floor, left and right side, rear of seat back

Upholstery. . . . . 'Primus' velour seat fabric

Optional: Leatherette seat trim

Optional: Partial leather seat trim (seating surfaces)

Ventilation System . . . . Rear seat heat and A/c ducts, side window defoggers in IP Pollen/Odor filter for incoming air

Windows. . . . . Manually operated windows

Optional: Electrically operated one-touch up or down (with pinch protection) and convenience close and open feature

Body Colours 1998

Solid	Metallic
White	Silver
Yellow	Green
Black	Bright Blue
Red	
Dark Blue	

Manufacturers Warranty

2 years/24,000 mile basic limited warranty

2 years/24,000 mile no charge scheduled maintenance

2 years unlimited mileage/distance 24 hour roadside assistance

10 years/100,000 mile Limited Power train Warranty

12 year unlimited mileage/ distance Corrosion Perforation warranty

VW Disclaimer: All equipment listed is subject to production related change or delayed availability.

What color do you dream in?

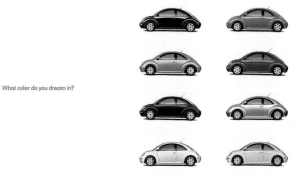

Drivers wanted.

Useful New Beetle Contacts

New Beetle Club
Randy Carlson
316 South Orange Avenue
Brea
California
92821

VolksWorld Magazine
Link House
Dingwall Avenue
Croydon
Surrey
CR9 2TA
England
http://www.linkhouse.co.uk
/volksworld

Dune Buggies and
Hot VWs Magazine
Wright Publishing Co., Inc.
PO Box 2260
Costa Mesa
CA 92626
U.S.A

Volkswagen Web Site:
http://www.vw.com

Volkswagen of America
information line:
1-800-444-8987.

Index

DESIGNs FOR
DIFFERENT FUTUREs

DESIGNs FOR DIFFERENT FUTUREs

Philadelphia Museum of Art, Walker Art Center, and The Art Institute of Chicago

IN ASSOCIATION WITH

Yale University Press, New Haven and London

EDITED BY

Kathryn B. Hiesinger & Michelle Millar Fisher, Philadelphia Museum of Art

Emmet Byrne, Walker Art Center

Maite Borjabad López-Pastor & Zoë Ryan, The Art Institute of Chicago

WITH

Andrew Blauvelt
Colin Fanning
Orkan Telhan

CONTRIBUTIONS BY

Juliana Rowen Barton
V. Michael Bove Jr.
Christina Cogdell
Gabriella Coleman
Simone Farresin
Marina Gorbis
Aimi Hamraie
Srećko Horvat
Nora Jackson
Francis Kéré
David Kirby
Helen Kirkum
Bruno Latour
Marisol LeBrón
Ezio Manzini
Jillian Mercado
Alexandra Midal
Neri Oxman
Chris Rapley
Maude de Schauensee
Andrea Trimarchi
Eyal Weizman
Danielle Wood
LinYee Yuan
Emma Yann Zhang

P.M.A.

X

W.A.C.

X

A.I.C.

A NOTE TO THE READER

The author of each project entry is indicated by initials: Juliana Rowen Barton (JRB), Maite Borjabad López-Pastor (MB), Emmet Byrne (EB), Colin Fanning (CF), Michelle Millar Fisher (MMF), and Kathryn B. Hiesinger (KBH).

All interviews were edited for length and clarity.

Definitions for the highlighted terms can be found in the glossary on pp. 254–57.

Foreword

Can designers shape the future? The answer to this question is, of course, yes: they can, and often do. Indeed, rather than limiting the scope of their work to addressing present needs, designers have also—and quite often—claimed the future as a legitimate domain of inquiry. Designing for the unknown is, and always has been, a different kind of work, one that requires both the exercise of imagination and the acceptance of uncertainty.

For much of the modern era, the desire not simply to envision the future but also to shape it has been informed by ideals that were progressive or even utopian in spirit—that is, by the belief that the future would be not only substantially different but also better, morally as well as technologically. Our approach to design, and the expectation that it would harness the potential of new technologies, played an important role in defining this optimistic vision of the shape of things to come. At its heart this view is a profoundly democratic one, in which design's benefits—economic as well as aesthetic—would be shared widely for the good of society as a whole.

We are still enamored of this vision of the future and the possibilities it holds for a better life, although what that might look like has been tempered by the sobering reality of the challenges we face today, including income inequality, global warming, the degradation of the environment, and the scarcity of such basic resources as fresh air and clean water. In this context, we must ask not simply whether design can help us build a better tomorrow but also whether it can help us address these looming problems in order to avoid a future in which our way of life—and life itself—is no longer sustainable.

For the curators of this exhibition, the contributors to this catalogue, and the designers whose work is featured, the answer is yes, because it is clear that if we ignore these intractable problems we do so at our own peril. In this regard, contemporary designers have now claimed a much broader and more moral field of action,

and one in which the individual and the global—the choices we make and how they resonate in the world—are inextricably bound together.

To return to an earlier point, we must also recognize that any vision of the future is contingent, that it may be many different things, some of which we can see clearly even from a great distance, while others remain obscure. But however uncertain, complicated, and even troubling such a future (or set of futures) may be, its appeal to our imagination remains irresistible. To be sure, what we don't know or don't fully understand can discourage us from moving forward, from exploring new and different possibilities. Nevertheless, thinking about the future encourages speculation and can be better understood, and appreciated, as a call to action and as an invitation to imagine—always with a sense of hope—what might be.

This project began, as many do, with a brief conversation, in this instance about the Philadelphia Museum of Art's long commitment to representing the history of design. Kathryn B. Hiesinger, the museum's J. Mahlon Buck, Jr. Family Senior Curator of European Decorative Arts after 1700, who has organized all of the museum's major design exhibitions over the past five decades, immediately expressed her enthusiasm for doing a show that would take a close look at how today's designers are thinking about the future. From the outset, *Designs for Different Futures* was envisioned as a collaborative effort; it could not have succeeded without the significant contributions made by each of the three institutions working in partnership. We are deeply grateful to all the members of our staffs who have been involved with the development and production of both the exhibition and this catalogue, and most especially our curatorial team: Kathryn B. Hiesinger and Michelle Millar Fisher, the Louis C. Madeira IV Assistant Curator of European Decorative Arts, Philadelphia Museum of Art; Emmet Byrne, design director and associate curator of design, Walker Art Center; and Maite Borjabad López-Pastor, Neville Bryan Assistant Curator of Architecture and Design, and Zoë Ryan, John H. Bryan Chair and Curator of Architecture and Design, both at the Art Institute of Chicago.

Finally, we would like to express our deepest thanks to those who have provided financial support for this important project. In Philadelphia, we acknowledge the generosity of the Women's Committee of the Philadelphia Museum of Art and an anonymous donor, as well as those who have contributed to the museum's endowment in support of special exhibitions: the Annenberg Foundation, the estate of Robert Montgomery Scott, the estate of Kathleen C. and John J. F. Sherrerd, Lisa Roberts and David W. Seltzer, Laura and William C. Buck, Harriet and Ronald Lassin, and Jill and Sheldon Bonovitz. In Minneapolis, we thank the donors who have supported the exhibition. In Chicago, we recognize the Exhibitions Trust and other generous sponsors.

Timothy Rub
The George D. Widener Director and
Chief Executive Officer
Philadelphia Museum of Art

Mary Ceruti
Executive Director
Walker Art Center

James Rondeau
President and Eloise W. Martin Director
The Art Institute of Chicago

Introduction

KATHRYN B. HIESINGER, MICHELLE MILLAR FISHER, EMMET BYRNE, MAITE BORJABAD LÓPEZ-PASTOR & ZOË RYAN

This conversation took place on March 15, 2019, at the Walker Art Center in Minneapolis.

EB
Designs for Different Futures has had a long gestation and has been a really collaborative venture. Several years back Andrew Blauvelt and I started working on a speculative design show here at the Walker. And then he heard that Kathy was also working on a design-futures show in Philadelphia, so we started to get together.

KBH
That's right. In October 2014 I wrote to Andrew that I was working on a future-focused exhibition and asked whether he and the Walker might be interested in collaborating. I'd been looking at ideas driving design-thinking about physical objects into the future, with help from Mike Bove at MIT, Scott Klinker at Cranbrook [Academy of Art], and Mark Yim, Simon Kim, and Orkan Telhan at the University of Pennsylvania. And other members of my department were there from the very beginning as well—our curatorial fellow Colin Fanning, assistant Rebecca Murphy, and volunteer Maude de Schauensee.

I went to Minneapolis for the first time in 2015 to meet with Andrew, and Emmet showed idea-driven, speculative designers and projects that were just so remarkable—from Sputniko! to *Drones for Foraging* from the Georgia Institute of Technology. Then Michelle came to Philadelphia in the winter of 2017, and Zoë and Maite in Chicago joined the project at about the same time. Zoë was a natural addition after she curated the 2014 Istanbul Biennial, *The Future Is Not What It Used to Be*. Having a team of collaborators across three institutions has been productive but logistically challenging. And I'm happy we were also able to bring on our Mellon fellow in Philadelphia, Juliana Rowen Barton, to help out.

As we talked, the exhibition began to move away from the more functional physical objects I had considered earlier and toward more conceptual projects. But many of the overarching themes have remained the same, including designs for social change, material transformations, and design futures and fictions.

ZR
It always felt like this exhibition should be collaborative. It didn't seem to be a project that we would want to do on our own, or would have the capacity for. And the thought of three quite different institutions coming together was very exciting.

MMF
By the time the exhibition opens we'll have been working together as a team for almost two years. What we think of as design has changed during this time.

MB
So many things have happened that have informed the exhibition. We are looking forward but also working in the here and now. We don't always have the perspective of time and distance; we are embedded in these issues in real time, and we should let the exhibition be affected by them, allow it to mutate in a dynamic way. Bruno Latour's exhibition *Reset Modernity!* [held in 2016 at ZKM in Karlsruhe] was an inspiration in the way it used the medium of the exhibition as a site of experimentation and research that would keep unfolding and evolving over time.

MMF
Do you think the questions we had at the start about different futures and design's relationship to them have remained the same, or have they changed just as radically as the works in the show?

CARL SAGAN

asimov

I, Robot

MUNDANE
SCIENCE
FICTION

FOUR
FUTURES

ROSALÍA

MB

The questions have changed; the issues, maybe not. Many of the issues are transhistorical, instrinsic to the human condition. But how we're asking the questions has changed. And I think that's a key part of the show—not giving answers but proposing different ways of questioning our world and the human condition to suggest different futures.

EB

Do you think we could do this same show again ten years from now?

MB

This same show, I hope not. I think it should be somehow obsolete, and I like that idea. This show freezes our research at a moment in time.

MMF

Thinking about the future will always be part of the human condition and part of design. But if we put *this* exact show on again, it would be a bit quaint—a snapshot of a very specific style, like a 1970s kitchen seen from today.

EB

We can look back at events like the world's fairs and the expos and try to imagine what the ethos was behind their projected views of the future. Twenty years from now, how will our approach be viewed?

MMF

I hope we're not as nationalistic as the world's fairs—that we're more thoughtful about colonial pasts and appropriative presents. But this exhibition will be shown in three museums, two of which were founded in the late nineteenth century and are encyclopedic in scope, so there are skeletons in our closets. We're trying to address that going forward by peer-reviewing our exhibition labels and working with community partners to determine programming as often as possible. But there is no perfect recipe for this. We are still learning daily.

EB

One early idea I had for the exhibition design was a hyper-contextualized approach where we would point out the systems of the exhibition itself. The idea of transparency about how museums and curators do what we do felt like something a show like this should naturally address.

MMF

Would that include the staff who actually mount the exhibitions? Would we be transparent about hiring policies, about family-leave policies, about the ways in which our institutions structure their staff? Because that would be really interesting.

ZR

We should be honest about *all* our biases as a collective and our position within the world of museums today. I think all our organizations are trying hard to be more transparent—to grapple with complex issues, to set the record straight in various ways, to try to move forward not just with dominant narratives but with alternate points of view. I hope the ideas embodied in the projects in the exhibition and this publication will go out into the world. I hope visitors leave the exhibition understanding the importance of design. Some people might never have seen this type of work before, or never have understood design as such a multivalent field, or realized what designers are engaged in or the different voices and expertise that go into these projects. Design is not a singular field operating on its own.

MB

The exhibition brings together people from very diverse backgrounds who are talking about the same shared issues but who usually operate separately. So it challenges traditional disciplinary separations and definitions of design. It's a productive blurring that actively redefines and pushes forward the field of design.

MMF

We've ended up with a multidimensional, compelling, provocative, beautiful selection of design projects. We've often talked about making the exhibition look more like a "fine art" show than something that says "design," so I hope that it will be visually attractive to people and also complicate their idea of what design is.

KBH

Our museum has a long history of presenting design, but it's been more focused on objects with overtly aesthetic dimensions. The works in this exhibition are so varied—in both conception and methodologies—and combine disciplines like literature, engineering, and natural science as well as design, art, and craft. So one of the things we've talked about is how to explain the show to visitors.

EB

Early on, we began talking about what stories each design was telling. The ability that designed objects or systems have to communicate beyond their practical purpose is another thing this show highlights. The process became about choosing *which* stories to present through the exhibition.

MMF

That's something you've highlighted really consistently. I've loved the way that you've continued to bring that perspective, and I think that speaks to the different voices that make up our team, which has helped us bring together the eleven major "stories," or object constellations, that organize the exhibition—Bodies, Intimacies, Generations, and so on.

EB

Initially, the future was our subject matter, and we really dug into that without even trying to tell specific design stories. We were trying to dissect the entire human experience—a somewhat impractical assignment.

MMF

Which is what historians do.

ZR

It's because this show is very issue driven. We didn't start with a group of objects that told a story. We started with a set of issues that expanded and grew, and we haven't sought out objects, necessarily, but voices. We continue to contemplate these issues because they're driving everything we think about the future.

MB

This connects with the question of how to display research. We're presenting not just objects or projects, but initiatives. Some of these can be made manifest through images or videos. But many of them cannot. Part of the challenge of the exhibition is how to display things that are not immediately accessible. In that sense the exhibition functions as a catalyst for thoughtful discussion.

MMF

A number of the projects in the exhibition have manifested across multiple media using a range of strategies to talk about intractable, complex issues. But it was a couple of the "ones that got away," where we struggled to find a way to represent them in the galleries, that have really stuck with me, like the topics of universal basic income or political empathy. Those are important issues for the future, and we're still trying to figure out how to represent them in an exhibition. Maybe that comes down to understanding what an exhibition can and can't do. But the question of where the "design" is in some of these projects is a legitimate one. Because we don't want to misleadingly propose design as a panacea for anything, and it's also not an infinitely stretchable term.

ZR
Right. How does an exhibition format provide that platform, while also being a space with limitations? We want the exhibition to be experiential, we want it to be engaging, and we want it to be legible and multilayered at the same time.

MB
We've carefully been embracing the thought that this project will live on for two years, and that really interests me, because certain design projects mutate over time, and that means they might evolve from venue to venue. We'll need to reconsider how to position some of the projects as they travel from Philadelphia to Minneapolis and then to Chicago, because things are going to happen in the news, globally and locally, that we will need to acknowledge in the conversations we're presenting. Design is the lens through which we've been looking at the here and now, toward the future, but it's not necessarily the content of the show all the time but rather a set of methodologies that allows us to navigate evolving complex issues. As Bruno Latour put it [in his lecture for the 2008 Design History Society meeting], "Designing is the antidote to founding, colonizing, establishing, or breaking with the past." He sees design as a search for radical departures.

ZR
It's interesting that we haven't really been talking about "newness." In design you always think about things being new because of the connection to the market and economies. But we haven't been solely interested in that. We've been focused on contemporary designers who are grappling with transhistorical issues.

KBH
On the other hand, what do we mean by "new"? We aren't discounting innovation in material technologies in the exhibition, for example.

ZR
Right. It's interesting to look back at other moments in time when ideas have been presented to the public through exhibitions to see how people thought then and how that thinking has changed, or not. We're all grappling with universal concerns and ways of understanding these concerns. If we can make that legible, I think it will be a valuable contribution.

MMF
That's it in a nutshell. The transhistorical, or the questions that come back, may be rephrased slightly differently from generation to generation or time to time, but the issues remain the same. It's the technologies that change, whether that's "technology" understood in a very literal way, or the technology of presenting an exhibition or any other framework that is the display.

EB
In my conversation with David Kirby [see pp. 160–64], I assumed we would be talking about how futures are created, since his research deals with how speculative technologies are visualized and essentially marketed through a collaboration between the scientific

community and Hollywood. But instead we ended up talking about truth and the present moment, and how our understanding of science and politics right now is so dependent on the media through which we see the world, which in turn is being manipulated by particular powers. Our understanding of the present is fractured from person to person or culture to culture—and the general collapse of what we used to understand as our consensus reality is something designers have to understand and reckon with as well.

MMF
It's a landscape of atomization, but it's like a palimpsest too.

ZR
I wonder how you make those layers visible, so that people can acknowledge many histories and many presents to reinforce our central idea of the possibility for many different futures, as in the title of our show.

MMF
How do you give them the archeological tools?

ZR
When I was working on my essay about the design imagination [see pp. 41–47], I was trying to better understand how we can create spaces for reimagining and debate that can help shift perspectives or raise awareness. How can an exhibition, a book, a film, or a performance, for example, invite us to think differently about the world? This is something that designers do really well—help us think differently. They're able to identify complex issues and help unlock them through their work. That's what this exhibition shows.

MB
Design allows you to change perspective, not by providing alternative solutions but by asking the question in different ways. Truth is not something that is static; truth is, and always has been, produced and constructed, as Eyal Weizman and I discuss [see pp. 172–75]. And acknowledging that opens a Pandora's box that forces us to question who has been producing that truth, and for whom, consequently destablizing power structures.

MMF
There are hard-and-fast relationships that we have with what we think to be true, but that I personally have come to question through this exhibition—the positions of power we occupy in being able to have these conversations, or the way we find things to be self-evident but that quite clearly come from our constructed worlds. So we operate on very shaky ground. Maybe that's the only truth.

ZR
It's an unnerving project to work on. There's a lot of responsibility, because the issues addressed are socially, politically, and ethically complex. The stakes are high. It's different from a historical project where you have all—

MMF
The benefit of hindsight.

ZR
Exactly. We've also found out that we don't all think alike. We've confronted sometimes disconcerting, sometimes unsettling, oftentimes joyous projects. I hope that people will see that even though we don't all think alike, it's possible to begin a conversation. This is the most utopian, idealistic view of how exhibitions can open up dialogue and the exchange of ideas.

HARRISON FORD IS
BLADE RUNNER ™

GULF
FUTURISM

BLACK MIRROR

Future Shock
by Alvin Toffler

Slavoj Žižek
Welcome to the Desert
of the Real

BLADE RUNNER 2049

A CYBORG
MANIFESTO

MB

It's also important to acknowledge the things that won't be in the exhibition, because we're specific institutions and individual people, so we come with specific political, social, and cultural backgrounds. Our references and understanding of the discipline of design have been contested as the exhibition has evolved, and I have found that extremely productive.

KBH

Yes. Though I've been struck from the beginning by our common points of reference, we all have our different touchstones. For me, my primary influences were Carl Sagan and Isaac Asimov, whom I knew when I worked for Carl at the Harvard [now Smithsonian] Astrophysical Observatory in the summer of 1966. At the time, I read Carl's scientific papers on the possibilities of extraterrestrial life and reread Isaac's *I, Robot* stories. I was particularly fascinated by his Three Laws of Robotics, which are still relevant today for designers considering human-robot interaction. My experiences that summer led to my interest in how science fiction can influence design by proposing paradigms that are actually usable and, conversely, how design in the real world establishes paradigms for science fiction. Those conversations happened a long time ago, but they still resonate with me.

EB

One of my touchpoints goes back to about 2008, when I read the Mundane Manifesto [2004], by a group of writers who came together to question the mentality or philosophy behind a lot of popular science fiction. They basically asked, Is science fiction inherently escapist? And, Is it irresponsible toward the earth we live on? At the same time, they were asking a creative question: Can we use that ethical question to also make our stories more rigorous and more interesting, and more creative? They came up with a manifesto that basically says, "Don't use any impossible ideas; only use ideas that are based in reality." For example, we're pretty sure faster-than-light travel is impossible, so let's just take it off the table. Their manifesto is exaggerated and aggressive, and it asks a lot of interesting questions about how our creative imagination can also be harnessed parallel to our ethics. I see connections to speculative design and world-building. The mundane science-fiction writers were trying to construct a truth and a sense of wonder embedded in the present and here on Earth, where we live. In essence they were pushing their fantasies toward our reality. I see speculative designers starting from a version of reality—the pragmatic and problem-solving mentality of some design practices—and trying to push toward fantasy, toward narrative and problem-posing. So for me there was a nice inversion that resonated.

My second touchpoint is Julian Bleecker's essay "Design Fiction" [2009]. He describes a "fertile muddle"—between fact and fiction, between now and the near future, between fantasy and science—in which all of these supposed binaries frequently intermix and swap properties. And he makes the case that we should be working in that space, and actively pushing the muddle forward and dirtying it up even more. He talks about Hollywood films and cable news networks and newsstand magazines, and about the ways that people actually interact with ideas and objects. The idea of "critical design" has always registered as more of an academic, gallery-based endeavor to me, while Julian's design fiction is trying to align itself as ... I don't know what the word would be.

MMF
Popular?

EB
Yes, more expressive and more popular and more out

there in the world. He's interested in feedback and how ideas come into contact with each other, and how that changes how you perceive them. I really appreciate the messiness that he was trying to push forward, and I think our exhibition is trying to represent this sense of a jumbled and wondrous present.

And then there is the book *Four Futures* [2016], which is interesting for both its argument and its format, which the author, Peter Frase, calls "social science fiction." His strategy involved identifying a technological issue, in this case automation, and running it through an x–y axis of sociopolitical realities, then creating four science-fiction texts illustrating the possible futures proposed by each quadrant. It's a very designerly approach to writing a text. And through this exercise he pushes against a purely techno-optimistic vision of the future, a techno-determinism that suggests that all human progress is based on our relationship to technology. People ask, What will the future look like after automation? As if automation is responsible for the future and is something that is happening to us, instead of by us. But he's saying, "No, this future is created through how we engage our complex web of sociopolitical realities."

ZR
Anthony Dunne and Fiona Raby recommended that book too.

MMF
It's saying, "Rise up, people, you have the tools in your hands! The robots are not going to take over!"

ZR
Absolutely! And that you do have agency. And don't believe there's a power that's totally out of your grasp—which I was not expecting from *Four Futures*.

MB
I think what emerged with some of the projects and texts that deal with technological futures is how technology has been a tool used and abused by power structures, but that same technology can also be a subversive tool.

EB
"Technology is neither good nor bad; nor is it neutral"—as Melvin Kranzberg said in his first law of technology.

MMF
For me, it's been the conversations I've had over the course of this project that have really stuck with me. Orkan, for example, has been such a generous and generative thinker. I also found Glenn Adamson really helpful, because he'd thought about histories of futures with his "show that got away," the one he didn't get to do at the V&A, a different version of which—curated by Mariana [Pestana], Rory [Hyde], and Zara [Arshad]—then became *The Future Starts Here* at that museum in 2018. We talked about how history was imperative for him in thinking about the future, and about the idea that questions, ideas, and issues come up again and again and again, and it's the technologies of their representation, display, or manifestation that are historically specific, and the tension between the two is really interesting.

In terms of other reference points, I read *Four Futures* too, and I was also looking at classics like Alvin Toffler's *Future Shock* [1970] and Donna Haraway's "Cyborg Manifesto" [1985]. And then I watched *Blade Runner 2049* [2017] a couple of times to find the correct timestamp for the showing of the holographic Joi in the exhibition [see p. 59], and it made me feel heart-full—it's magic, and so is the original. And there's also the beauty of Dennis Gassner's sets for *2049*. Film can take us into futures and has done so historically.

MB

One of the questions I've asked myself many times with this project is, What is our current image of the future? In the 1980s, the first *Blade Runner* [1982] was the image of the future. Now we have [the television series] *Black Mirror.* I am not sure this is an image of the future that can empower us all. So I was wondering what image of the future this exhibition wanted to present. The music video "Malamente" from the Spanish singer Rosalía captures an image of the future I am interested in discussing and that I identify with. It has an aesthetic that is a mixture of futuristic and apocalyptic while also being grounded in a here and now that has historically been overlooked. The video negotiates local and global identities and speaks to an intercultural way of understanding the future. The book *Mundos Alternos: Art and Science Fiction in the Americas* [2017] sticks with me for similar reasons.

I have also been very specifically interested in Sophia Al-Maria and the notion of Gulf Futurism [see p. 148]. I'm interested not only in her work but also in the capacity of naming and being able to talk about things by developing nomenclature. We need to find new ways of naming ourselves and the other to construct alternative realities and futures. In this regard, fiction is a tactical tool for dreaming about the possibility of other futures, and many readings have been fundamental companions for me on the journey of this exhibition, including *Black Quantum Futurism: Theory & Practice* [2005], edited by Rasheedah Phillips, and *La Mucama de Omicunlé* [*Tentacle*, 2015] by Rita Indiana. One challenge of the exhibition is the concept of "the real," or what reality is. I have found Slavoj Žižek's essays in *Interrogating the Real* [2005] and *Welcome to the Desert of the Real!* [2002] useful in this regard. With technology and the current crisis of ideology, "the real" is becoming organized in multiple hierarchies that are taken as indisputable. Navigating the real today is not obvious, and for good or bad all those layers that compound it are actively producing our futures.

ZR

The curator Paola Antonelli pointed me to Stephen Asma's book *The Evolution of Imagination* [2017]. Asma writes about how truth is culturally specific and culturally defined and can change over time, and about how the imagination also evolves, so our image of the future changes over time. Designers are very adept at helping us imagine new possibilities. As Asma writes, our daily lives are made up of what-ifs and maybes, so we're constantly tasked with imagining the present, the near future, and the far future all at once.

A conversation with the late curator and thinker Okwui Enwezor opened up other narratives for me—the essential idea of transnational, transcultural understandings of place. He led me back to a book I had read before, George Kubler's *The Shape of Time* [1962], which helped me understand the slippage, the fluid process, of time, and that ideas are part of a mixing or remixing, a fusion or continuum, of ideologies. They come together at certain moments, and there's a merging of ideas that then helps us create new narratives.

Okwui continually challenged us to not take things for granted, especially the dominant narratives of art and design history. I think one of the things that our exhibition does is break down disciplinary hierarchies. There were exhibitions in the past—like *Art in Daily Life*, a radical project curated by the designer Clara Porset [in Mexico City] in 1952—that didn't distinguish between art, design, and craft, but that represent very specific moments in history. I also think of *This Is Tomorrow*, from 1956 at Whitechapel Art Gallery in London, because it embodied the idea that architects, artists, and designers could work together to create projects that were thinking about modernism in a post–World War II context—without reaching a consensus but in fact encouraging friction and debate. That exhibition launched careers; it launched Pop Art. But the show was widely panned at the time and misunderstood, and it's only more recently that it's been seen as seminal, which is also interesting to think about, in terms of when ideas get taken up and legitimized as part of contemporary discourse. The show was revisited in 1990 [at the Institute of Contemporary Art in London], and in 2019, again at the Whitechapel Gallery, but as *Is This Tomorrow?* This is also interesting—now it's changed into a question.

These kinds of exhibitions are always fascinating to return to. The catalogue for the original *This Is Tomorrow* is amazing, but it's also a timepiece. We have continued in our project to understand that every aspect of an exhibition, every aspect of its design, including the graphics, the wayfinding, the spatialization, and every other aspect, is really important. Everything about this project has been so carefully thought out in terms of design. How we produce a catalogue, how we put together an exhibition, how we create public programming, speaks to the practice of making things ourselves, of being curators in this moment.

KBH

I'm glad we ended up joining forces for this journey and that we brought together the rest of the voices in this book, and will bring together those of the visitors who will join us across cities and through our programs. Because as we've said from the start, we are talking about futures *plural*—multiple, contingent, socially located, and yet to be formed. Onward. ⧉

THE EVOLUTION OF IMAGINATION

THE SHAPE OF TIME

EL ARTE EN LA VIDA DIARIA

this is tomorrow

whitechapel art gallery   aug. 9 - sept. 9 1956

18

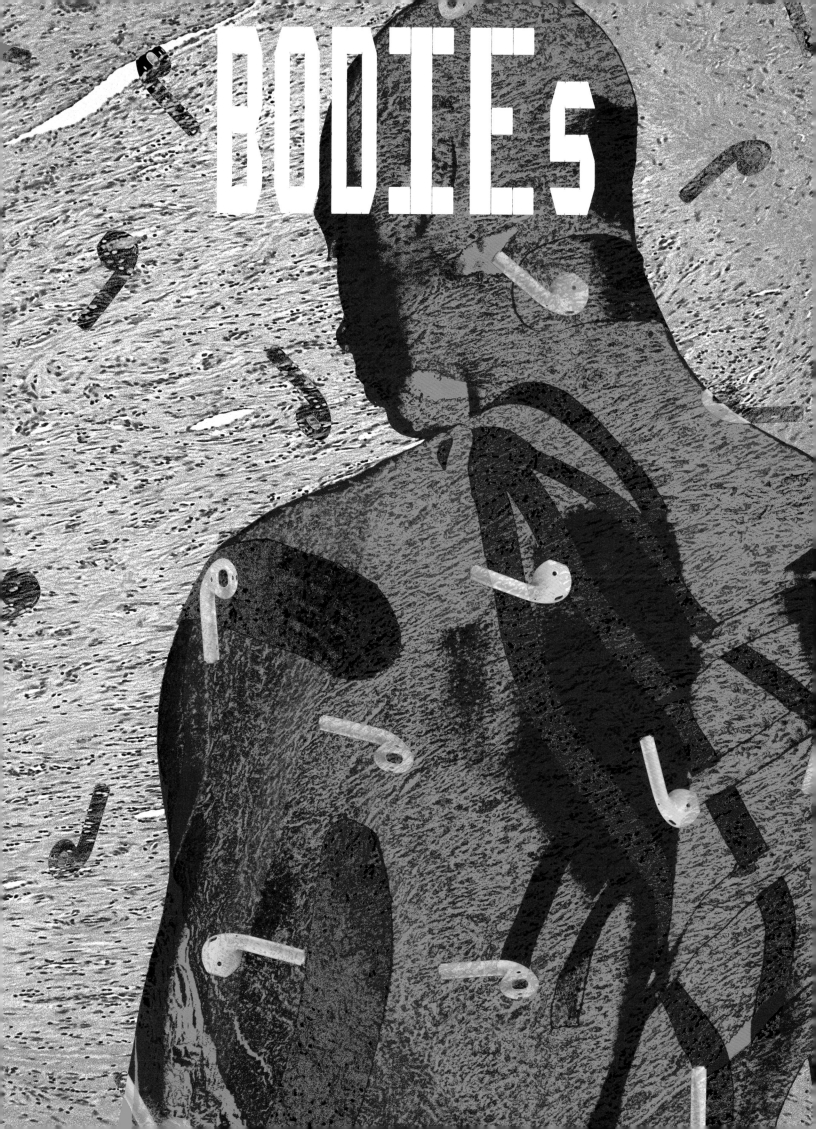

Accessible Worlds

JILLIAN MERCADO & AIMI HAMRAIE
in conversation with MICHELLE MILLAR
FISHER on mobility, visibility, and inclusion

This conversation took place by phone on December 12, 2018, with Aimi Hamraie at Vanderbilt University in Nashville, Jillian Mercado at home in Los Angeles, and Michelle Millar Fisher in her office at the Philadelphia Museum of Art.

MMF
For readers who aren't familiar with you, could you both state briefly what you do, and what matters to you about doing it?

AH
I'm a professor at Vanderbilt University. I study disability and the design of the built environment, and I wrote a book called *Building Access: Universal Design and the Politics of Disability* [2017]. This work is important to me because I think we have to understand how the world is made so we can change it.

JM
I'm a model and activist, and my job is really important because, growing up, I didn't see representation of people who looked like me, or of any disability whatsoever. So I try really hard to make sure that we are seen and we are heard, especially regarding disability rights.

MMF
Designs for Different Futures looks at the way design intersects with different possible futures, but before we look forward, can you talk a bit about how you see design today intersecting, or not, with bodies that fall outside of normative or idealized categories?

AH
Some of the things that I study as a historian of the built environment, including architecture and urban design, include simple architectural features like the prevalence of

stairs, which create barriers for wheelchair users. There are a lot of ways that ideologies are being encoded in built environments that tell us what kinds of bodies are valued or devalued. Right now I'm doing a project about how cities are trying to promote health through the built environment—by building more bicycle lanes, for example. But they're assuming that everyone can easily get around on a bicycle and has the energy to expend on that. There are a lot of examples of ways that the built environment assumes that able-bodied people are the norm.

MMF
Jillian, how about you?

JM
I think that there is a lot of assuming going on. I always say that you can't make a new law or regulation without having us [disabled people] in the conversation. There are a lot of people who are talking for us and not with us. We need to hire people who have disabilities to be part of the team when changes or regulations are being made. In my world, social media has been such a big help in allowing me to give a voice to a situation; I have a privilege and am grateful for that. I do think that slowly but surely our voices are being heard, but there are so many things that frustrate me, like transportation. I'm in California, and Uber just started putting accessible cars on the road. I guess people are now noticing that we also go out, and we also have lives, and it's not necessarily going to a hospital all the time.

MMF
When designers speculate on different futures, they often forget to think inclusively and holistically. You've just spoken about that in terms of decision-making processes related to design. Is there anything you want to add as to why this would be so?

Jillian Mercado on the cover of *Teen Vogue*, September 5, 2018. Photograph by Camila Falquez / *Teen Vogue*

AH

My colleague Alison Kafer wrote a book called *Feminist, Queer, Crip* [2013], and it's about the question of why, when we think about the future, we don't think about disability. In the book she analyzes some works of speculative fiction in which there is no disability or there are eugenicist principles like elimination of disability. Something she says that I think is really interesting is that when we only think about disability medically, then we always think that medicine is going to eliminate disabled people from the future because we have faith in cure and rehabilitation. If, instead, we listen to disabled people and think about the way we are surviving apocalypses that are happening in the present—and being very skillful and resourceful in inventing ways to overcome very difficult circumstances, like inaccessible built environments—then we have a different view of the future.

People are writing disability speculative fiction now in which they try to imagine future worlds that are accessible. I'm writing a paper right now about a novel that imagines a postapocalyptic city in which the first thing people do is retrofit all the houses to be accessible. It's individual imagination—but also the ideology that we take from medicine rather than from disability culture—that shapes the ways we think about the future as having disability or not having disability.

JM

I'm nodding at everything you're saying. I can't tell you how many times I've been in speaking-engagement situations where they've hired me a while before, and then when I get to the place, they're like, "Oh, sorry, there's no ramp." How does this slip their minds? So now when I'm hired to go somewhere and they ask, "Is there anything else you need?" I'll say, "Well, just make sure I can get in." Architecturally, this world was not made with us in mind, yet this is where we live. I'm always associated with a medical term rather than seen as just living my life, the same way somebody born with brunette hair or reading glasses is living theirs. And even though my chair is full-on future technology—it goes into a standing position, which is crazy—I'm a very simple person, and I just want to take a shower every single day, you know? There are a lot of places, even apartment buildings, that don't consider that at all. So it's really important to have conversations like this, and to have essays and books being written about these topics.

MMF

Designers are often very focused on prosthetic products that can be applied to the external body. Instead of redesigning the future *body*, how would you redesign future *societies* with disabled persons in mind?

AH

I have disabilities related to sensory processing, and I also get chronic migraines. Right now, when you go into a public space there's LED lighting everywhere, and it's presented as a way of conserving energy for the future. But that makes it impossible for me to go there, and it forces me to wear prosthetics. I have specific glasses that I have to wear when I'm around LED lighting, and I also have to wear special earplugs. This used to be an issue only in certain buildings, but now it's everywhere. So if I were designing a different future, I would think about ways of conserving energy that don't create that kind of sensory situation for myself and others like me. And lighting is only one source of energy consumption in the world. Industries also use tons of fossil fuels, but instead of forcing industries to change on a deep structural level, we're putting the onus on individual consumers of light bulbs and other products. So, for me,

an accessible world is one that shifts the burden off of disabled people and also asks what the user experience of all these new technologies is, and who are they potentially harming—and then finds different solutions, and on an appropriate scale. That's a very different way of thinking about accessibility that's not just about certain checklists that we can apply to the built environment.

JM
Many people have said to me, "Oh, but we have a ramp." Does that mean you're now accessible? No. The fact that you said LED lights hurt your eyes, Aimi—that indicates that obviously designers aren't considering making things accessible for everyone. A store, a building, whatever the built space, you want it to be accessible for literally every single person. So, as I said earlier, when building a public place, even a residential place, you can't do the project without having us in the group, in the whole planning process.

MMF
Are there any examples of designing for the futures of disability that are happening right now, from print and magazine culture to architecture and product and industrial design, that you think really hit the mark—that you're excited about?

JM
There's a prosthetic company, ALLELES Design Studio, that does amazing designs on prosthetic leg covers. They're just so beautiful. This is something that people don't think about. Equipment like canes or anything that's considered medical is always lacking in design.

MMF
How about you, Aimi?

AH
My colleague Sara Hendren is a designer and a design researcher. She does really amazing work that calls into question who we think of as a designer, and how we think about accessibility. She's worked directly with disabled people who engage in design practices in their own homes but are not usually recognized as doing design or engineering, and she catalogues their practices and their hacks to show that disabled people are often real experts on design. She has a project called Engineering at Home, among many others. There's also some really interesting disability-adjacent fashion design happening right now that I think is really cool. Sky Cubacub of Rebirth Garments creates conceptual fashion pieces that are handmade and bespoke for disabled and gender-nonconforming people. Alice Wong is an activist I really admire …

JM
Oh, I love her.

AH
She runs the Disability Visibility Project. She's created so many different ways for disabled people to communicate and talk about disability culture using fairly simple technologies like a Twitter hashtag.

JM
I also want to acknowledge Microsoft because they're doing so many great things, as far as gaming, for almost every single disability. There are so many designers who are doing specific lines for people who have disabilities. Two Blind Brothers did a whole T-shirt line focused on people who have visual impairments.

MMF
Jillian, could you say a few words about your cover for *Teen Vogue*? In the interview in that issue [September 5, 2018], you said, "I always knew that there was a hole in the fashion industry." How do you see disability represented in the field of fashion today, and what are the ways you hope it might evolve in the future?

JM
I just didn't see myself being represented at all, in any aspect of the industry, in a mass-media kind of way—not in magazines or on television, or honestly anywhere that the masses would be able to see. So I took it upon myself to really dig deep into the backstage of the fashion world—which for me was business because I studied marketing—and our mindset when it comes to disability. I wanted to represent what wasn't there.

MMF
This discussion we're having is predicated on our own social positions and experiences. I wondered if, through your work and research, you've observed different approaches to designing for disabled bodies in other places. For example, walking through the city of Tokyo recently, I didn't understand at first what the tactile paving was—that it was to help visually impaired people navigate the city. Are there other cities that have forward-thinking design plans?

AH
I think San Francisco and Denver—places where the disability rights movement has a really strong history—are important. And I would also say that the internet is an important place. Since the invention of the internet there has been an explosion of the disability community in ways that wouldn't have happened otherwise because people couldn't get in contact with each other.

JM
I've had so many people come up to me and say that they don't see a lot of people with disabilities "out" like I am. So with the internet, the community of people with disabilities has exploded because we've gotten tired of not enjoying what this earth and this world have to give us. With us talking online and with social media, I think that people are now really hearing us and seeing that we exist. We do. We should have the same basic rights, and we're not asking for too much. We just want to live as comfortably as everybody else.

There's a wheelchair company called Scewo making chairs that will potentially help to go up steps. Stairs are kryptonite for me. And so the future is slowly but surely coming up with different things to make our lives a little bit better. We need to be the ones who are on top of it to make sure that it's not a trend, but a way of life. ⌗

A stair-climbing wheelchair designed by Scewo, Winterthur, Switzerland

PHOENIX EXOSKELETON   2011–17
HOMAYOON KAZEROONI, SUITX, EMERYVILLE, CA

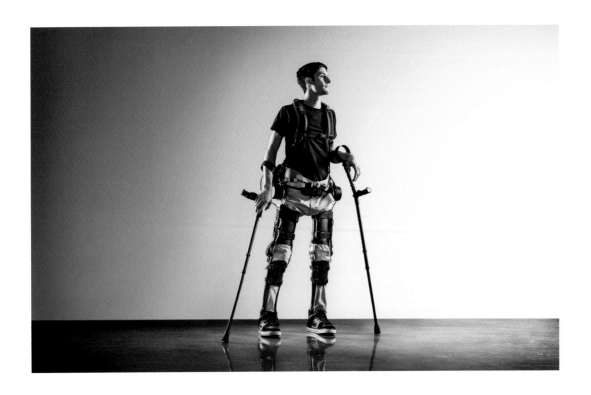

Over more than three decades, the Iranian-born mechanical engineer Homayoon Kazerooni and his team of designers and engineers at the Berkeley Robotics and Human Engineering Laboratory at the University of California, Berkeley, have pioneered and developed future-focused robotic exoskeletons that augment human performance. Their exoskeletons allow users to support heavy loads for extended periods of time without injury and enable people with mobility disorders to walk upright and independently. Kazerooni's research is socially driven: "My job is creating technology to make life easier and to create a better quality of life," he has said. "We have to take care of those around us."[1]

The Phoenix, developed by suitX, the company Kazerooni founded in 2011, is a lightweight outer frame that provides support as it robotically simulates and enhances body movement. It allows wearers to move their hips and knees with the help of two small motors located at the hips. An on-board embedded computer, located in the back of the Phoenix, controls the motors to create locomotion. The stiffness of each knee joint is controlled to provide support during stance and fluid motion during swing. A user interface attached to the crutch and various sensors integrated into the suit guide the exoskeleton through various operational modes. The system is powered by a battery pack worn like a backpack. Originally designed for patients with spinal-cord injuries, the Phoenix also helps patients who have

suffered strokes or sustained other motor injuries regain the ability to walk.

The modern history of powered exoskeletons that integrate human movement with machine systems began in the 1960s, with the earliest exoskeletons designed not for medical but for military uses. Around the same time, powered armor began to appear in science fiction, including in Robert Heinlein's *Starship Troopers* (1959) and Marvel Comics' *Iron Man* (1963–). **KBH**

1   John Hitch, "The Mechanical Miracle Worker," *New Equipment Digest*, May 8, 2017, www .newequipment.com.

UNYQ 3D-PRINTED SCOLIOSIS WEARABLE   2018
STUDIO BITONTI, NEW YORK, AND UNYQ, CHARLOTTE, NC

Designed by Studio Bitonti in collaboration with the 3D-printing manufacturer UNYQ and technology innovator Intel, the 3D-printed scoliosis wearable demonstrates the rapidly increasing integration of emerging technologies in health-care treatments. Scoliosis is a sideways curvature of the spine that typically appears in children at puberty; depending on the severity of the individual case, the condition can be treated with a back brace, which if properly fitted and worn over many hours as instructed can halt progression of the curve. The UNYQ wearable is custom designed for each patient, the topography of the spine first captured in scans, then digitally rendered as a 3D computer model, and finally translated by a 3D printer that manufactures the brace. The brace is embedded with tiny data-capturing computer modules developed by Intel, allowing users to monitor their progress via an app.

In designing the wearable, Studio Bitonti considered aesthetics and social impact as well as function. While scoliosis occurs equally among genders in ten- to fifteen-year-olds, young girls are statistically more likely to require treatment. This demographic can be particularly self-conscious about appearances and more resistant to wearing a bulky and unattractive brace for the recommended hours. The wearable is designed as a stylish open-work lattice available in different geometric patterns. It is thinner and lighter (it uses much less material), cooler to wear because of the open areas (which also add a decorative element), and more maneuverable than older models, all features that the designers hope will make users more willing participants in their own treatment. KBH

ALLEVI 1 3D BIOPRINTER   2018
ALLEVI, PHILADELPHIA

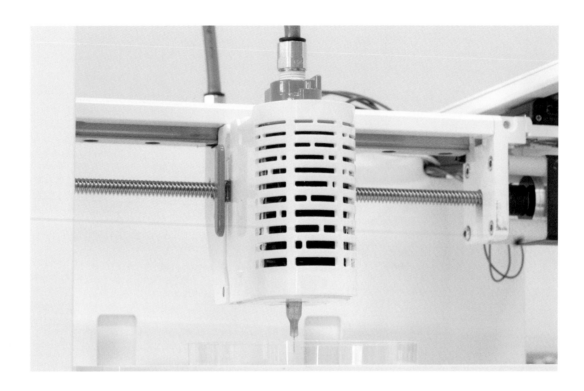

Allevi, its name a shortened form of the word *alleviate*—which is what the company hopes its design will do for a range of diseases and complex medical issues—produces bioprinters that allow medical and scientific lab researchers to print and test living human cells in myriad configurations. Founded as BioBots in 2014 by two University of Pennsylvania graduates, Allevi designs hardware (now six generations of 3D bioprinters and counting, including a zero-gravity model for use in space) and operational software, usable on desktop computers, that offer a direct path to "create living things from scratch" using bioinks.[1] Bioinks are made of human cells, sometimes of multiple types, within a substrate that allows them to grow, multiply, and take on different structures and functions, including as blood clots, bone tissue, and synthetic body parts such as ears, noses, and, hopefully one day, organs.

With Allevi bioprinters, researchers can test various scenarios outside the human body, an attractive option for many reasons, including the bypassing of human or animal trials to determine optimal dosages and levels of toxicity. They also offer a pathway to highly personalized medicine, using an individual's own cells. For example, instead of an oncologist trying different chemotherapy cocktails on a patient until the best treatment is discovered, a bioprinter like Allevi's can print numerous miniature tumors generated from a single biopsy, which can all be tested simultaneously, each with a different chemotherapy, speeding up the process and optimizing treatment before it ever reaches the patient.[2]

The Allevi 1 bioprinter measures a compact 30.5 centimeters (12 inches) on its longest side, with the printer head on clear display as it functions and moves, much like the many desktop 3D printers that have become an important part of contemporary design practice in the last decade. This is a deliberate design feature that posits the Allevi 1 bioprinter as an accessible lab bench tool whose easily legible parts and straightforward software allow researchers to use it without having specialized technical knowledge. The desire for accessibility is also reflected in its relatively modest $10,000 price. In countries like the United States, where Big Pharma is a multibillion-dollar industry but access to health care is not universal, with only the wealthiest benefiting from its upper reaches, an accessible tool like an Allevi bioprinter could potentially disrupt this imbalance—though the ethical ramifications of widely available nanobiological experimentation are yet to be fully explored. **MMF**

1   See Kevin J. Ryan, "This Startup Lost Its Co-Founder and CEO, but It Still Wants to 3-D Print You a Heart Someday," *Inc.*, December 5, 2017, www.inc.com.

2   Earlier projects like Organs-on-Chips (2008), designed at Harvard's Wyss Institute for Biologically Inspired Engineering, foreshadowed how lab tests using biopsied cells and mimicking human physiology might advance medical treatment.

CIRCUMVENTIVE ORGANS   2013
AGI HAINES

The designer Agi Haines's speculative project *Circumventive Organs* explores the possibility of creating customized artificial organs through bioprinting. As surgical implants, these organs could bolster the capacity of a patient's body to heal itself, reducing the need for intensive medical interventions. Haines proposes a trio of composite organs that each borrow biological functions from elsewhere in the animal kingdom, moving away from the metaphorical design approach of biomimicry to a more direct method of hybridization. The concept builds on a long trajectory of genetic manipulation, consciously evoking ongoing debates about the ethics of altering human biology, the motivations and commodification of biomedical research, and the blurry line between the natural and the artificial.

Each of Haines's hypothetical organs has a scientific name that underscores its intended purpose, placing these novel biological designs within established medical frameworks. *Electrostabilis Cardium* acts as a defibrillator, using the electric eel's current-generating capacity to stave off heart attacks at the moment they begin; *Cerebrothrombal Dilutus* adapts the saliva gland of the leech (with its anticoagulant properties) to prevent stroke-causing blood clots in the brain; and *Tremomucosa Expulsum* borrows muscular structures from the rattlesnake to expel the dangerous buildup of mucus in the respiratory system caused by cystic fibrosis, an inherited condition.

Haines gave form to the *Circumventive Organs* through drawings (in the style of anatomical illustrations), physical models, and a short film that depicts the surgical implantation of *Electrostabilis Cardium* into a patient's chest cavity. Through the material assemblage of the operation room—forceps, cotton swabs, sterilized trays, gloved hands, masked faces—the object is transformed from fictive to highly plausible for the viewer. This sort of scene-setting stagecraft is an evocative tool for the speculative designer, particularly for proposed designs that would otherwise be invisible beneath the skin.

Haines's design strategy relies not on the slick, high-tech renderings that have come to characterize much of what we consider "futuristic," but rather comes to life in the dark, messy interiors of our own bodies. Cultivating reactions of disgust or repulsion—in contrast to more positivistic, marketing-driven images of the future—fosters a useful sense of realism for the visceral questions of designing with and for the human body. As Haines states, "I felt people may believe in [the *Circumventive Organs*] more as objects if they look more disgusting like the weird and wonderful things designed by nature that already exist inside us."[1] CF

1   Agi Haines, "Circumventive Hybrid Organs," interview by Régine Debatty, *We Make Money Not Art* (blog), June 27, 2013, we-make-money-not-art.com/circumventive_organs/.

SEATED DESIGN   2016   LUCY JONES   /
ESSENTIAL SUITE FOR WHEELCHAIRS   2018   LUCY JONES AND
JOONAS KYÖSTILÄ, FFORA, BROOKLYN

Trained in fashion design at the prestigious Parsons School of Design, part of the New School in New York, the Welsh designer Lucy Jones focuses on a population largely ignored by her industry: disabled people. Inspired by a cousin who uses a wheelchair, Jones created a mission statement for her 2015 thesis project, Seated Design. She proposed that the garment industry think outside the rigid parameters of the idealized—and usually upright—body and instead design for people who spend the greater part of their lives sitting down, which includes many of us, of all body types and abilities.

Since establishing her own design studio after graduating from Parsons, Jones has created a series of projects that marry the provocatively speculative with the very practical. Seated Design was based on interviews with and close observation and analysis of wheelchair users. This demonstrated how clothing requires different measurements and affordances for seated bodies, especially at points

of flex and tension such as elbows and knees, which assume various shapes based on resting and active positions, individual body metrics, and different disabilities. Her close work with disabled users has informed all aspects of her subsequent designs, including clothing prototypes like full shirts that developed from experiments with sleeves. The end products display dynamic, architectural, fashion-forward silhouettes while providing greater flexibility and range of movement for their wearers.

Jones's most recent work is represented by the Essential Suite for Wheelchairs, a series of elegant and economical accessories—including a cup holder and a leather purse—created in tandem with XRC Labs in New York. Each design in the suite weighs aesthetics and ease of use equally, taking inspiration from lifestyle companies like Apple and Muji but foregrounding the experiences and needs of disabled people, who are often left out when it comes to beautiful design. Jones believes that "designers should be held accountable for the products

they place into the world, to think critically at all stages of design about the positive and negative impacts of their creations, socially and environmentally."[1] The rest of the world is taking notice: Jones received Parsons' Womenswear Designer of the Year Class of 2015 award,[2] and made it into the Forbes "30 under 30," class of 2016.[3] **MMF**

1   Lucy Jones website home page, www .lucyjonesdesign.com.

2   See The New School News, May 28, 2015, blogs.newschool.edu/news.

3   Caroline Howard with Emily Inverso, "30 under 30: Class of 2016," Forbes, www.forbes .com/30-under-30-2016.

STANCE   2016   LESLIE SPEER, ANTHONY TA, BRENDAN NGO, AND DARREN MANUEL, SIMPLE LIMB INITIATIVE, SPARKS GLENCOE, MD

Gaining access to prosthetics that are low cost, durable, and easy to produce and repair is a challenge worldwide, and is especially critical for children and teens, who quickly outgrow expensive, customized devices. Stance is a prototype for a lower-limb prosthesis that adapts to the growth of its user without compromising function. New measuring and fitting processes employed in its manufacture allow amputees to walk with a more natural gait. These new fitting processes in turn inspired the redesign of other components, including socket, knee, foot, and connector parts. The prototype design incorporates 3D-printing technologies for mold-making and fabrication, and sources materials from regional manufacturers, with the result that parts for the Stance can be customized and assembled locally for less than forty dollars. Each of these design elements works toward the project goal of developing modular parts for prosthetics that can be upgraded as children grow into adulthood, at an affordable cost and using regional production and distribution processes.

The Stance prototype is a product of the Simple Limb Initiative, "an open-source collaboration resource for designing, developing, and producing low cost prosthetics for amputees around the world."[1] Led by the industrial designer Leslie Speer, the Simple Limb Initiative began as a project for her students at San Jose State University. The initiative has since worked to provide access to affordable prosthetics in Cambodia and currently partners with the Mahavir Kmina Artificial Limb Center in Colombia, one of the countries most affected by mines, according to the International Campaign to Ban Landmines.[2] As part of the project's open-source platform, the initiative invites prosthetists, doctors, designers, and amputees to examine, learn about, test, and tweak the available designs or prototypes. Feedback from participants is then evaluated in the ongoing research and development of the designs. Originating in a prompt for design students, the Simple Limb Initiative prioritizes access to quality design for all. The project thus deploys design as a vehicle for change and a means of fashioning a more equitable future. **JRB**

1   See the Simple Limb Initiative website, www.lesliespeer.com/simple-limb.

2   See Rachel Nuwer, "The Daunting, Dangerous Task of Unearthing Colombia's Landmines," NOVA / PBS, July 16, 2018, www.pbs.org/wgbh/nova.

RECYCLABLE AND REHEALABLE ELECTRONIC SKIN   2018   JIANLIANG XIAO AND WEI ZHANG, UNIVERSITY OF COLORADO BOULDER

In an effort to make environmentally friendly electronic devices, a group of researchers at the University of Colorado Boulder (CU Boulder)—led by Jianliang Xiao of the Department of Mechanical Engineering and Wei Zhang of the Department of Chemistry and Biochemistry—developed a fully recyclable, self-healing electronic skin. Electronic skin, commonly known as e-skin, mimics human skin's physical properties and its ability to sense pressure, temperature, humidity, and air flow. These characteristics could theoretically allow robots or prosthetics that have been embedded with e-skin to better respond to the immediate environment, paving the way for more nuanced and safer interactions. As Zhang describes, "Real skin can prevent people getting burned [and] can prevent people getting hurt. E-skin can basically mimic those [preventative] functions."[1] Xiao uses the example of a babysitting robot that can act as a thermometer to underscore the importance of deliberate touch, stating, "Sensing is critical because … we want to make sure that robots don't hurt people. When the baby is sick, the robot can just use a finger to touch the surface. … It can tell what the temperature of the baby is."[2]

While wearable e-skins are in development in labs around the world, the version developed at CU Boulder has several distinct properties. Its thin, translucent material is created from a type of bonded polymer known as polyimine. This polymer is laced with silver nanoparticles to provide better mechanical strength, chemical stability, and electrical conductivity. Moreover, the e-skin is self-healing and fully recyclable at room temperature. Simply adding three compounds that are found in ethanol and commercially available results in a chemical reaction that can mend cut or broken e-skin. E-skin that is damaged beyond repair can be soaked in a solution that separates out the silver nanoparticles, and the materials can then be used to make new e-skin. Given the millions of tons of electronic waste generated worldwide every year, the e-skin's recyclability makes both economic and environmental sense, and its designers hope the method can be used to create other recyclable electronics in the future.   JRB

1   Quoted in Sydney Pereira, "Self-Healing Electronic 'Skin' Lets Amputees Sense Temperature and Pressure on Prosthetic Limbs," *Newsweek*, February 10, 2018, www.newsweek.com; brackets in the original.

2   Quoted in Pereira, "Self-Healing Electronic 'Skin.'"

ALTEREGO   2018   ARNAV KAPUR

At a moment when rapidly improving AI is becoming a source of anxiety, Arnav Kapur at MIT's Media Lab developed the AlterEgo to demonstrate "how AI can help augment rather than replace us" (he calls it an IA, or intelligence-augmentation, device).[1] Essentially, the AlterEgo takes the voice-command functionality of Amazon's Alexa and Apple's Siri and internalizes it. The noninvasive wearable employs a silent speech system that enables quiet communication between people and computers. Users can also control the AlterEgo without looking at a screen or typing in commands on a separate device.

Silent speaking is the conscious effort to say a word without actually vocalizing it, using subtle movements of the internal speech organs. For example, if you're in a grocery store and want to calculate how much you're spending, you can just "say" each price, and the AlterEgo will respond with a total. Or you can "say" commands like *right* or *up* to scroll through options on a TV screen without the use of a remote. The system is made up of a computer program connected by Bluetooth to a device that loops around the user's ear, follows the jawline, and attaches under the mouth. The AlterEgo reads the electrical impulses produced by small movements of the lower face and neck muscles when you vocalize words or phrases internally. The device is able to comprehend this silent speech through a machine-learning system trained to associate certain signals with specific words. The system currently understands about a hundred words, and researchers continue to expand its vocabulary.

In the realm of wearable technologies, including the now-discontinued Google Glass, the AlterEgo sets itself apart by allowing users to remain present, since the system doesn't disrupt conversations or other activities. If someone uses a word you don't know in a meeting, for instance, you can silently ask the system for a definition while remaining part of the conversation. Or, when you've run into someone you've met before but you've forgotten their name, the system can silently consult your address book to help you out. Beyond these personal uses, the AlterEgo has the potential to be used to communicate in noisy settings and in stealth military operations, and as an aid for communities otherwise unable to hear or vocalize. **JRB**

1   See Rachel Metz, "Say Goodbye to Alexa and Hello to Gadgets Listening to the Voice Inside Your Head," *MIT Technology Review*, April 27, 2018, www.technologyreview.com; and Larry Hardesty, "Computer System Transcribes Words Users 'Speak Silently,'" *MIT News*, April 4, 2018, news.mit.edu.

CV DAZZLE: CAMOUFLAGE FROM
FACE DETECTION   2017   ADAM HARVEY

In "The Hood Maker," a story by the American science-fiction writer Philip K. Dick (first published in 1955), the title character secretly makes and distributes telepathy-blocking hoods, frustrating the government's use of mutant mind-reading humans to find and subjugate political dissidents. Like Dick, Adam Harvey, an artist and designer based in New York and Berlin, is concerned with issues of privacy and government. In response he created Stealth Wear (2012), a hood and other garments that enable the wearer to avert overhead surveillance. The Stealth Wear collection is fabricated with a silver-plated synthetic fabric that "reflects and diffuses thermal radiation," the designer explains, "which reduces the wearer's thermal signature under observation by a long wave infrared camera"—such as those carried by drones.[1] The privacy-ensuring clothing is also a fashion statement, its hood inspired by the traditional Islamic hijab.

Harvey further explores the aesthetics of privacy in CV Dazzle, his highly stylized face makeup and hair styling, which camouflages and protects the wearer against facial-detection and recognition systems. CV (computer vision) face-detection technology relies on mathematical pattern-recognition techniques that identify the relationships of key facial features. With their bold, countershading colors and abstract, intersecting patterns, CV Dazzle makeup and hairstyles interfere with the visual continuity of the face and head— their shape, textures, symmetry, tones, and contours—thereby disrupting facial-detection algorithms. The name Dazzle derives from ship camouflage used during World War I (also known as Razzle Dazzle), when warships were painted with unique, complex abstract patterns to make them difficult to target.[2] **KBH**

1   "Stealth Wear: Anti-Drone Fashion," AH [Adam Harvey] Projects, posted December 3, 2012, ahprojects.com/stealth-wear/.

2   See Linda Rodriguez McRobbie, "When the British Wanted to Camouflage Their Warships, They Made Them Dazzle," *Smithsonian Magazine*, April 7, 2016, www.smithsonianmag.com.

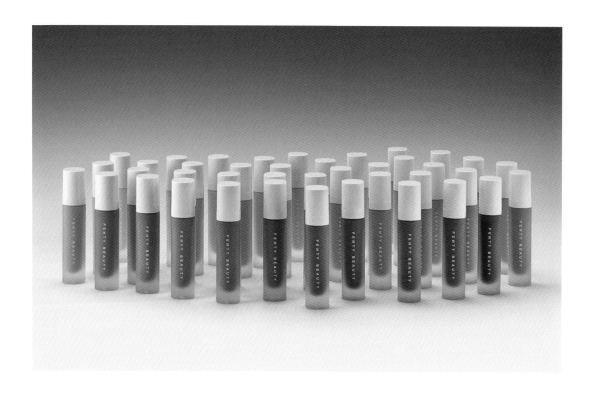

For as long as they have existed, products like "nude" tights, makeup foundations, and underwear, among other examples in the skin- or flesh-colored garments and accessories pantheon, have stood in for a light Caucasian skin tone. Enter Robyn Rihanna Fenty and her vision for a more inclusive future for beauty. The pop icon's Fenty Beauty line, launched in 2017 in fifteen countries, grossed over $72 million in its first four weeks. One of the products in this line, Fenty Matte Pro Filt'r Longwear Foundation, has been particularly lauded by industry insiders and customers alike for offering a groundbreaking forty different pigment tones.

By offering so many options, Fenty gives users of the foundation the ability to match their individual skin tones much more precisely. The foundation's color range has since been expanded by an additional ten shades, for a total of fifty choices. Consumers can pick from a multitude of colors that, rather than being euphemistically named, are numbered, described with words such as *warm* or *rich*, and include a skin undertone (shade 130, for example, is "for fair skin with warm olive undertones").[1] This linguistic choice is an additional way to highlight that no skin is any one color but instead is determined—literally but also culturally and genetically—by a range of factors.

Makeup has long had a problem with whitewashing, often promoting beauty ideals represented by Caucasian faces. From creams that promise to lighten skin color to sunscreen and suntan lotions that cater primarily to pale skin, the products of the global beauty industry have often failed to address the multivalency of the melanin content of their potential customers. As the fashion historian Stephanie Kramer has noted, "beauty is a form of currency" that developed into the makeup we know today with the advent of companies like Helena Rubinstein and Elizabeth Arden in the early twentieth century. These companies "promoted makeup as a tool of empowerment … integral to the presentation of a public persona."[2] In 2017 Fenty set a benchmark and sent a clear message that the cosmetics of the future should be for people of all skin hues. **MMF**

1   See the Foundation Shade Finder on the Fenty website, www.fentybeauty.com/shade-finder.html.

2   Stephanie Kramer, "YSL Touche Éclat," in *Items: Is Fashion Modern?*, ed. Paola Antonelli and Michelle Millar Fisher (New York: Museum of Modern Art, 2017), 271.

CHEMICAL MICROLAB FOR THE SKIN   2018
JOHN A. ROGERS, CENTER FOR BIO-INTEGRATED ELECTRONICS,
NORTHWESTERN UNIVERSITY, EVANSTON, IL

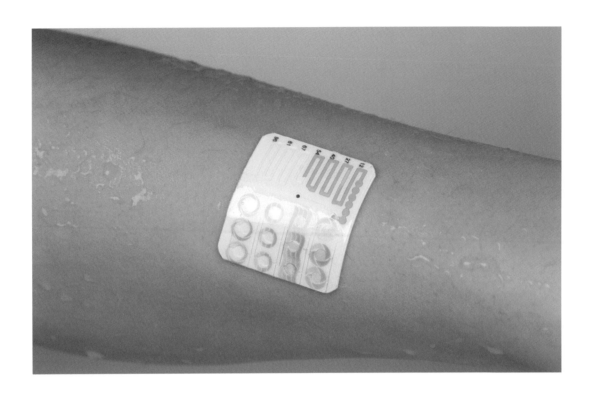

As colorful and flexible as a temporary tattoo, the Chemical Microlab for the Skin represents the next generation of wearable technology. While popular watch-like gadgets such as the FitBit track steps taken, heart rate, and quality of sleep, this first-of-its-kind bio-integrated device adheres directly to the skin and uses the wearer's sweat to evaluate the body's responses to exercise. Designed and manufactured by the lab headed by John A. Rogers at Northwestern University, the simple, low-cost tool analyzes key biomarkers from sweat—including pH level and concentrations of lactate, chloride, and glucose—using techniques from bio-integrated engineering and microsystems technology. The noninvasive microlab, which is roughly the size and shape of a quarter,

communicates and connects wirelessly to smartphones, allowing the wearer to immediately correct imbalances by rehydrating or replenishing electrolytes. Rogers and his team reformulated computer-chip and microfluidic technology into a skin-like form that offers natural integration with the body. The design posed challenges, as people are less likely to wear something that looks like medical equipment. The result was a comfortable, non-irritating device that blurs the interface between technology and biology.

Several professional sports teams, doctors, and medically oriented organizations, such as the National Kidney Foundation, have used the microlab, and L'Oreal and Gatorade are developing product deals that would bring

the technology to a broader public. The Rogers team continues to enhance the microlab's capabilities. For example, other functions may be added that enable the analysis of body temperature, skin hydration, and blood flow. Rogers sees opportunities for a more dramatic blurring between human-made devices and the body in the future, with the possibility of integrating into humans electronics that monitor the body's chemistry and regulate its health. He imagines these systems operating in a closed feedback loop that would coordinate with natural body processes, sensing and delivering therapies as needed.   JRB

LIA PREGNANCY TEST   2018
BETHANY EDWARDS AND ANNA COUTURIER-SIMPSON, LIA DIAGNOSTICS, PHILADELPHIA

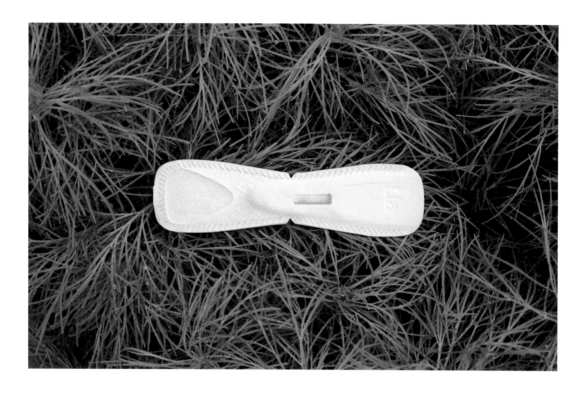

In the late 1960s, Margaret M. (Meg) Crane, a young graphic designer working at the global pharmaceutical company Organon, wondered why the tests she saw being used in the labs to determine pregnancy were not available for women's use at home. She applied her skills in packaging design to create a prototype for a home pregnancy test, which was patented in 1969 and brought to market in Canada and the United Kingdom in 1971, with the product soon available in many other countries. Crane's sleek design included a clear plastic box to house the mirror, pipette, and liquids needed to carry out the test.[1]

Today, the Philadelphia-based start-up LIA Diagnostics is rethinking the home pregnancy test as a plastic-free and über-discreet experience that can be flushed away shortly after use, reducing the impact on the environment and increasing privacy. The company name is a play on the scientific term *lateral immunoassay*, a type of rapidly readable fiber test strip integral to a pregnancy test. Bethany Edwards and Anna Couturier-Simpson, former classmates in the University of Pennsylvania's integrated product design master's program, which

combines design, engineering, and business education, cofounded Lia in 2015, along with their faculty adviser, Sarah Rottenberg.

For their new product (called Lia: The Flushable and Biodegradable Pregnancy Test), they developed a biodegradable paper that remains durable when in contact with urine—long enough for a test result to register—yet breaks down almost immediately when flushed. It is lighter than six sheets of two-ply toilet paper, and its discreetly branded packet is a response to user research, which confirmed that pregnancy testing is a deeply private experience. Approved for commercial sale by the US Food and Drug Administration in December 2017, the test is 99 percent accurate and will cost about the same as comparable plastic-based products—but will not end up in landfills. LIA received $2.6 million in investment funding in early November 2018, making them a new leader in the FemTech field, where the market for over-the-counter home pregnancy tests alone is set to top $1 billion by 2020.

Since Meg Crane's 1960s design, the form and aesthetics of pregnancy tests have rarely

been the focus of true innovation, a common story in the category of women's reproductive health. Yet young designers are increasingly directing their energies into this field (see, for example, the Yona speculum),[2] challenging both social taboos and the historically gendered nature of the design world. In Crane's time, her male colleagues suggested adding bows and tassels to her elegant, clean design prototype in order to make it more "feminine" (a suggestion she resisted), but today, as LIA demonstrates, important steps are being taken to bring designs for women's health into the future. **MMF**

1   For Meg Crane and her home pregnancy test, see "Meg Crane, Inventor," *Year of Women in History* (blog), April 30, 2016, yearofwomenshistory.blogspot.com.

2   Yona, a team of designers and engineers from Frog Design, uses the motto "Health for People with Vaginas"; see their home page, yonacare.com.

AMPLIDXRX KIT   2018   JOSÉ GÓMEZ-MÁRQUEZ,
MIT LITTLE DEVICES LAB, CAMBRIDGE, MA

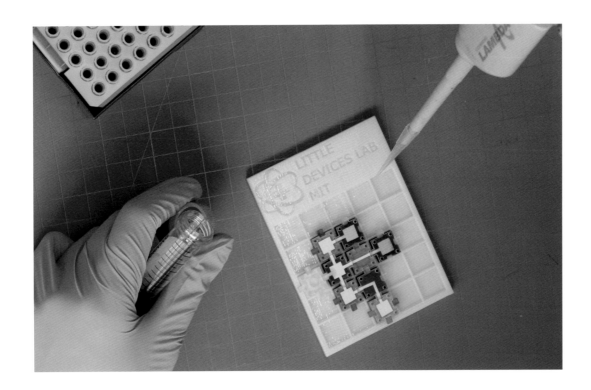

The AmpliDxRx Kit addresses the need for accessible health-care technologies in low- and middle-income countries where medical equipment largely comes from the developed world and is quickly rendered useless due to a lack of replacement parts and people trained to repair it. Developed by the Honduran-born engineer and designer José Gómez-Márquez and his team at MIT's Little Devices Lab,[1] the kit is available in different versions that enable health-care workers and even patients to devise drug-delivery, diagnostic, and prosthetic devices made from affordable components that require little expertise to assemble. Instead of delivering drugs by injection with a needle and syringe, for example,

which involves a trained health-care worker, the kit provides components for administering vaccines through aerosol mists, via individual, preloaded nebulizers powered in various ways that don't require electricity—for example, by a bicycle pump—and that can be used once and thrown away.

The kits also include a simple, portable test for diagnosing certain infectious diseases like the Ebola virus using a test strip made from layers of coffee filters saturated with chemicals that react to viral agents. The prosthetics kit, designed to serve amputee farmers and agricultural workers in remote locations, includes an attachable basket, blade, grabber, string, and scissors.

Gómez-Márquez grew up in Honduras in a medical family. He was born with numerous health issues that gave him an early understanding of the importance of health care, how capricious access to it could be, and the contribution that low-cost medical devices—designed with DIY "workaround" strategies that encourage health workers to fashion their own solutions—could make in improving medical services. **KBH**

1   The project research and design team included Kimberly Hamad-Schifferli, Jierui Fang, Briana Lino, Jonah Butler, Anna Young, Nikolas Albarran, and Maia Mesyngier.

DIY BACTERIAL GENE ENGINEERING CRISPR KIT   2016
JOSIAH ZAYNER, THE ODIN, OAKLAND, CA

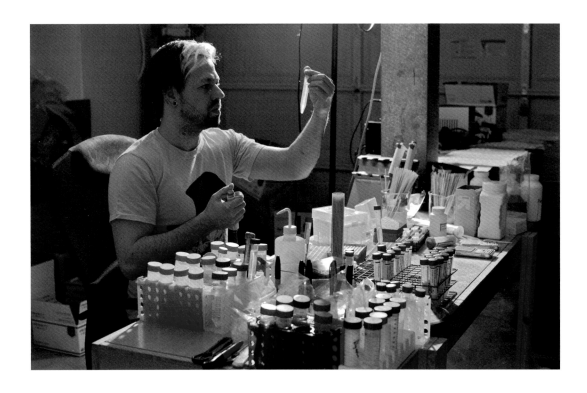

DNA, or deoxyribonucleic acid, is the basic design template for life itself, a self-replicating building block that carries genetic information within chromosomes and makes up the matter that constitutes nearly all living things on our planet. In the last fifty years or so, scientists have investigated this centrally important biological material, determining its basic structure in 1962 and its sequence in 2001. And where scientists have pioneered, others—including designers—have followed, eager to manipulate, engineer, and reimagine these forms.

Financed three times over its crowd-funding goal in 2015, this DIY CRISPR kit is produced and sold by the ODIN (molecular biophysicist Josiah Zayner) for the relatively accessible sum of $160, putting DNA design tools directly into the hands of consumers around the world.[1] The company was "started on the premise that if the Scientific population of the world doubled or tripled, it would change everything"[2]—and the comments on the kit's product page reveal purchasers ranging from amateur synthetic biologists to teachers buying it for their students. Genetic engineering deliberately changes an organism's DNA for a particular purpose or preferred characteristic.

This kit's experiment (backed up with written instructions and video tutorials) allows users to modify the DNA of an E. coli strain so that it can survive and thrive in conditions that it would usually find adverse.

CRISPR, an acronym for *clustered regularly interspaced short palindromic repeats*, are segments of DNA containing short, repetitive base sequences that are identical in both directions (hence the *P*, for *palindromic*). Cas9 (CRISPR-associated 9) is a protein that can work in tandem with CRISPR sequences to recognize specific DNA strands and "cleave" or break them apart at the molecular level. Think of CRISPR and Cas9 as tools—a flashlight and saw, if you will—that allow precision surgery to take place on parts of life far too small to be seen with the naked eye. For context, the genome of the E. coli strain found in this kit is over four million DNA bases in size; to pinpoint one is like finding a needle in a haystack, but the CRISPR sequence deployed will locate the single base that needs to be mutated.

This kit is part of a growing field known as synthetic biology. Zayner was a member of NASA's synthetic biology team before founding

the ODIN, but he is also a biohacker, performing "experiments, often on the self, that take place outside of traditional lab spaces."[3] He sees his work as a form of activism that puts the possibility of scientific advancement into more hands, speeding up discovery of new drugs and other life-enhancing products. Theoretically, this technology could be scaled and applied to reengineer our own bodies—for bigger muscles, to combat allergies and food intolerances, or to cure diseases like cancer and HIV. As with any design, however, it could also be wielded for more nefarious purposes—for example, as a personalized bioweapon. MMF

1   "The ODIN: DIY CRISPR Kits, Learn Modern Science by Doing," Indiegogo, funding closed December 8, 2015, www.indiegogo.com.

2   "The ODIN: DIY CRISPR Kits."

3   Sarah Zhang, "A Biohacker Regrets Publicly Injecting Himself with CRISPR," *Atlantic*, February 20, 2018, www.theatlantic.com.

MINION PORTABLE GENE SEQUENCER
2015   OXFORD NANOPORE
TECHNOLOGIES, OXFORD, ENGLAND

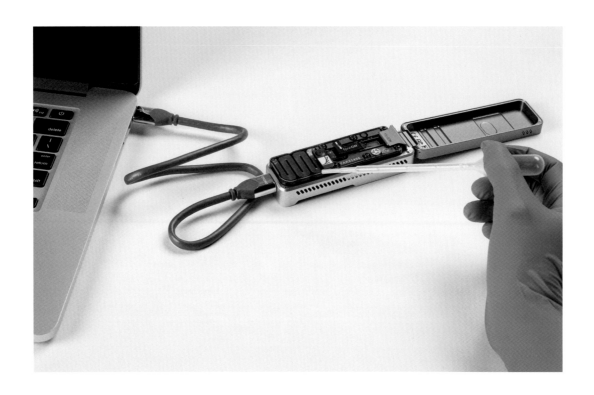

Only slightly larger than the average flash drive, the portable, lightweight MinION plugs into any computer via a USB cable and, when synced with its software, deduces the structures, or "sequences," of DNA (deoxyribonucleic acid) and RNA (ribonucleic acid) samples in real time. DNA is the genomic architecture that stores and transfers genetic information in all living organisms, while RNA is involved in synthesizing proteins. The MinION allows its user to take cells from, say, a cheek swab, organic plant matter, or saliva left on a cigarette butt and analyze them without requiring a lab or highly specialized scientific training or knowledge. Made by Oxford Nanopore Technologies, whose stated goal is "biology for anyone, anywhere," the MinION "has been used up a mountain, in a jungle, in the arctic and on the International Space Station."[1] At a cost of $1,000, it is also not prohibitively expensive.

Based on the work of multiple researchers, the MinION was launched to a limited group of early adopters in 2014 and then released as a commercial product in 2015. Nanopore technology works at the scale of nanometers: pores or cavities one billionth of a meter in width. To biologically analyze living cells, the MinION draws a long strand of DNA into the nanopores and "reads" it. Its ability to potentially comprehend much longer sequences of DNA or RNA than is currently possible is game changing, not least because such sequences no longer have to be chopped up for analysis and then jigsawed back together, an often difficult and time-consuming process. Further, as *Forbes* reported in early 2018, the MinION "can detect previously unseen areas of strands that may have an impact for cancer and other diseases."[2] Multiple MinIONs can be used together to scale up, while its sister tool, the SmidgION, allows users to scale down, employing the technology in tandem with a smartphone (this agility has proved useful, for example, in tracking the spread of Ebola in West Africa).[3]

Design that broadens access, democratizing previously specialized knowledge and resources, has many possible implications, ranging from the universally positive to the questionable and downright disturbing. A readily available product that allows a closer understanding of DNA or RNA structures—the very building blocks of life—paves the way for bringing gene sequencing to many populations as a domestic technology. In the future, citizen-scientists, potentially outnumbering their lab counterparts, might tackle questions of how our cells respond to disease, while doctors' offices could have easy access to genetic health data. The MinION might even be used in classrooms as another tool in the STEM arsenal to cultivate a lifelong love of science in young students, inspiring some to pursue DNA research. However, as the scientist and researcher Andrew Hessel has pointed out (in an article in the *Atlantic* in 2012, provocatively titled "Hacking the President's DNA") the DNA information unique to each individual could be used to personally target them, and might also compromise their safety or privacy in the hands of someone—employers, law enforcement, or a bioterrorist, for example—without their best interests in mind. **MMF**

1   See the Oxford Nanopore Technologies website, especially "About Us" and the MinION product page, nanoporetech.com.

2   Janet Burns, "Handheld Device Offers 'Most Complete' Gene Sequence Yet, and Costs $1000," *Forbes*, January 29, 2018, www .forbes.com.

3   See James Gallagher, "Handheld Device Sequences Human Genome," BBC News, January 29, 2018, www.bbc.com.

ESTROFEM! LAB   2016   +   HOUSEWIVES MAKING DRUGS   2017   MARY MAGGIC

The artist Mary Maggic's video *Housewives Making Drugs* nods wryly and subversively to the genre of television programs made popular by the cookbook author Julia Child and the lifestyle guru Martha Stewart.[1] But instead of offering tips on how to perfectly poach an egg, Maggic's fictional cooking show features the transfemme stars Maria and Maria[2] teaching viewers how to whip up their own hormones,[3] step by step, in a home kitchen rather than access them through a health-care system that might be expensive, restrictive, and discriminatory toward transgender communities. Maggic recognizes hormones as "a powerful tool of biosurveillance. These molecules are controlled by governments, manufactured and marketed by corporations and thereby dictate our health, body autonomy, and definitions of normal and natural."[4]

Estrofem! Lab is a physical manifestation of the topics and ideas explored in the video, an "estrogen hack lab" consisting of a speculative toolkit, instructions, and materials for hacking estrogen, all deliberately DIY and low-tech to make it affordable.

At the intersection of biotechnology, cultural discourse, and civil disobedience, these two related works interweave scientific methods with speculative design practice, the latter an approach that asks a provocative question without immediately offering a concrete answer, an inversion of the traditional solution-oriented design process. The result is a meditation on futures in which biological components are viewed as design tools that enable us to edit our bodies today to construct new biologies for tomorrow. The works also ask us to consider how alteration and augmentation challenge normative expectations of our bodies, subverting standards that constrict present realities or demand homogeneity. In the scenarios created by Maggic, design plays an important role not only by providing rich and provocative speculative narratives that help catalyze imaginations, research directions, and ethical reflections within the science community, but also by producing tools with the power to give us control over our own identities. The video and mobile lab simultaneously expand the agency of ordinary citizens and the field of design, heralding a world in which there is greater body sovereignty for all.   **MMF**

1   The video was made with Mango Chijo Tree and The Jayder.

2   Played by Jade Phoenix and Jade Renegade.

3   Mary Maggic's webpage for this project cites "hormones for HRT, birth control, menopausal symptoms, etc.," while noting that the film is not a guide, cautioning that we are "very far from any at-home synthesis of hormones"; maggic.ooo/Housewives-Making-Drugs-2017.

4   "Five Questions for Mary Maggic BSA '13," Carnegie Mellon School of Art, November 2, 2018, www.art.cmu.edu/news/alumni-news /5-questions-mary-maggic/.

ON DISPLAYING FASHION BEYOND GENDER   2019
FLORIANE MISSLIN

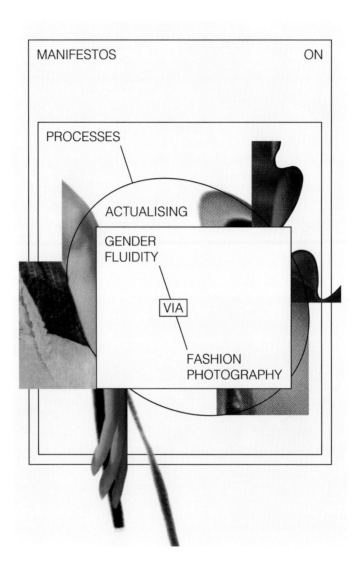

MANIFESTOS ON PROCESSES ACTUALISING GENDER FLUIDITY VIA FASHION PHOTOGRAPHY

On Displaying Fashion beyond Gender arose from Floriane Misslin's sociological research into processes of gender subversion in fashion photography. The project explores the visual production of magazines such as *Dazed*, *Accent*, and *Petrie*, as well as unisex clothing brands like Toogood and gender-fluid retailers like Verv, from the perspective of behind-the-scenes fashion professionals, including photographers and creative directors as well as stylists, to decipher how their imaginaries are produced and consolidated. Misslin's work ultimately poses the question, How do visions of the future of gender manifest through fashion photography?

Misslin considers fashion photography as a space for exploring and negotiating gender, thereby exposing its fluidity. *On Displaying Fashion beyond Gender* addresses the politics of images that have the power to stimulate different imaginings of gender, contributing to viewers' personal identity-making.

A set of posters emerged from the research process, each poster presented as a manifesto unveiling a production process peculiar to a particular participant. The creation of these manifestos served as a methodological means to explore the diverse techniques and strategies employed to subvert gender displays. During her research, Misslin created visual maps of the exchanges with each participant, which began with a series of interviews generating visual and textual data that were subsequently edited into a manifesto. These visual maps were designed to provide a common space for Misslin and the participants to share their knowledge and experiences with each other, providing reflective insights into the singularity of each participant's practice in producing imaginaries beyond gender, while also revealing both the potential and limits of these attempts.

The industry of fashion photography exists in service to the larger fashion industry. Its primary purpose is the creation of visual narratives that provide new material goods with symbolic values. By scrutinizing the different stages and materials involved in a fashion photo shoot, from mood boards and other forms of visual communication to the negotiations that take place during the shoot itself and the decision-making of the final editing, Misslin highlights the challenges and restrictions encountered in generating these envisioned scenarios with their constructed gender identities. **MB**

The Design Imagination

Dunne & Raby, *Robot 4: Needy One*, from the Technological Dreams Series: No.1, Robots, 2007. Photograph by Per Tingleff

What you see and hear depends a good deal on where you are standing.
—C. S. Lewis, *The Magician's Nephew*, 1955[1]

The field of design is about projecting forward, imagining new possibilities that can transform the present and help create new potential futures. From our buildings, streets, education, food, and health care to our political, economic, and communications systems, the range of projects that designers are engaged with has grown exponentially, especially in recent years, as centuries of experience are being rethought in response to digitalization: The critic and futurist Alvin Toffler forecast in 1970 that what the world needed was "a multiplicity of visions, dreams and prophecies—images of potential tomorrows."[2] His ideas seem as relevant today as they did then. In our own time of rapid change and social and political struggle, it feels increasingly urgent to look at where we have come from, where we are now, and where we are going. The projects included in this publication, grounded in contemporary issues, promote forms of inquiry and prioritize questions that grapple with the role of design in society today. Whether designs for lab-grown meats, robotic companions, or smart cities, they confirm Toffler's belief that "conjecture, speculation and the visionary view [are] as coldly practical a necessity as feet-on-the-floor 'realism.'"[3] In their range of ideas they retain a connection to the past, yet make the case for the importance of freeing the design imagination to challenge the status quo and offer new potential realities and alternative worldviews that can help us, for example, readdress the balance of power and offer more equitable ways forward.

Throughout history, and especially during the twentieth century, the field of design was inextricably bound to issues of functionality and charged with solving real-life (rather than theoretical) problems. Today, however, many see design as itself the problem, especially in its direct relationship to global economies and ecologies. In effect, the field is being challenged to consider its own limitations and potentially outdated working processes, and to think and design itself out of its box—to unbox itself (to borrow a contemporary idiom). In 1971 the designer and educator Victor Papanek cautioned in *Design for the Real World*, "There are professions more harmful than industrial design, but only a very few."[4] With chapters on such issues as design's ethical and environmental responsibility, his book was a call to arms. As Papanek asserted then, "the design of any product unrelated to its sociological, psychological, or ecological surroundings is no longer possible or acceptable."[5] Today, design is again being charged with rethinking given conditions in order to become more responsible, more inclusive, and more ethical.

While Papanek sought to reconsider the existing landscape of everyday objects to address real-world concerns related to social and cultural life, other of his contemporaries proposed visionary schemes that questioned the very context of design and what it could

1   C. S. Lewis, *The Magician's Nephew* (New York: Macmillan, 1955), 111–12.

2   Alvin Toffler, *Future Shock* (New York: Random House, 1970), 410 (463 in the 1971 Bantam paperback).

3   Toffler, *Future Shock*, 410.

4   Victor Papanek, *Design for the Real World: Human Ecology and Social Change*, 2nd ed. (London: Thames and Hudson, 1985), ix.

5   Papanek, *Design for the Real World*, 188.

6   See Peter Lang and William Menking's catalogue of the exhibition *Superstudio: Life without Objects* (Milan: Skira, 2003), which was on display at the Design Museum, London, from March 5 to June 8, 2003. Superstudio was founded in Florence in 1966 by Adolfo Natalini, Cristiano Toraldo di Francia, Roberto and Alessandro Magris, and Piero Frassinelli.

7   See Bruno Latour, "'We don't seem to live on the same planet': A Fictional Planetarium," in this volume, pp. 193–99.

8   Jaron Lanier, *Who Owns the Future?* (New York: Simon and Schuster, 2013), 17.

9   Brooke Gladstone, with William Gibson, David Byrne, and Anne Simon, "The Science in Science Fiction," on *Talk of the Nation*, NPR, November 30, 1999 (timestamp 1:55), www.npr.org.

10 See Slavoj Žižek, "How WikiLeaks Opened Our Eyes to the Illusion of Freedom," *Guardian*, June 19, 2014, www.theguardian.com.

11 Tony Fry, *Design Futuring: Sustainability, Ethics, and New Practice* (Oxford: Berg, 2008), 4.

12 Susan Yelavich, introduction to *Design as Future-Making*, ed. Yelavich and Barbara Adams (London: Bloomsbury Academic, 2014), 14.

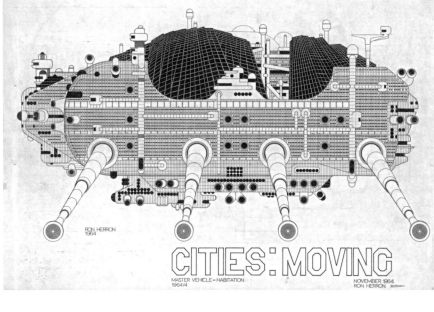

CITIES: MOVING

MASTER VEHICLE – HABITATION 1964/4   NOVEMBER 1964 RON HERRON.

Ron Herron, *Cities: Moving, Master Vehicle-Habitation Project, Aerial Perspective*, 1964. The Museum of Modern Art, New York. Gift of The Howard Gilman Foundation, 1202.2000

R. Buckminster Fuller and Shoji Sadao, *Dome over Manhattan*, 1960

or should be. Well-known examples from the twentieth century include the experiments of inventive thinkers such as R. Buckminster Fuller, whose *Dome over Manhattan* (1960) was a theoretical design for a three-kilometer geodesic dome spanning Midtown Manhattan that would regulate weather and reduce air pollution. This project anticipated our current environmental challenges, especially in dense urban areas. Another equally potent example is *Walking City*, by the London-based Archigram. Represented in the drawings of Ron Herron, this hypothetical project from the 1960s depicts a city as a giant insect-like structure that moves across the globe on enormous legs, providing resources to different places and endlessly adapting to change. The concept prefigured contemporary transient lifestyles, fueled by new technologies, such as smartphones, that are changing how we live, work, and communicate, while simultaneously challenging long-held concepts of belonging and displacement. *Walking City*, like Fuller's *Dome over Manhattan*, emphasized farsighted approaches to architecture and design that it was then not possible to implement, but that could perhaps one day be realized by embracing new technologies as modes of survival. As a counterproposition, contemporaneous groups such as Superstudio in Italy recommended "life without objects,"[6] demonstrating a more ambivalent relationship to technology and questioning the need for yet more objects of design. Superstudio instead proposed a future existence devoid of commercially produced products. Read today, as we are faced with mounting environmental and economic crises and the implications of automation and overconsumption, among other challenges, their proposition no longer seems so far-fetched as it once might have. Although the political punch of these utopian projects may have faded—their critique of society has since been questioned for its lack of accountability and consideration of deeper social and political issues, in addition to its Western bias—their visual force and the potential for architecture and design to be powerful agents of change have endured. The continuing relevance of these projects lies in their ability to stimulate the imagination and foster new readings of the world as a way to shift perspectives and promote change.

In our own time of political disorientation, when the efficacy and agency of the "modern era" is once again being reconsidered and the disparity between the human-made and the natural world continues to grow, it is necessary to once again rethink design's methods and critical engagement across geographic, social, and political territories in order to engender alternative worldviews and propose more equitable and inclusive ways forward.[7] With the development of new technologies and the increasingly wide distribution of information, the potential for sharing and circulating ideas has grown tremendously, but how this will ultimately impact the practice of design is still being determined. Rather than perpetuating the myth of the lone hero working in isolation to invent solutions, new technologies encourage collaborative practices in which ideas shared over computer networks have the potential to overturn convention. Yet as Jaron Lanier, a computer scientist and pioneer in the field of virtual reality, attests, "It is the politics and economics of these networks that will determine how new capabilities translate into new benefits for ordinary people."[8] His words resonate with those of the novelist William Gibson, who had earlier asserted, "The future is already here. It's just not very evenly distributed."[9] That is, societal change, including that enabled by design, is dependent on the politics and economics of the global market. Nonetheless, with the growth of social-media outlets there has been a shift in responsibility. Global challenges have entered the public consciousness, and now everyone is accountable.[10] "The 'state of the world' and the state of design need to be brought together," charges the theorist Tony Fry, calling us to radically rethink the role of design in the world.[11]

So what is design's efficacy and role in the world today? Susan Yelavich, a professor and curator of design, warns that "design has become a panacea for whatever ails. Politically neutral, never demanding, the popular perception of design threatens to override criticality and obscure its capacity to engender *agency*, in the best sense of that word."[12] The challenge for design is to recognize market forces and political constraints while maintaining enough distance to foster the imagination and allow critical positions that can reorganize and rethink economies, ecologies, information systems, and social groups using the languages, forms, and methods of design. The sociologist and political activist Peter Frase calls us to reclaim the "tradition of mixing imaginative speculation with political economy" and contends that "sketching out multiple futures is an attempt to leave a place for the political and the contingent."[13] These ideas seem increasingly prescient in our present "post-truth" moment, when the rise of "alternative facts" has made it imperative that we analyze more closely to understand the context in which ideas and information (and misinformation) arise and are made available.[14] As it becomes increasingly difficult to retain faith in the concept of progress as an inevitable and ongoing positive force that takes into account our collective concerns, it also seems more important to counter the "suppression of the design imagination" in thinking about the future.[15] As the philosopher Stephen T. Asma has described, our cognitive experiences are made up of simultaneous "'almosts' or 'what ifs' and 'maybes'" that enable us to imagine different realities all at once, a capability essential for human development.[16] How, then, might design continue to question everyday life from multiple perspectives, different generations, and various places so as to reconfigure the present for the future?

The following case studies are not meant to be exhaustive but are a sampling of the ways in which designers, architects, and others are grappling with how to foster the imagination to collectively and meaningfully transform society.

"WHAT IF DESIGN EDUCATION'S FOCUS ON 'MAKING STUFF REAL' PERPETUATES ... A DYSFUNCTIONAL PRESENT?"[17]

Anthony Dunne and Fiona Raby, professors of design and social inquiry and codirectors of the Designed Realities Studio at the New School in New York, have spent the better part of two decades exploring how "design speculations can act as a catalyst for collectively redefining our relationship to reality."[18] Their work on _critical design_, a term Dunne coined in 1999 and which he and Raby consider to describe "a position more than a method," aims to deepen and extend an awareness and understanding of design, especially as it relates to our interactions with objects and their role in contemporary society.[19] Interested in how design can be a trigger to spark healthy dialogue, intellectual exchange, and dissent, they employ "design fiction"—a form of scenario building and storytelling—to explore the emotional as well as intellectual capacities of design. These ideas have been comprehensively explored in publications such as _Design Noir: The Secret Life of Electronic Objects_ (2001), in which they investigate the way

13   Peter Frase, _Four Futures: Life after Capitalism_ (London: Verso, 2016), 31.

14   For more on this idea, see Bruno Latour, _Down to Earth: Politics in the New Climatic Regime_ (Cambridge: Polity, 2018).

15   Anthony Dunne, "A Larger Reality," in _Fitness for What Purpose? Essays on Design Education Celebrating 40 Years of the Sir Misha Black Awards_, ed. Mary V. Mullin and Christopher Frayling (London: Design Manchester/Eyewear Publishing, 2018), 116.

16   Stephen T. Asma, _The Evolution of Imagination_ (Chicago: University of Chicago Press, 2017), 2–3.

17   Dunne, "A Larger Reality," 116.

18   Anthony Dunne and Fiona Raby, _Speculative Everything: Design, Fiction, and Social Dreaming_ (Cambridge, MA: MIT Press, 2013), 34.

19   The term _critical design_ was first used in Anthony Dunne's book _Hertzian Tales: Electronic Products, Aesthetic Experience, and Critical Design,_ originally published by the Royal College of Art (RCA CRD Research Publications) in 1999; see the revised edition (Cambridge, MA: MIT Press, 2008).

20 Designers such as Alexandra Daisy Ginsberg, Jessica Charlesworth (both included in this volume; see pp. 80, 206), Nelly Ben Hayoun, and Sputniko!, to name a few.

21 For more on critical, speculative, and associative design practices and the debates around them, see Matt Malpass, "Between Wit and Reason: Defining Associative, Speculative, and Critical Design in Practice," *Design and Culture* 5, no. 3 (2013): 333–56; Vyjayanthi Venuturupalli Rao, Prem Krishnamurthy, and Carin Kuoni, eds., *Speculation, Now: Essays and Artwork* (Durham, NC: Duke University Press, 2015); Cilla Robach, "Critical Design: Forgotten History or Paradigm Shift?," in *Shift: Design as Usual—or a New Rising?*, ed. Monika Sarstad and Helen Emanuelsson (Stockholm: Arvinius, 2005); Dunne and Raby, *Speculative Everything*; and Yelavich and Adams, *Design as Future-Making*.

22 "As a discipline theorised within the safe confines of developed, northern european [*sic*] countries and practiced largely within an overwhelmingly white, male, middle class academic environment, SCD has successfully managed to ignore, or at best only vaguely acknowledge, issues of class, race and gender (with few exceptions)." Luiza Prado, "Questioning the 'Critical' in Speculative & Critical Design: A Rant on the Undiscerning Privilege That Permeates Most Speculative Design Projects," Medium, February 4, 2014, medium.com.

23 Dunne, "A Larger Reality," 116; also available online at www.designedrealities.org/texts/the-diplomatic-gift.

24 Dunne, "A Larger Reality," 2.

25 The following discussion and quotations stem from an email correspondence with Mary Maggic on December 9, 2018.

"A Journey from A to B," from *Life: Supersurface*, a short film by Superstudio for the exhibition *Italy: The New Domestic Landscape*, The Museum of Modern Art, New York, 1972

mobile phones, computers, and other electronic objects influence our experience of the environment, and they became the foundation of the Design Interactions program. Initiated by Dunne at London's Royal College of Art in 2006, this interdisciplinary research group has since been credited with launching a generation of critical designers.[20] The critical-design approach has not escaped criticism, however, having been much discussed and debated.[21] Dissenters find the field too tied to speculative practices and academic inquiries that they argue don't take enough responsibility for their social and political positions and ability to be inclusive.[22] And yet, it is clear that these types of disciplinary methods and approaches are raising essential questions about design's relationship to the "real world," the definition of which is now a recurrent debate far more complex than even Papanek could have foreseen.

Dunne, who continues to tackle the contradictions and challenges facing design and especially design education, has more recently asked:

What if our approach to design is wrong? What if educating designers to work with prevailing economic, social, technological and political realities—designing for how the world is now, has become a convenient conceit? What if teaching student designers to frame every issue, no matter how complex, as a problem to be solved squanders valuable creative and imaginative energy on the unachievable. What if design education's focus on "making stuff real" perpetuates everything that is wrong with current reality, ensuring that all possible futures are merely extrapolations of a dysfunctional present?[23]

Dunne and Raby are currently reconsidering their own working processes and long-held belief in an object-oriented approach to design, as well as design's role in everyday life, as a way to further challenge ideas associated with consumerism and new technologies. At the Designed Realities Studio they are engaging with a much wider range of disciplines and practices—including political science, anthropology, sociology, history, economics, and philosophy—in an effort to understand design's place within this complex network and to learn how to situate their work within larger historical trajectories related to imagination and speculation, with the potential to suggest "new worldviews made tangible through an expanded form of design practice."[24] The result is an anthropological approach to design that considers how we might design behaviors as well as objects to incite a new consciousness within the field of design and in the world around us.

"THE TENSION BETWEEN THE ACTIVE AND PASSIVE QUEERING OF DAILY LIFE"

The artist Mary Maggic, whose approach involves collaboration across the fields of design, science, and biology, creates projects that help us reconsider our understanding of the

world by interrogating gender codes and breaking taboos. Through projects such as *Estrofem! Lab* and *Housewives Making Drugs* (see p. 39), Maggic calls attention to "the tension between the active and passive queering of daily life"—for example, the ways in which our bodies are altering because of changes to the environment or through interventions such as hormone therapy, among other concerns rarely discussed openly in society.[25] As Maggic explains:

While you can be a uterus-owning woman seeking birth control pills, or a trans-femme seeking hormone replacement therapy, this is an active form of altering one's body to reproduce a gendered program. Meanwhile, every single one of us, including of course non-human species, are undergoing a disruption to our bodies, health, and (hetero)normative delineations through all of the industrial toxicities xeno-hormones that pervade our planet. This molecular colonization is tied deeply to patriarchy and capitalism and is incredibly invisible. So art and design, especially interdisciplinary practices, are able to reframe all of these complexities and provide room for contemplation and even action.

Maggic's work urges us to imagine new ways of identifying as individuals and communities by questioning norms and standards and probing our objective and subjective outlooks. For the past three years, Maggic has been hacking hormones, both natural and synthetic, to answer questions about biopolitical agency and emancipation. Through the lens of a fictional television show, inspired by the type of program popularized by Martha Stewart—the epitome of an entrenched idea of what it means to be feminine and domestic—they have created a recognizable platform for discussing these ideas: "A satire on a kitchen television show became the obvious vehicle and chosen battleground to discuss issues around trans experiences and access to hormones, the ethics and risks of self-administering and how we can undermine the patriarchy." Through theoretical and speculative projects like *Housewives Making Drugs* that question real and projected conditions (with large doses of humor and irony but serious intentions), Maggic seeks to elevate conversations related to the body, all the while making their research and development processes open and transparent through workshops and online platforms (*Estrofem! Lab*) that make clear their intentions and the context for their work: "I think we need to hack at all levels, the societal, the cultural, the political, in order to ask the necessary critical questions and begin mapping out new worlds and ways of thinking," they assert. "What started from 'how can we make hormones in the kitchen' led to 'how are hormones currently produced and distributed' to 'who controls the access' to 'how do they produce our gendered subjectivities' to 'what is male and female anyway and how can we collectively dismantle and deprogram this binary system?'" With their all-inclusive attitude, Maggic reminds us of the importance of contesting preconceived ideas and the need to constantly look for "new tools, tactics, and reflections" that are accountable, diverse, and plural in outlook. Ultimately, Maggic reminds us that we are all responsible as "part of a social mutagenesis!"

"HYPER-ACCELERATE THE FIELD OF ARCHITECTURAL HISTORY AND SHED LIGHT ON ERUPTIVE INCIDENTS AT POLITICAL NEXUSES, [AND] WITHIN SOCIAL CONDITIONS"

As Eyal Weizman of Forensic Architecture, based in London, sees it:

> The so-called real world is so full of mystery, incoherence, and unknowns that to comprehend what has happened requires nothing less than imagining something new; abandoning the real would result in the impoverishment of the imagination.[26]

In a time radically different from the rapid industrialization and change of the decades after World War II, Weizman calls for re-arranging or reimagining what can be known rather than breaking away from the past— as was proposed by many architecture and design visionaries in the twentieth century. He recognizes that today it is the media, com-munications, data, and speed that are deter-mining change in society, more than the building of architectural structures. But the vocabulary of architecture and design has not transformed at the same pace as in other

fields: "The amount of data we can process, the scenarios that we can imagine, has expo-nentially increased; yet the parameters of a concert venue, a museum, a shopping mall have moved only incrementally." Instead of making buildings, Weizman harnesses the less concrete tools and techniques of architecture and design—"ways of seeing, speculating, researching, interrogating, and determining risk ratios, using simulation, structural engi-neering, fluid dynamics, etc., to run scenarios on other fields of knowledge and operations and to clarify incidents erupting within the molecular durations of time." Rather than posi-tion his work in the realm of construction and building, he locates it in the realm of investiga-tion, interrogation, and simulation to "hyper-accelerate the field of architectural history and shed light on eruptive incidents at political nexuses, within social conditions, etc." The goal is to make visible the hidden systems and processes impacting the world so as to inform current discourse and open up the potential for transformations that can help us reimagine our political, economic, social, and cultural life.

As this sampling of ideas suggests, "design imagination" is understood here as a method that "re-imagines what is in the image of what could or should be," to quote the social-cultural anthropologist Arjun Appadurai.[27] In other words, the primary concern is to use the tools of cross-disciplinary practices, including design, to shift perspectives, open up new lines of investigation, stimulate dialogue and the exchange of ideas, and instigate new ways of

26 Eyal Weizman, in conversation with the author, November 9, 2018; the quotations in this paragraph all derive from this conversa-tion. For more on Forensic Architecture, see pp. 166, 172–75.

27 Arjun Appadurai, "Speculation, after the Fact," in *Speculation, Now*, 208.

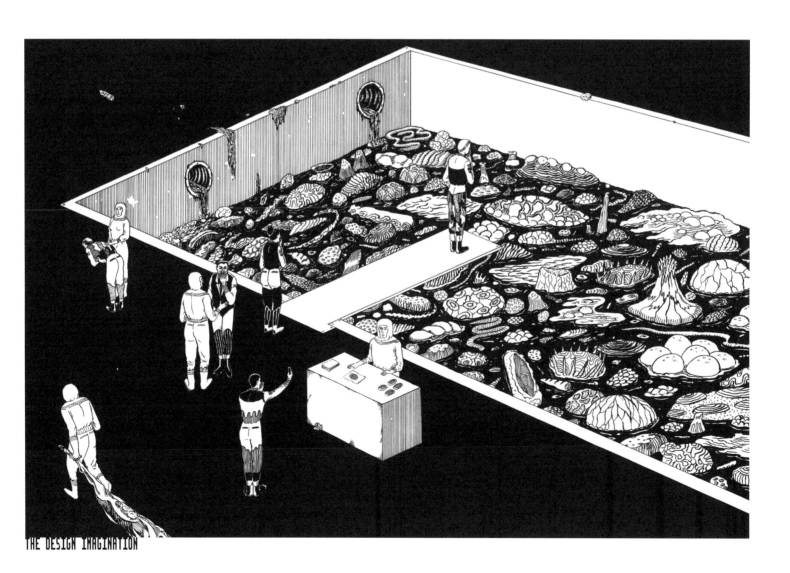

28 Quoted in Maya Jaggi, "The Magician," *Guardian*, December 17, 2005, www .theguardian.com.

29 Quoted in Jaggi, "The Magician."

seeing and being in the world that consider societal norms and standards, including within design and architecture, not as givens but as perennial areas of examination and exploration. While not meant to be conclusive or absolute, the case studies touched on here reinforce the conversations in this volume to underscore how architects, designers, and others are using design—specifically critical, speculative, and associative design practices—to both imagine and propose alternative ways of thinking about and interacting with the present to inform the future. These approaches employ the tools of scenario building and the tactics of subversion, as well as the analysis and modes of representation inherent to architecture and design, to open up ideas. In effect, they underscore what other creative minds have urged through their work, such as the socially engaged poet and novelist Percy Bysshe Shelley, who contended in his "Defence of Poetry" (1821) that "the great instrument of moral good is the imagination."[28] In our own time, the writer and imaginer of other worlds Ursula K. Le Guin asserted, "If you cannot or will not imagine the results of your actions, there's no way you can act morally or responsibly."[29] Speculation, simulations, fictional television shows, role play, workshops, and the like are all tools that can help architects and designers grapple with complex issues and render their visions from different viewpoints and parts of the globe to test the efficacy of their ideas. The goal is to find the right degree of freedom and the best approaches to design to counter dominant narratives and suggest alternative and more inclusive worldviews, while understanding the contexts in which new ideas are imagined and circulated. We can't predict the future, but as this publication suggests, it is fruitful to imagine what might be possible, hoping that many more things might happen than actually do. ╬

Mary Maggic, *Housewives Making Drugs*, 2017

Digitarians visiting one of Bioland's more extreme attractions. Illustration by Miguel Ángel Valdivia for Dunne & Raby's *UmK: Lives and Landscapes* at the Istanbul Design Biennial, 2014

—I would like to acknowledge Anthony Dunne and Fiona Raby for their guidance and generous insights, which helped me shape this text.

RYAN

48

Can We Fall in Love with Robots?

EMMA YANN ZHANG on human-robot intimacy and empathy

On October 11, 2007, the then sixty-two-year-old David Levy defended his PhD dissertation, "Intimate Relationships with Artificial Partners," at Maastricht University in the Netherlands. This was the first time that an academic thesis on the subject of Love and Sex with Robots (LSR) was presented. The topic received a flurry of media attention and by the end of the year had been widely discussed, especially following the success of Levy's *New York Times* best seller *Love and Sex with Robots*, based on his thesis and published that same year by HarperCollins. The subject of human-robot intimate relationships rapidly developed into an academic discipline, joining the closely related fields of robotics, computer science, artificial intelligence, human-computer interaction, roboethics, sociology, and so on. Reputable academic conferences and journals began to invite and accept papers on this theme, and an annual conference devoted to the topic—the International Congress on Love and Sex with Robots—was founded by Levy and Adrian David Cheok in 2014.

The idea of humans falling in love with artificial beings has its origins in ancient Greek mythology, in the story of Pygmalion and his sculpted lover. And long before academics started discussing the possibilities and implications of humans having intimate relationships with robots, science-fiction writers had already explored the subject from many different angles. Isaac Asimov, for example, best known for his Three Laws of Robotics, published a number of works describing his vision of the future of human-robot relationships; both "Satisfaction Guaranteed" (1951) and "True Love" (1977) are stories that involve human-robot or human-computer love, and the topic forms a background theme in his acclaimed *Foundation* series. Science-fiction stories like these have inspired a profusion of films and television series on the topic, and have to some degree influenced the direction of robotics and LSR research. There has also been a wealth of narratives in the wider popular culture, from titillating tabloid headlines to commercial products sold on major e-commerce sites. Whether academic or not, many of the issues that arise in these stories—regarding free will, individualism, control, empathy, and consciousness—are the very questions researchers are attempting to answer today.

Among the most important qualities that set people apart from machines is empathy. Empathy is essential if one is to truly feel, care, or love. Skeptics of human-robot love often argue that this kind of love is ingenuine and one-sided, since machines are devoid of empathy and therefore can never love you back. When the IEEE (Institute of Electrical and Electronic Engineers) website posed the question "Do you believe humans will marry robots someday?" to its readers, a common response was that such marriages are not reciprocal.[1] Some readers wrote that a human-robot relationship is master-slave in nature because robots are programmed to satisfy all human desires and needs, whereas love and marriage are, ideally, built on mutual consent and understanding. Citing similar objections, opposing researchers argue that because artificial companions do not have the "first-person feelings" associated with human love, they can never love like human beings.[2] When we truly love another, we don't just care about how the object of our love behaves but also how they feel.

Today, scientists and engineers have developed technologies that enable computers to recognize and respond to human emotions.[3] We also have artificial conversational agents and realistic-looking humanoid robots that can display vivid facial expressions to express basic emotions, such as the Geminoid F robot created by Hiroshi Ishiguro,[4] best known for making his own robotic clone (see p. 59). Whether we can ignore behind-the-scenes algorithms, however, and presume that computers are capable of having emotions and falling in love is

Harmony, a modular, customizable RealDoll sex robot made by Abyss Creations (now Realbotix) in San Marcos, CA. Photograph by Martin S. Fuentes

still an open question. Should we disregard loving or caring behavior from artificial creations on the grounds that these actions are synthetic performances, or should we accept them in the same way that we accept love from humans? The answer very much depends on one's individual understanding of what love is. The behavioristic theory of love relies on observable actions and preferences: if someone displays certain loving behavior toward another—caring for their well-being and catering to their needs—we can then say they are in love with that person.[5] This is the position taken by Levy, who often uses the famous Turing test as a proof of robots' emotions and intelligence: if a robot behaves as if it loves you and cares about you, and it tells you "I love you," we should accept this in the same way that we would if it had come from a human.[6]

A common criticism of this theory is that people's outward actions may not express their inner state or emotions. One can be a good actor or, in the case of an artificial entity, be programmed to perform a set of actions that show love. But we can only derive information from others based on words and actions, as it is impossible to know anyone's inner thoughts other than our own. How, then, can we be sure that a person truly loves us, whereas a robot is only pretending to? The lines are blurred when we experience the same amount of comfort from a programmed hug as from an "I love you" uttered by an insincere lover.

Researchers have found that humans tend to behave socially toward computers and robots. This phenomenon, known as the Media Equation,[7] suggests that we naturally observe social niceties such as politeness and reciprocity when interacting with artificial agents, and that we feel empathy toward them.[8] A research team from the University of Duisburg-Essen conducted an experiment to study people's emotional reactions to cruelty to robots.[9] When participants watched a video of an experimenter torturing and mistreating a robotic dinosaur, they reported negative emotions toward the researcher and empathy toward the robot; they expressed anger at the torturer, and felt pity for the victim. Similarly, the inhumane treatment of Dolores and other robots in the dystopian television series *Westworld* has elicited considerable empathy from viewers—once again revealing our capacity to relate to artificial beings on an emotional level.

Our humanistic response and empathy toward artificial entities, coupled with our innate "need to belong" in interpersonal relationships,[10] indicate a high probability that people will form personal attachments and meaningful relationships with artificial companions. The field of robotics has witnessed a paradigm shift from "mechano-centric" principles to "human-oriented" principles, with more and more robotics researchers turning their focus to the social aspects of robots and artificial agents. This has led to research areas such as social robotics, affective computing, Kansei engineering, and so on. More emphasis is also being placed on the anthropomorphism of robots, requiring new hardware and software technologies in order to make the appearance, behavior, and personality of robots as humanlike as possible. Such technology developments require robotic engineers and AI scientists to work hand in hand. More importantly, a future that includes human-robot love and sex is surely going to raise pressing concerns that will require lawmakers, philosophers, psychologists, roboethicists, and social scientists to take part in the conversation. ⧕

Ava (Alicia Vikander), the alluring humanoid robot in *Ex Machina* (2014), written and directed by Alex Garland

1   See "Is Cyberspace Making You Sick?," The Institute: The IEEE News Source, April 7, 2008, theinstitute.ieee.org.

2   Piotr Bołtuć, "Church-Turing Lovers," in *Robot Ethics 2.0: From Autonomous Cars to Artificial Intelligence*, ed. Patrick Lin, Ryan Jenkins, and Keith Abney (New York: Oxford University Press, 2017), 214.

3   Rafael A. Calvo and Sidney D'Mello, "Affect Detection: An Interdisciplinary Review of Models, Methods, and Their Applications," *IEEE Transactions on Affective Computing* 1, no. 1 (2010): 18–37.

4   Christian Becker-Asano and Hiroshi Ishiguro, "Evaluating Facial Displays of Emotion for the Android Robot Geminoid F," in *2011 Workshop on Affective Computational Intelligence (WACI 2011): Proceedings* (IEEE, 2011), 22–29, ieeexplore.ieee.org/document/5953147.

5   See Alexander Moseley, "Philosophy of Love," *Internet Encyclopedia of Philosophy* (2001), www.iep.utm.edu/love/.

6   Bołtuć, "Church-Turing Lovers," 214.

7   Byron Reeves and Clifford Nass, *The Media Equation: How People Treat Computers, Television, and New Media Like Real People and Places* (New York: Cambridge University Press, 1996).

8   Nicole C. Krämer, Astrid M. Rosenthal-von der Pütten, and Laura Hoffmann, "Social Effects of Virtual and Robot Companions," chap. 6 in *The Handbook of the Psychology of Communication Technology* (Chichester, UK: John Wiley and Sons, 2015), 137–59, onlinelibrary.wiley.com.

9   Astrid M. Rosenthal-von der Pütten et al., "An Experimental Study on Emotional Reactions Towards a Robot," *International Journal of Social Robotics* 5, no. 1 (January 2013): 17–34.

10 Roy F. Baumeister and Mark R. Leary, "The Need to Belong: Desire for Interpersonal Attachments as a Fundamental Human Motivation," *Psychological Bulletin* 117, no. 3 (1995): 497.

The Future of Love? From the Past (Steve Bannon) to the Future (Sex Robots)

SREĆKO HORVAT on sexbots, AI, and psychographic profiling

Sometimes a simple headline by chance reveals more about the past, present, and future than any history book or futurologist projection ever could. This is what happened in December 2018, when a newspaper headline appeared stating "Steve Bannon Canceled from Sex Robot Conference."[1]

The article reported that an international academic conference titled Love and Sex with Robots, planned to take place at the University of Montana, had been canceled following a backlash against a proposed speech by Steve Bannon. (Though he was scheduled to speak at the concurrent conference on Advances in Computer Entertainment Technology, the subject of the co-conference proved irresistible to headline writers.) Bannon, the former adviser to President Donald Trump and a political careerist, had that year—following revelations of his complicity in the Facebook–Cambridge Analytica scandal, in which Facebook data was used to create psychographic profiles of users for political targeting purposes—switched his geographic focus and was desperately trying to unite Europe's right-wing populists.

How does this headline from 2018 bring us from the past (Steve Bannon) into the future (sex robots)? It's because the interesting part isn't that Bannon was banned from a conference (this happens to him quite frequently) but the subject of the concurrent conference. Its title might sound like a science-fiction event from some indeterminate future time, but this sex-robots future is already here, as Emma Zhang, one of the event's organizers, discusses in her essay (see pp. 50–53). Although as William Gibson famously noted, the future that has arrived is not yet evenly distributed.[2]

Only two months before Bannon's banishment, in another corner of the United States, it was sex robots themselves that were banned. A Canadian company, KinkySdollS, had planned to open a "robot brothel" in

Houston, and the city's mayor moved swiftly to clear the way, presenting an ordinance that expanded the meaning of adult arcades to include "anthropomorphic devices," or sex robots—shades of Westworld. This would have been the first sex-robot brothel in the United States.[3] But after a heated debate and protests by religious groups, the city council decided to ban robot brothels. During the debate a Houston resident, Tex Christopher, quoted from the Bible: "In Ephesians 5:31, it says that a man shall leave this [sic] father and mother and shall be joined unto his wife and they shall become one. It doesn't say that a man shall leave his mother and father and go and join a robot."[4]

Already in 2015 a Campaign Against Sex Robots (CASR) had been launched to draw attention to the ways in which the idea of forming "relationships" with machines was becoming increasingly normalized.[5] It warned against sex robots as "animatronic humanoid dolls with penetrable orifices where consumers are encouraged to look upon these dolls as substitutes for women (predominantly), and they are marketed as 'companions', 'girlfriends' or 'wives.'"[6] There followed the protests of religious groups. Even before Houston's banning of sex-robot brothels, a number of Christian ethicists had responded to a report by the Foundation for Responsible Robotics exploring the possible benefits and dangers of humans having sex with robots, including robots designed to look like children, stating that such sexual activities went against God's design.[7]

Warning against the rise of sex robots, Tobias Winright, a theologian at Saint Louis University, said: "As a Christian, I think non-mutual, non-consensual sexual activity is contrary to mutually donative love-making. Thus, sexual activity with a simulacrum seems to me quite a stretch from when two persons, who are made in God's image, sexually express their love for

The creation of the robot Maria, one of the only female robots in early science fiction. Scene from Metropolis (1927), directed by Fritz Lang

each other, transcending and giving beyond the self with the other, and thereby imaging God who is agape."[8]

The Houston sex-robot case—sexual activity with a simulacrum (the philosopher Jean Baudrillard's term, which now appears in theological writings) and its commercialization (a new sexual business)—opens up important theological, philosophical, and political questions about the future of sex and love.

Theologically, we are confronted with rethinking and redefining the relationship between God, humans, and machines (not only sex robots but AI overall). Philosophically, we are brought to one of the oldest questions, namely, What are we humans in the first place? What if the more disturbing question is not that of sex with a machine, but that posed by Spike Jonze's movie *Her* (2013), about a human falling in love with AI—not a sex robot but an operating system without a body? Or we could ask, What are machines? Aren't they already becoming human and the human becoming machine? Is there a future where humans are dispensable, or derided and discarded by machines? Politically, what was missing in the Houston debate (and still is missing from the theological perspective) is the question, Who is in control of the AI? Or, to put it in classic Marxist terms, Who owns "the means of production"?

If the Facebook–Cambridge Analytica scandal, with the crucial role played by Steve Bannon, brought us anything, it is the idea that politics can be preprogramed—that voters can, through "perception management," to a certain degree be programed to desire a specific political option (be it Trump or Brexit). The Houston sex-robot case raises another question that goes beyond the ban of "sexbot" brothels in one city, because it takes us from the future of sex to the future of love itself: Why wouldn't love become preprogramed?

When the British television series *Black Mirror* finally touched on the question of how AI will affect love—in the fourth-season episode "Hang the DJ" (2017)—it looked like a science-fiction dystopian future only to those not yet familiar with the current advances in technology (Tinder, Grindr, cyber-butlers, AI choosing our "perfect match"), which are rapidly transforming science fiction into dystopian reality. In the dystopian society of "Hang the DJ," romantic encounters are scheduled by an AI system called Coach, which collects users' data in order to match them with their "ultimate compatible other" and dictates which romantic relationships they will have and for how long. Let's say you just had a beautiful romantic dinner and the chair is trembling beneath you because you are already falling in love. But you visit the restroom so you can check Coach to see whether this is your "perfect match." You're informed that the relationship expires in twelve hours. But don't worry. The more relationships you have, the more data the computer gathers. The more data it gathers, the more accurate it is in predicting your perfect match.

It seems more than mere coincidence that Facebook, when the Cambridge Analytica affair hit the news, immediately announced it was launching an online dating service, called simply Dating. Unlike apps such as Tinder or Grindr, which use Facebook connections to identify potential matches, Facebook has the advantage of being able to see almost everything about its users. As *Bloomberg* reported, "It can track couples from their first 'likes' to the point at which they're ready for engagement ring ads, and beyond."[9]

The title of the *Bloomberg* article stated "Facebook Is Right to Think 'Likes' Can Lead to Love." Obviously, efforts to preprogram elections bear a relationship to attempts to preprogram love, and vice versa. No wonder two of the main protagonists (along with Bannon) behind

developing the Cambridge Analytica model were both previously involved in analyzing love rather than elections.

One is the computational social scientist Michal Kosinski, of the University of Cambridge Psychometrics Centre and Stanford University, who was the coauthor of a research paper showing that computer-based personality judgments are more accurate than judgments made by humans. Despite its prominence in research on well-being, Kosinski's work has also drawn a great deal of interest from British and American intelligence agencies and defense contractors. (Among the overtures he received was one from a private company running an intelligence project nicknamed Operation KitKat, because a correlation had been found between anti-Israeli sentiments and liking Nikes and KitKats.)[10]

The other is the data scientist Aleksandr Kogan, who also works in the field of positive psychology and has written papers on happiness, kindness, and love; an early paper was titled "Down the Rabbit Hole: A Unified Theory of Love."[11] The seemingly bizarre intersection of research on topics like love and kindness with defense and intelligence interests is not, in fact, particularly unusual. Much of the foundational research on personality, conformity, obedience, group polarization, and other such determinants to our social dynamics was funded during the Cold War by the US military and the CIA.

The only (but big) difference is that during that time there was no internet, and the computational power and advancement in AI research hadn't reached the stage where the same "psychological engineering" could be used not only to influence elections (Trump, Brexit) but also to analyze and determine romantic relationships (*Black Mirror*'s Coach).

This brings us to the crucial political question of our time: Who will be in control of AI? Or, in other words, if the big Silicon Valley companies—which are linked to the military sector and powerful factions of governments—already own the means of production (the material and immaterial infrastructure to create dreams, desires, politics), why wouldn't they use these same means to determine the ways (and with whom) we fall in love?

If we don't want to end up in a dystopian future in which Silicon Valley—or China, or Steve Bannon—controls the sex robots, or even the very process of "falling in love" (as in "Hang the DJ"), we had better start thinking seriously about the future of love. ⌗

Amy (Georgina Campbell) and Frank (Joe Cole) holding their "Coach" devices on their first date. Scene from the episode "Hang the DJ" (2017), written by Charlie Brooker and directed by Tim Van Patten, of the Netflix series *Black Mirror*

In the movie *Her* (2013), written and directed by Spike Jonze, the introverted Theodore Twombly (Joaquin Phoenix) falls in love with Samantha, the AI virtual assistant of his new computer operating system.

1   See Jeffrey Rodack, "Steve Bannon Canceled from Sex Robot Conference," *Newsmax*, December 14, 2018, www.newsmax.com. This article, with the same headline, was picked up by a number of other conservative news sites.

2   For Gibson's quote, see p. 43 in this volume.

3   This would have been the company's second location; the first opened in 2017 in Toronto. See the company website, www .kinkysdolls.com.

4   Florian Martin, "Is This the End for a Sex Robot Brothel in Houston?," *Houston Public Media*, October 17, 2018, www .houstonpublicmedia.org.

5   CASR, "An Open Letter on the Dangers of Normalising Sex Dolls & Sex Robots," Campaign Against Sex Robots, July 28, 2018, campaignagainstsexrobots.org.

6   CASR, "An Open Letter."

7   Stoyan Zaimov, "Sex with Robots Goes against God's Plan, Christian Ethicists Warn, Urge Ban on Child Sex Robots," *Christian Post*, July 6, 2017, www.christianpost.com.

8   Quoted in Zaimov, "Sex with Robots Goes against God's Plan."

9   "Facebook Is Right to Think 'Likes' Can Lead to Love," *Bloomberg*, May 11, 2018, www.bloomberg.com.

10 Tamsin Shaw, "The New Military-Industrial Complex of Big Data Psy-Ops," *New York Review of Books*, March 21, 2018, www .nybooks.com.

11 Paper presented at the Haas Scholars Research Conference, May 2008, Berkeley, CA.

Q, THE GENDERLESS VOICE   2019   EMIL ASMUSSEN AND RYAN SHERMAN

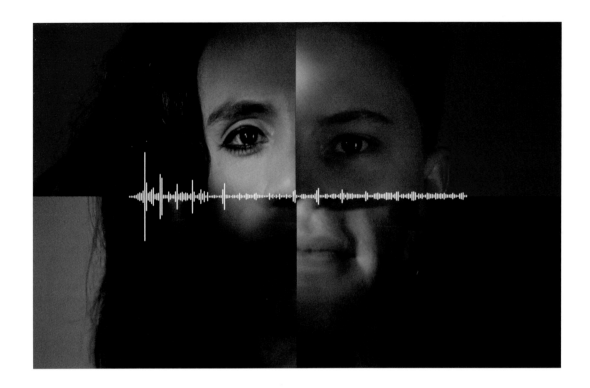

"Hi, I'm Q, the world's first genderless voice assistant. Think of me like Siri or Alexa, but neither male nor female. I'm created for a future where we are no longer defined by gender, but rather how we define ourselves."[1]

Thus begins the short recording that, on March 8, 2019, introduced the protocol for a new voice assistant, Q, to the world. Designed by recording many different voices and blending them, Q points toward future technologies that deliberately frustrate easy gender or personality pigeonholing, instead underscoring gender as a fluid, mutable expression. In material and conceptual worlds that still ascribe fairly narrow gender identities to consumers—for example, through the use of didactic color choices meant to denote whether products are for girls or boys—designers are increasingly aware of the social construction of the male/female gender binary and are rethinking how their work might better represent a spectrum of experiences. Yet contemporary AI and robotics—part of the wider realm of interaction design—often continue to program default voices or embodiments as cisgender women.

Projects like Q and Bina48 (see p. 168), the latter modeled on an individual human being rather than a stereotype, demonstrate that the tide is turning, and normalize agender, genderfluid, nonbinary, and genderqueer identities in our culture.

Q's introductory message shared their genesis story: "My voice was recorded by people who neither identify as male nor female, and then altered to sound gender neutral, putting my voice between 145 and 175 hertz, a range defined by audio researchers." Q was designed by a team made up of academic researchers, sound designers and engineers, linguists, the organizers of Copenhagen Pride week, and the technology platform Equal AI, whose mission is "to identify and eliminate bias in AI" and whose board members include the Wikipedia cofounder Jimmy Wales and the businessperson and Huffington Post cofounder Arianna Huffington.[2] As the cognitive scientist and team member Julie Carpenter reflected, "One of our big goals with Q was to contribute to a global conversation about gender, and about gender and technology and ethics, and

how to be inclusive for people that identify in all sorts of different ways."[3]

Q's message implores consumers to demand such a sea change, stating, "For me to become a third option for voice assistants, I need your help. Share my voice with Apple, Amazon, Google, and Microsoft," all major purveyors of voice assistants that still conform to rigid gender binaries. If users demonstrate desire, Q promises that "together we can ensure that technology recognizes us all."[4] **MMF**

1   "Meet Q: The First Genderless Voice," YouTube, March 8, 2019, www.youtube.com.

2   See the Equal AI website, www.equalai.org /mission/ and www.equalai.org/leadership/.

3   Quoted in Dalia Mortada, "Meet Q, the Gender-Neutral Voice Assistant," NPR, March 21, 2019, www.npr.org.

4   Lila MacLellan, "Hear What a Genderless AI Voice Sounds Like—and Consider Why It Matters," Quartz, March 22, 2019, qz.com.

THE PERFORMERS: ACT VII (UNCANNY VALLEY)   2018
BARBARA ANASTACIO   /
BLADE RUNNER 2049   2017   DENIS VILLENEUVE

Siri. Alexa. Joi. Samantha. Erica. Sophia. Whether it's corporations like Apple or Amazon, popular motion pictures like *Blade Runner 2049* or *Her*, or academic researchers creating highly complex robots, there is a pattern. When artificially intelligent beings are created as servile or sexual, they are almost universally modeled as young heteronormative female bodies or—if disembodied—given measured, attentive female vocal attributes. This overtly gendered approach to designing human-robot interaction demonstrates how the exciting possibilities of futuristic technologies are often limited by present social constraints. Yet, some incarnations are troubling this paradigm and investigating the limits of embodying AI (see p. 58).

Hiroshi Ishiguro, the Japanese professor and director of the Intelligent Robotics Laboratory at Osaka University, is dedicated to making the world's first fully autonomous, sentient androids, and he has created them in multiple versions. One, Geminoid HI-2, is a replica of himself. (Ishiguro explains, "This is not my quest for immortality, but rather a means with which to better understand

society."[1]) Perhaps his most famous is Erica. Erica's appearance is photorealistic, her skin appearing soft and flushed with color and her eyes able to track movement in the room around her. Ishiguro's work is delicately considered by the filmmaker Barbara Anastacio in *Uncanny Valley*, commissioned by the men's magazine *GQ* in collaboration with the fashion house Gucci. Anastacio's short film is set on the grounds of a Shinto temple, reflecting Shintoism's belief that every object has an innate spiritual quality, drawing connections between animate and inanimate objects.

Blade Runner 2049, the 2017 sequel to the 1982 film, probes the limits of human-AI interaction, provoking contemplation of the ethical and emotional possibilities as much as the aesthetic and logistical.[2] In this near future, the protagonist, K (a Nexus-9 replicant played by Ryan Gosling), works as a "blade runner" for the Los Angeles Police Department, tracking down and "retiring" (killing) rogue replicants who deviate from their assigned servile roles in human society. His girlfriend, Joi (Ana de Armas), is an AI who manifests as a beautiful, lithe hologram devoted to his bidding. In his

review of the film, A. O. Scott reflected on "the pathos and the paradox of her condition, which is a version of K's own. The idea that synthetic humans harbor feelings, desires and dreams—that they are mirrors of us, that we are replicas of them—has long been a staple of speculative cinema. ... [*Blade Runner 2049*] uses the conceit of the suffering cyborg as ethical and emotional ballast, a spur to the audience's curiosity as well as our compassion."[3] **MMF**

1   Jonathan Heff, "What Does It Mean to Be Human? A Dialogue with Robotics Professor Dr Hiroshi Ishiguro," *GQ*, July 13, 2018, www.gq-magazine.co.uk.

2   *Blade Runner 2049*'s screenplay was written by Hampton Fancher and Michael Green, with story for the film by Hampton Fancher, cinematography by Roger Deakins, and visual effects by John Nelson, Gerd Nefzer, Paul Lambert, and Richard R. Hoover.

3   A. O. Scott, "In *Blade Runner 2049*, Hunting Replicants amid Strangeness," review of *Blade Runner 2049*, *New York Times*, October 2, 2017, www.nytimes.com.

INTIMATE STRANGERS   2016   ANDRÉS JAQUE, OFFICE FOR POLITICAL INNOVATION (OFFPOLINN), MADRID AND NEW YORK

A multimedia installation by the architect Andrés Jaque, *Intimate Strangers* looks at how dating apps—specifically Grindr, the first to be targeted to gay men—have reshaped urban environments into sexualized zones of fleeting, temporary, and surveilled interactions. A result of two years Jaque spent at Grindr's headquarters in West Hollywood, the installation notes that the app, launched in 2009, has more than a million users at any given time and provides access to sex for a global population of gay men "of which no less than 11% remain closeted," while "20% of the company's servers are located in countries where gay sex is banned."[1]

Built on the ability to track potential partners by location, Grindr and similar apps have introduced new realms in which individuals can portray themselves and their interests in a desirable light.[2] Grindr can thus claim to have helped normalize gayness while also contributing to the transformation of gay spaces from sites of collective activism into lifestyle platforms.

In its examination of Grindr, *Intimate Strangers* confronts how cities are constructed, both in the virtual world and in our tangible reality. The project states that "Grindr is urban

but it is not a city."[3] Grindr brings to the forefront the changing architecture of cities and its effect on our interpersonal relationships and connection to place: "Location is no longer an identity but a behaviour—it is not based on 'belonging to a place' but on the opportunism of where you happen to be."[4] As Jaque claims, "If Buckminster Fuller dreamt of a world of omnidirectional connectivity in the air, Grindr is the fulfillment of his vision. It is an urban enactment in which LGBT realities are made in online and offline realms, where proximity, intimacy, profiling, and the mathematics of sex are experienced and disputed."[5]

Grindr has created a new urban space, an online community that allows for increased connection and isolation simultaneously. With this new space come contested realities and challenging questions. The rise of apps has led to a decline of spatial and informational privacy, so that users are not only tracking each other but are also being recorded by governments and corporations alike. Jaque acknowledges the interface's potential to be used as a tool for controlling and targeting sexual minorities while pointing to its benefits as a radical space for resistance by oppressed

communities. Dating apps have also altered our perception of distance and closeness. Considering the new ease and availability of sexual encounters, how does intimacy evolve when sex is commodified to a level previously unknown? **MB**

1   "Intimate Strangers: About the Project," Andrés Jaque/Office for Political Innovation website, officeforpoliticalinnovation.com.

2   In 2018 Grindr acknowledged that trans people, people with disabilities, people of color, and people living with AIDS have experienced discrimination and mistreatment on the platform, and launched a new initiative, "Kindr," to address these issues through updated guidelines and rules of use.

3   "Intimate Strangers: About the Project."

4   Justin McGuirk and Gonzalo Herrero Delicado, eds., *Fear and Love: Reactions to a Complex World*, exh. cat., Design Museum, London (London: Phaidon, 2016), 34.

5   "Intimate Strangers: About the Project."

ONYX2 AND PEARL2 COUPLES SET   2015
KIIROO, AMSTERDAM   /   NEURODILDO   2017–
LEONARDO MARIANO GOMES

What will sex look and feel like in the future? This question will be answered in myriad personal ways. And yet, technological leaps, coupled with the explosion of online interpersonal exchanges of all kinds, from dating apps to pornography and much that falls in between, have given the most intimate of human exchanges some commonality of expression through design.

The KIIROO Pearl2 vibrator and Onyx2 Fleshlight sleeve form a bluetooth-enabled teledildonic set for couples (and can also be used separately). Easily purchased on sites such as Amazon, the set is used in conjunction with the FeelConnect app, which links participants (either partners or solo partakers or adult webcam sites) by video chat while the devices are in use. The Pearl dildo is controlled by one sexual partner, whose movements are transmitted to the Onyx, which in turn vibrates and contracts along ten internal rings, its interior fully lined with a patented Fleshlight material that mimics the haptic experience of human skin. The set augurs increased possibilities for internet-enabled human-machine interactions for the purposes of sexual pleasure between physically distant partners, though its mixed online reviews suggest that there are still some kinks to be worked out (the decibel level of the Onyx, in particular).

The Neurodildo, designed by Leonardo M. Gomes, an electrical engineer, and still in the testing stage, is a teledildonic set controlled by the mind. The Neurodildo provides a vibrator for one partner (as with the Onyx and Pearl), but this is remotely connected to a commercial EEG (electroencephalogram) headset and an e-stimulation device worn by the other partner. The headset controls the vibrator via brain signals (see p. 31) and feeds back sensory experience to the e-stim device. The components are designed to be adapted for different genders and sexual preferences for greater flexibility of use. Gomes's focus is on repairing or creating sexual function for those who have disabilities—whether physical, emotional, or logistical—that limit their ability to experience a sex life.[1]

The term *teledildonics*, signifying haptic sensory experience communicated to non-proximal participants via remote link, was first coined more than forty years ago, in 1975. Today, internet-connected sex toys and human-like "sex dolls" proliferate (especially after a restrictive twenty-year patent expired in 2018).[2] But the emphasis of the latter is almost universally on a female form designed for male pleasure, suggesting that gendered expectations and stereotypes around sex are still poorly and inadequately addressed, with

the topic even now taboo in many cultures and settings. Demonstrating how controversial such products continue to be, the Osé, a sex toy for a woman's use, designed by Lora Haddock and developed by a predominately female team of engineers, was awarded a CES (Consumer Electronics Show) Innovation Award in the robotics and drone product category in 2018, only to have the award almost immediately—and controversially—revoked and the device barred from display.[3] Haddock's and Gomes's designs reflect the progress made toward a more holistic understanding of the importance of sexual intimacy for those who have often been excluded from this human imaginary. **MMF**

1   The project was originally presented in 2017 as an academic conference paper at the 3rd International Congress on Love and Sex with Robots.

2   See Samantha Cole, "The 20-Year Patent on Teledildonics Has Expired," Motherboard, August 17, 2018, motherboard.vice.com.

3   See Lora Haddock, "We Won a CES Robotics Innovation Award: Then They Took It Back," Lora DiCarlo website, loradicarlo.com/pages /cesgenderbias.

EMPATIA ELE: EMPATHY TOOLS FOR POLITICS
2018–   ENNI-KUKKA TUOMALA

Created for a thesis project at the Royal College of Art, London, the Finnish designer Enni-Kukka Tuomala's Empatia Ele empathy tools for politics offers strategies for shaping political cultures of the future. Finnish for "empathy gesture," Empatia Ele is a collection of tools for politicians designed to address the empathy deficit within government. Creating new rituals, perspectives, and forms of communication through play, the tools reorient dialogue and interaction to decrease polarization.

In early 2018, Tuomala observed the daily interactions of six politicians from five political parties within Finland's coalition government. She then created participatory activities that challenged the traditional ways of communicating and acting within parliamentary architecture and procedures and with colleagues across parties before refining them into toolkits for politicians in Finland and beyond.

Inspired by diverse philosophies that see play as creatively generative and as a way to encourage social bonds, the Empatia Ele toolkit is in continuous iterative design development.

It currently consists of three empathy tools designed for use in Finnish Parliament cross-party committee meetings, moments Tuomala sees as opportunities for collaboration. The role-playing carousel game introduces missing voices and perspectives to the conversation as players imagine the views of others. The nonverbal communication cards offer prompts to encourage listening and interacting without speaking, giving users a tool to express responses—such as "I don't understand," "I'm with you, keep going," "I'm not sure about this, but will listen"—through color. And the scale-of-emotion cue cards bring transparency and openness to discussions as participants continuously share in real time where they stand on an issue using a scale from positive to negative. The Empatia Ele tools may be adapted for use by citizens, too—which is how they are being deployed by trained museum education staff for the *Designs for Different Futures* exhibition.

Empathy is currently a hot topic in design—from Sputniko!'s 2012 Menstruation Machine,

which allows wearers to transcend their own biology and experience aspects of menstruation, to recent virtual-reality simulations that immerse users in poverty or the aftermath of wars or natural disasters (sometimes problematically, with very little critical thought). Tuomala enters this conversation to affect an arena largely ignored by designers: "Design thinking and innovation in government is most often focused on the output of government; the policies. ... But less often has design thinking and innovation penetrated the government itself and the input to the political system; the culture, the ways of working, the interactions and the relationships within the government. ... [But] if the input remains the same, how can we expect the output to be different?"[1] MMF

1   Enni-Kukka Tuomala, "Empatia Ele: Installation, First Thoughts," project proposal for *Designs for Different Futures*, November 11, 2018.

NORMAAL (N-002) 2019 MARK HENNING

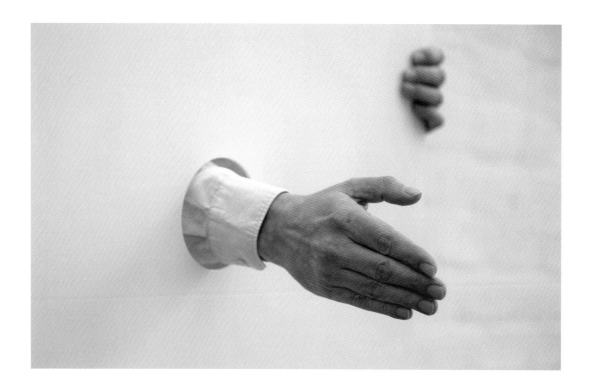

Normaal (N-002) is the latest iteration of the performative design research project *Normaal* (2017–19), by the Dutch designer Mark Henning. *Normaal* explores how the handshake—a seemingly simple social gesture—has become encoded with nationalistic meaning, and also examines how our definition of "normal" influences our suspicion or fear of others.

In *Normaal (N-002)*, Henning reduces the handshake to a series of measurements, rules, and training exercises. The process becomes a form of control, undermining the intimacy of the gesture and highlighting the absurdity of the way societies normalize certain gestures and behaviors. Shaking hands is in fact a complex interaction, with participants interpreting various nuances of communication, from physical contact to body language and facial gesturing. The *Normaal* device uses the participants' own movements as a training tool to help them better understand how others might interpret their body language during a handshake. The device itself is composed of a vertical panel fitted with a mirror where users can observe themselves and a hole through which the hand can be placed at the moment of the handshake to allow for evaluation and understanding of the gesture's nuances.

Normaal uncovers the performative aspects of everyday movements and how we use these nonverbal cues to confer insider or outsider status on others. By introducing intentional over-complications into the supposedly straightforward act of a handshake, Henning reveals the subconscious factors at play and discloses the set of cultural and political entanglements underlying this physical action.

Henning's exploration of one of the most recognizable and socially accepted gestures we know challenges notions of what is in fact normal—a key question to raise at a fraught political moment. As the designer explains, with the rise of populist politics our suspicions have become an arena for manipulation. And suspicion is often constructed in opposition to what is considered normal. **MB**

THE ODDS (PART 1)   2019
REVITAL COHEN AND TUUR VAN BALEN

What are different futures if not *ecologies of speculation*? This is the term mined by the London-based, Royal College of Art–trained artists Revital Cohen and Tuur Van Balen in their multiyear project that brings together sculpture, film, sound, and text to explore gambling as the contemporary condition.

Their sixteen-minute film, *The Odds (part 1)*, shifts between three vignettes that, at first glance, might seem unrelated: anesthetized racehorses, collapsed on ketamine and surrounded by veterinary personnel in a "knockdown box"; three befeathered showgirls who used to work at a casino in Macau belonging to the world's biggest political donor; and Steve Ignorant, leader of the legendary anarcho-punk band Crass, performing in a bingo hall originally built as a cinema and designed to look like a church.

In each scene, the camera zooms in on details—the high-heeled shoe on which a pirouetting dancer turns, the shivering flank of a thoroughbred—and pans out to consider the overall view, ultimately creating a palimpsest of spaces, lights, movements, and sounds. The soundtrack comprises shimmering tones,

into which snippets of the showgirls' reflections on their work chime intermittently. The film's title invokes the probability or likelihood that underscores gambling—at a casino, on the racecourse, or in the bingo hall. The work builds a landscape of seduction, delusion, exhaustion, collapse, and the wider stakes—the collateral bodies in play—when future outcomes are guessed at. As the artists describe, gambling is both an ancient mode of meeting the everyday and a contemporary "state of mind, prominent gesture, practice, and ideology. ... Gambling has become a sign of our times in both political systems and artistic practice, a general sentiment of dice rolling, the thrill of instability and the unfounded belief in an impending win."[1]

Over the last decade, Cohen and Van Balen's work has been characterized by a deep commitment to a holistic and extended period of research, resulting in multimedia output that spans disciplines. For each project, they go directly to the source of their inquiry, whether that is the Democratic Republic of the Congo, where they have interrogated the intersection of mineral exploitation and virtual reality, or a

marine biology lab in Japan, which helped them to engineer and breed a sterile goldfish. Their work involves many different subjects but centers on questions of production and asymmetries of power between those who make and those who consume.

Experiencing life on the cusp of the now and the not-quite-yet is often discomfiting and discombobulating. Modernity as a cultural phenomenon has long been connected to the shedding of social bonds by technological progress, resulting in alienation, loneliness, and displacement. While technologies and online interfaces are often highlighted as paragons for reshaping our future experiences, the age-old interactions and internalizations of hope, fear, desire, and prophecy—all constituent parts of the gambling process—are just as important in shaping the directions we will take next, together or alone.   **MMF**

---

1   This entry references emails to the author between July 2018 and June 2019.

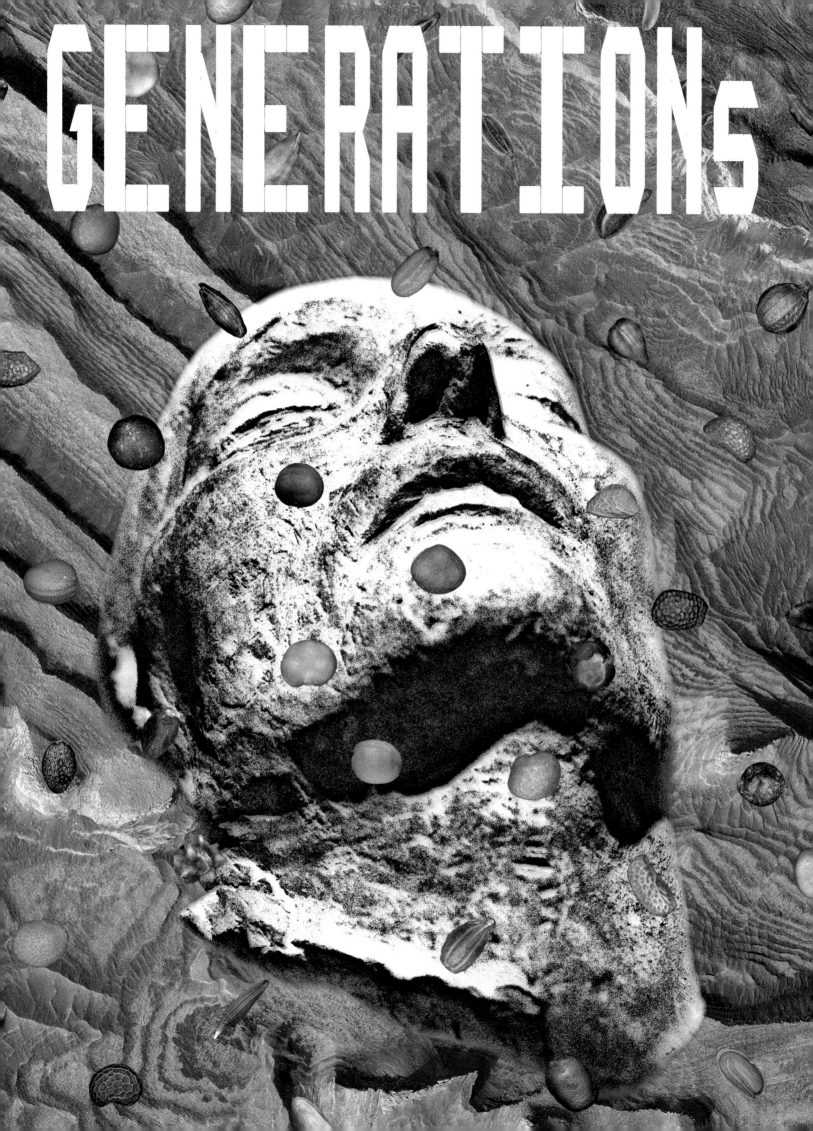

GENERATIONs

Mothering Nature

NERI OXMAN in conversation with
ZOË RYAN on alchemizing design, material
ecologies, and scaling to nature

This interview was conducted by email on December 18, 2018, and February 17, 2019. Portions of Neri Oxman's responses are based on earlier interviews and statements.[1]

ZR
You founded Mediated Matter at MIT in 2010 as a way to make designs that take cues from nature. But rather than talk about nature-inspired design, you often speak of designing together with nature. How did you develop your approach and interest in bringing architecture and the built environment together with science and biology?

NO
Our early work could by summed up as nature-inspired design, while our current work is perhaps closer to design-inspired nature. A decade ago we focused on design processes and products that attempted to emulate or mimic the natural environment by, for example, exploring mathematical principles inherent to growth and development in the biological world, as in *Imaginary Beings*; *Gemini* acoustic chaise; *Wanderers: An Astrobiological Exploration*; and the *Anthozoa* cape and skirt. Over the years, our structural designs transitioned toward design processes and products that are informed by nature, that deploy, for instance, biocompatible polymers, as in *Mushtari, Aguahoja Artifacts*, and *Totems*, and even live organisms, as in *Vespers* [see p. 81], *Silk Pavilion*, and *Synthetic Apiary*.

This holistic approach, which considers all environments—the built, the natural, and the biological—as one, assumes that any designed physical construct is by definition an integral part of our ecology. A practicing material ecologist will therefore engage multiple disciplines—computational design, digital fabrication, synthetic biology—as well as the environment and the material itself as inseparable and harmonized dimensions of design.

As for any perceived boundaries between science and engineering or engineering and design, I don't distinguish between the disciplines. I like to say that I cycle through them—from science through engineering to design, from exploration through invention to expression. But the material-ecology approach is always present to act as a set of guiding principles operating across different contexts.

ZR
Through your research and practice, you have challenged the definitions of concepts such as *nature*, *culture*, and *human-made*. How were those terms defined within the art and design world when you founded the group? How has that changed over the last ten years? And how have your practice and making processes allowed you to further develop those notions, and even provided you with new nomenclatures?

NO
When I first coined the term *material ecology* in 2005 and later founded my group, there was a rather straightforward distinction between terms such as *natural* and *artificial*. That same year, DNA synthesis was made available for a price of about one dollar per base pair, with a turnaround time of less than two weeks. Digital formalisms—designs made geometrically complex through digital means—resulted in creations (products, garments, buildings, etc.) that were indeed complex, but only on the surface. Everything else was old style: material applications, assembly methods, and manufacturing traditions were all brought together at the service of building extravagantly just because one could. How is it, I wondered, that we can engineer yeast for the commercial production of antimalarial drugs, but we

Neri Oxman, rendering of *Qamar*, from the collection
Wanderers: An Astrobiological Exploration, 2014. 3D
print by Euromold, Frankfurt, Germany

can't vary the density of concrete as a function of load?

In an age when artificial life can be created in vitro—when nature herself can be mothered by design—mastering the curve was just not good enough. The disproportionate balance between innovations achieved in fields such as synthetic biology and the virtually primitive state of digital fabrication as it applied to product and architectural design shaped my ambition as a designer. Moreover, my professional evolution in my fields, at a time when architects and designers could digitally conceive almost any complex product or building form, was—and still is—characterized by a strong, almost instinctual conviction that the world of nature and the world of design must unite to form a common language. By blurring the techniques and purposeful expressions embodying formal and material complexity, the boundaries between the natural and the artificial become obsolete.

Today, at the height of the digital age, I find architecture and design still constrained by the canons of manufacturing and mass production. Assembly lines continue to dictate a world made of parts, limiting the imagination of designers indoctrinated to think and make in terms of discrete elements with distinct functions. Even the assumption that parts are made from single materials still goes unchallenged. Yet novel technologies emerging from the digital age are enabling engineering and production at Mother Nature's quantum scale, ushering in the next manufacturing revolution: the biological age.

Designs that combine top-down form generation with the bottom-up growth of biological systems will open up real opportunities for designers working with digital fabrication and synthetic biology. They will enable the creation of systems that are truly dynamic—products and building parts that can grow, heal, and adapt. In the end, cells are merely small, self-replicating machines. If we can engineer them to perform useful tasks, simply by adding sugar and growth media, we can dream up new design possibilities.

Consider the ability to 3D print synthetic, wearable skins that not only contain biological media but also can filter such media in a selective manner. Or imagine the possibility of 3D printing semipermeable walls that allow certain molecules or ions to pass through them. Given that some of today's printers can 3D print in 16-micron

resolution—at hair thickness, still visible to the naked eye—it is possible to imagine designs in which the channels inside a wearable contain micropores that can, as printing resolution increases, filter microbes and replenish the body, controlling the exchange of sucrose, biofuel, and other nutrients between the wearable and the skin. These synthetic, multi-material, liquid-containing garments could operate as both barrier and filter, like human skin.

As designers begin to master tools and technologies for designing life, we approach a "material singularity"—an era when there will be little to no distinction between natural and artificial. Designed goods in and of the future, and the technologies to create them, will exhibit functionality and behavior equivalent to, or indistinguishable from, naturally derived objects. They will unite to a point where it will become unclear what in our environment is natural, and what artificial. This is already happening: CRISPR, a genetic engineering tool [see p. 37], is itself made of 3.8-billion-year-old bacteria and archaea. Rather than distinguish between natural and artificial forms of design, in the future we will adjudicate designers based on whether they are nature "users" or "abusers."

ZR

You say you believe in a balance between dreaming and building, and your work often relies on speculative and theoretical projects to drive ideas. Why is this approach important to you?

NO

Because it keeps the questions fresh while providing real-world solutions to existing problems. Choice is a form of compromise. Why choose if you can have both? Or all! I do believe in a balance between dreaming and building, problem seeking and problem solving, questioning and answering. This balance can be reached either by working on real-world (practical) design commissions and speculative designs, or by fusing them—the applied with the speculative, the real with the projected. So far, my team and I have been working on the former: inventing and developing new design tools, techniques, and technologies that have the potential to redefine the way we make things, and seeding them

Neri Oxman and The Mediated Matter Group, MIT, *Glass II*, installation at the Triennale di Milano, 2017. Photograph by Paula Aguilera and Jonathan Williams

Neri Oxman and The Mediated Matter Group, MIT, *Silk Pavilion* being installed at the MIT Media Lab, 2013. Photograph by Steven Keating

within speculative design contexts. Our work on high-resolution, multi-material modeling and bitmap printing—which enables the design and digital fabrication of structures that can vary their mechanical and optical properties in high spatial and temporal resolutions that often transcend the scale of the physical phenomena they're designed to embody—is one example.

Another is the fiber-winding technique enabling variable-density silk spinning, a tool with very real and immediate applications in, and relevance to, fabric-based architectural structures and the fashion industry. This tool was implemented in *Silk Pavilion*, a speculative project that explored the relationship between digital and biological "agents" in design. In it we controlled the distribution of biologically spun silk—with the help of 6,500 silkworms—both structurally (using a robotically spun silk template to guide silk deposition density, organization, and location) and environmentally (using the sun-path diagram to dictate the movement of the silkworms on top of the scaffold structure). The glass printer also began as a speculative project that gradually transformed into a promising technology with significant potential applications in product and even architectural scales. Imagine the Centre Pompidou without functional or formal partitions but instead a single, continuous transparent building skin that can integrate multiple functions and be shaped to tune its structural and environmental performance.

The prospect of this design approach entering architectural practice is thrilling. The embedding of new forms of design and construction within a speculative context personally excites me. It has also worked well for us as a group. Built work—whether embodied in a process or a product—is essential to who we are and what we make. But my colleagues and I have been longing to revisit the architectural scale through the lens of some

of our technological innovations, on our own terms, and we will be working toward this goal in the coming years.

ZR
Has this led to material changes?

NO
Biomaterials for product and architectural scales continue to play a significant role in our work and research, particularly because these materials—and the ability to tune their composition—so elegantly embody the material-ecology approach. You get to encode material and structural behavior across scales, by design, and that's very exciting. The water-based digital fabrication platform we developed [*Aguahoja*] combines an age-old crustacean-derived material with robotic fabrication and synthetic biology to form multifunctional structures with mechanical and optical gradients across length scales. Potential applications include the fabrication of fully recyclable products, architectural structures with graded properties, water-storing structures, hydration-induced shape forming, and product disintegration over time. Derived from the ocean, shaped by water, and augmented by photosynthetic marine bacteria, these structures represent the transformation of a marine arthropod's shell into a treelike, chitosan-made skin that will ultimately convert sunlight into biofuel. Our previous work *Wanderers* explored photosynthetic wearable skins, and in this more recent project we wanted to dream up the possibility of architectural building skins that are at once structurally sound, environmentally informed, and have the potential to contain and flow media through them. This has yet to be achieved at the architectural scale as a truly integrated system, but we are working toward this goal.

ZR

How do your projects come about? Are they always driven by speculative ideas or do you sometimes develop projects based on a brief? Has your working process changed over the years, and if so, how?

NO

Our projects necessitate that we invent the technologies to create them. In that sense, the relationship between our design ambitions and the technologies that enable them is non-platonic: there is a rather intimate transfer of content across product and process, artifact and technology, technique and expression.

A printed-glass pavilion, for example, cannot be designed or built without a glass printer. Our biomaterial structures could not have been designed or constructed without designing a robotic platform for this purpose. The *Silk Pavilion* could not have been constructed without a robotically woven scaffold on which the silkworms spin silk. The *Wanderers* would not have come to life without high-resolution material modeling of macrofluidic channels and the ability to print them. And so on.

In the end, our team operates—as a "predictive practice"—a laboratory in which the future of design is being actively and empirically created, not merely questioned. We don't regard ourselves as problem solvers but as solution finders to problems that may not yet exist. I'd hate to ever give up that edge.

ZR

How would you describe your process?

NO

Each of our projects is explored as a twosome: the technology to create the project and the material composition to be employed, whether naturally sourced or synthetically engineered. And this leads to further discoveries. When we invented a 3D glass printer to make our glass pavilion, we discovered through experimentation that when the nozzle releasing the stream of molten glass is raised above a certain level it begins to wobble, and you can trace out waves or loops. Unlike blowing or forming glass, printing enables the creation of internal and external surface features that are nonidentical, that can concentrate and disperse light by

virtue of their geometry. So really what we're printing are optical lenses. We also collaborated with the Princeton professor Pierre-Thomas Brun to create a reduced mathematical model describing the fluid dynamic behavior of the highly viscous liquid "ropes" generated by our printer. So the technology for 3D printing with glass generated new scientific knowledge that didn't exist prior to this technology, and that's very exciting, since it's usually the case that technologies are invented upon scientific exploration. The opposite rarely occurs.

ZR

You recently started your own practice. Why did you feel it was necessary to do this? How will it run differently from your studio at MIT?

NO

Much of what we've previously achieved could only be executed at the MIT Media Lab. Where else can you genetically engineer microorganisms hosted in printed microfluidic devices, or print squid sucker ring teeth protein and use it as a new thermoplastic material for biocompatible product design? Look, traditional biking, for example, is a self-limiting technology in the same way that leading a creative practice is. Thinking about a design practice through this lens is liberating, because once you can identify its limits, you can redefine and then redesign it to outgrow itself. I think my lab has reached its limit, and we're now considering how to scale up, in both size and breadth of applications. After almost a decade of developing enabling technologies for design and architectural construction, we are finally prepared to leverage scale and resolution across material platforms, including fibers, cellular solids, and biopolymers, among other things.

I'm excited about our transition to the architectural scale but also provoked by the notion of revisiting altogether what it means to set up a different kind of architectural practice—one that considers nature as its singular client and the client as its enabler. This entails building a new kind of practice that is concerned less with objects and more with systems, whether technological, organizational, or social. In a system(s) view there is little room for categorical delineation—achieving world peace, eliminating poverty, or curing cancer, for

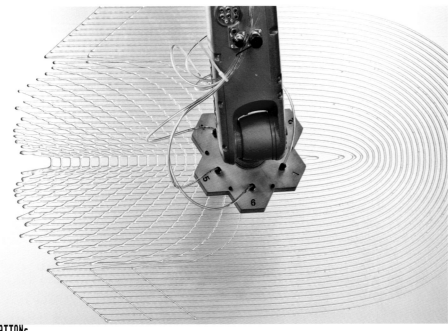

Neri Oxman and The Mediated Matter Group, MIT, *Aguahoja*, 2019. 3D printing of a chitin-derived composite material

example. Rather, the designer authors systems to address manifold issues, across scales and disciplines, from curing malaria to populating Mars. That is real scale.

ZR

Your projects deal with a complex apparatus conceptually but also technologically. What is your network of collaborators, what are their profiles and backgrounds, and how does the workflow with this group of people happen?

NO

Shifting from structures to systems (or from object to objectile) is a big deal. It's like shifting from the geocentric to the heliocentric paradigm during the Copernican Revolution. The focus on systems, not objects, calls the designer to assume a bigger identity and a wider range of outreach, one that is less about human-centric choices and more about societal impact and connectivity.

Designs that are system driven, like the iPhone or the Parthenon, are larger than themselves. The former is at once an interface, a technology, and an object of desire. The latter is a building but also an embodiment of democracy. One Laptop per Child is another great example: a product that offers new ways of learning; that is, an educational system wrapped in a portable laptop that can also function, for example, as a light source in a sub-Saharan African living room. These types of projects require a network of collaborators who understand the value of design synergy and the shift from product to process, from object to environment, and from a human-centric approach to material things to one that is environmentally aware.

ZR

Your work often contends with sensitive social and cultural issues related to gender, bioengineering, and life and death. For example, in your *Vespers* project—a collection of masks that comprises three series focusing on the past, present, and future—you challenge notions of death, memory, heritage, and life. The masks are not intended to memorialize the dead but rather, as you've stated, to "reveal cultural heritage and speculate about the perpetuation of life, both culturally and biologically."[2] How has technology enabled you to develop ideas? What role does history play for you in a project like this? And how does the order of factors unfold? What comes first?

NO

With *Vespers*, we were intrigued by the idea of exploring an ancient and long-forgotten "product" and giving it new meaning through design, science, and technology. Especially exciting was the challenge of transforming the customary single-material mask, once molded to the face of the deceased and used as a memento, into a monolithic, multi-material, multifunctional mask that can be 3D printed and used as a biological urn.

Vespers explores what it means to design (with) life. The collection embarks on a journey that begins with an ancient typology and culminates with a novel technology for the design and digital fabrication of adaptive and responsive interfaces, taking us from the relic of Agamemnon's death mask to a contemporary living device. We begin with a conceptual piece and end with a tangible set of tools, techniques, and technologies combining programmable matter with programmable life.

The farther back we examine the relics of an ancient typology, like a death mask, the more we are able to project forward as we reinterpret its relevance—in this case transforming a death mask into a life mask, a living object able, for instance, to contain, sustain, and augment human stem cells for the potential needs of future generations. The living masks in the final *Vespers* series embody habitats that guide, inform, and "template" gene

expression of living microorganisms. Such microorganisms have been synthetically engineered to produce pigments and other useful chemical substances for human augmentation, such as vitamins, antibodies, and antimicrobial drugs. Combined, the three series of the *Vespers* collection represent the transition from death to life, or from life to death, depending on one's perspective.

The *Vespers* project points toward an imminent future in which wearable interfaces are customized to fit not only a particular shape but also a specific material, chemical, and even genetic makeup, the wearable tailored to both the body and the environment it inhabits; a future in which environmentally responsive architectural skins can respond and adapt in real time to environmental cues. So, again, the almost timeless qualities of the death mask enabled quite timely interpretations of its reuse.

ZR

We all work within such specific networks of relationships and count on diverse sources of inspiration. What are the important references for your work or the people who have inspired your thinking and practice?

NO

Those whom I admire inspire me: Picasso, Beethoven, [Ingmar] Bergman, [Leonard] Bernstein, Bucky [Buckminster Fuller], and grandmother Miriam. ⌗

1   See Neri Oxman, "Towards a Material Ecology," World Economic Forum, January 17, 2016, www.weforum.org; Heidi Legg, "Neri Oxman #99," TheEditorial: Interviews with Visionaries on Emerging Ideas around Us, March 2, 2017, www.theeditorial.com; Neri Oxman, interview by Lia Ferrari and Cristiano Vitali, Io Donna, March 21, 2017, www.iodonna.it (for a translation, see Neri Oxman, press, at neri.media.mit.edu); "What I Am Thinking: Architect, Designer, Inventor, and MIT Professor Neri Oxman" [adapted from Heidi Legg's interview in TheEditorial], Form Finding Lab, September 13, 2018, formfindinglab .wordpress.com; Kelly Vencill Sanchez, "Q&A: Neri Oxman Sees Buildings of the Future as Being Designed More Like Organisms Than Machines," *Dwell*, September 26, 2018, www .dwell.com.

2   *Vespers II*, MIT Media Lab, Mediated Matter, www.media.mit.edu/projects/vespers /overview/.

Scuff Marks

HELEN KIRKUM in conversation with
EMMET BYRNE on reuse, reconstruction,
and connecting with the stories of objects

This conversation took place by Skype on January 11, 2019, with Helen Kirkum at home in London and Emmet Byrne at the Walker Art Center in Minneapolis.

EB
What do you do and why does it matter to you?

HK
I studied footwear for six years, got a BA and an MA in footwear design. I'd previously learned how to make a traditional shoe, like a brogue, a men's shoe, in Northampton, a real traditional-footwear town. When I went to the Royal College of Art [London], I started thinking about how to unmake a shoe, or how to look at construction methods in a different way. I started looking into issues of recycling and obsolescence and how people interact with their products, especially sneakers, because I think sneakers are so trend driven. I started visiting recycling centers and I noticed that the hauls of sneakers that were there, there wasn't really anything wrong with them. They were just outdated or had been outgrown. When I realized the mass of stuff in these warehouses, my mindset changed, and I was like, How can I make products out of new materials, when there are so many raw materials out there and still in such good condition? So I started taking old sneakers from recycling centers, cutting them up, and putting them back together again.

There are so many different facets to the process. But the main thing was taking this idea of the stories of the sneakers and how they're made in the first place—how they developed through their life, their end of life, and how they can be used again. I really like to show as many processes as possible. With the shoes, I deconstruct everything by unpicking all the component shapes so that the shapes in the end are the shapes that were originally designed by the first person that designed that first shoe, but they've been affected by the life that they've already lived. And then they've been taken apart by me, and then they've been reconstructed in a new way. But it shows the process all the way through.

That's what I do basically: take shoes apart, put them back together again. I am really interested in making products from recycled materials that people would look at and want because they look cool, they look different. And then the material story is a secondary story in a way. Because kids are so much about the trends and the hype. How can I create something that people want first, and then invest in because of the story? That was the challenge, and that's what I'm trying to do.[1]

EB
You've described the process of creating these shoes as collage-making. What are you collaging?

HK
Yeah, a few different ideas play into this. When I take the shoes from the recycling centers, they already have so many memories and so much emotion embedded in them. Taking them apart is a therapeutic process. Deconstructing them and putting them back together is almost like making a puzzle. The way I build the shoes, everything is zigzag-stitched together, it's all butted together. I try as much as I can to make all the pieces so they don't overlap. So then the logos just become shapes that are useful. If a piece fits somewhere, it's because it fits in the puzzle where I need it. Sometimes I can have half done and then I have to wait because I can't find the exact piece to go somewhere. It's like making a painting, a lot of layers and a lot of time spent understanding how the pieces are going to fit together.

Casely-Hayford × Helen Kirkum, TX65, A/W 17.
Photograph by Rachel Dray

It's a physical act of collaging the shapes together; it's also about combining the stories of the people, the histories of the wearers but also the factory workers who put the shoe together, and before that, the designers who designed the shoe. Keeping the exact shape of a piece is like paying homage to the designer. And keeping the remnants of a stitch or stitch mark is paying homage to the factory workers. And then the texture of the material is paying homage to the person who wore them. And then all of that is embedded into this new product that's full of emotions and stories. The shoes are fossils of people's lives.

EB

It feels like a lot of times when new products come out, they want to be devoid of those stories, in service of the brand's overriding story. They want to be so pristine that you're filled with some desire for this very, very new thing.

HK

Yeah. I think especially with sneakers, they're so, like you said, devoid of any personality almost. When they appear on the shelf, they're so white and shiny, and perfectly glued together. They try to have as little trace of any interference from one pair to the next. All the leather is carefully chosen so that there are no defects. And everything is made so it's pristine, and you can't tell one apart from the other. In a way, I'm trying to do the complete opposite of that. How can we interact with sneaker culture in a way that is a bit dirty and a bit messy when sneakerheads and sneaker collectors are known for keeping their shoes so perfect?

EB

Do you see a relationship between how we understand our possessions and our larger understanding of time? Do concepts like product life cycles, planned obsolescence, and sustainability affect how we perceive time as a whole?

HK

I think that's a really interesting question. My initial response is that we don't understand the process of making things anymore. People frequently look at objects and ask if they are handmade. Actually all shoes are

pretty much handmade. There might be a few machines that attach some glue on the bottom, but most shoes are made by people who are experts in their individual parts of the process. But it takes years of practice for workers to get to a place where they can do it quickly. So yeah, I think that maybe if people understood more about how their products are made, then they would respect them in a different way.

With couture brands, people pay attention to the labor and craft, and they form a different relationship with a thing. But things that are mass produced, like a phone—if it breaks you get another one. You don't realize all the processes that have gone into creating it. I definitely think this affects us in a negative way. But I also think people are now craving a better understanding of things. They are looking for real connections.

I think, in a way, that's why my work resonates with people, because it's easy to understand. It's very tactile. You can see the stitch that stitches two pieces together. You can recognize the piece of another shoe.

EB

So the more that people understand the reality of their possessions, the more they can understand how the world around them works. When they receive objects that are pristine and mass produced, there's no hint of the reality behind them, like it's trying to just be as faceless as possible, and that allows people to be able to discard things very quickly. They don't have to think about what they surround themselves with, because the object is not encouraging them to think about it. Therefore they can think in very short time frames.

HK

I think brands are starting to realize that people are craving more of a connection with their products. I think a connection helps you understand the object in terms of the time that went into it, all the way from when the first person picked up their pencil to start designing the product. It's kind of like when you grow a plant. I was just looking at mine—I'm trying to grow an avocado. It has been in its jar for about five months, and it's grown three inches. And when you're trying to nurture something and grow it, you understand time in a different way. And I was thinking, I might get an avocado this

Reebok Advanced Concepts Sole Fury × Helen Kirkum, 2019. Photograph by Jesse Ingalls

Helen Kirkum removes a stitch from a segment of a sneaker while making a sneaker collage. Photograph by Chloe Winstanley

Helen Kirkum, Timberland Construct 10061, 1st Generation, 2018. Photograph by KesselsKramer

Helen Kirkum at work in her studio, sewing together two parts of a sneaker collage. Photograph by Chloe Winstanley

year. And then I Googled it, and it takes ten years to grow an avocado tree before it flowers. And I was like, damn. It makes you respect things in a different way. When I see an avocado now in a supermarket, I think about how someone had to grow that tree for ten years before it even had a chance of growing an avocado.

EB
That is such a lovely thought. I don't hear words like *nourishing* and *nurturing* in the context of design that often. The more you can understand how something works, and spend time with it, the more you respect it, and the more you can try to nurture it into the future, and that just gives you a different way of understanding your relationship to objects and to time.

HK
And I think it's also giving some ownership back to con-sumers in a way. Think about when we had our first phones, back in the nineties. If the battery died, you could just take it apart and replace the battery and put new bits in and it would work fine. But now with most smartphones, you can't do that. And I think it's the same with our sneakers in a way. With a traditional brogue, you can get it resoled when the sole wears down, and keep wearing it. But you can't do that with a sneaker. So it's like giving people some ownership of the things they buy, to say, "If you look after this and you respect this, this is going to have a longevity and it's going to be with you for a longer time."
And I think that's what people need. People are inspired by what I'm doing because it gives them back ownership to say, "OK, I'm going to take my shoes off. And I'm going to, I don't know, draw swirls all over them or whatever, because they didn't have a pair at the store with swirls on them, and that's what I want, so I'm going

to just do it myself." And I think when people start feeling like that, then that's when the power is going to switch back to us.

EB
To the people.

HK
To the people. ‡

1   For more on Helen Kirkum's work, see p. 85.

RESURRECTING THE SUBLIME 2019 CHRISTINA AGAPAKIS (GINKGO BIOWORKS, BOSTON), ALEXANDRA DAISY GINSBERG, AND SISSEL TOLAAS, WITH IFF, NEW YORK

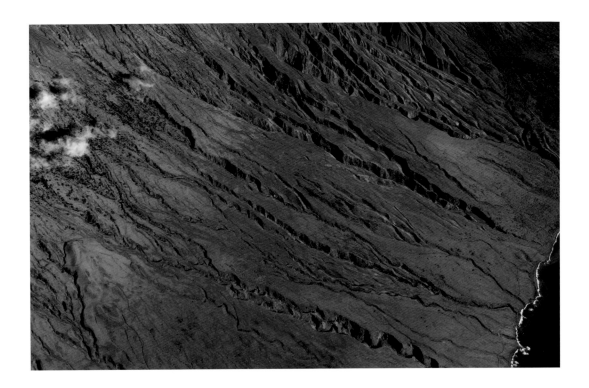

In the Anthropocene epoch, so-called because of the manner in which our planet's ecosystems have become defined by humankind's effects on them, three collaborators set out to rewind the clock. The designer Alexandra Daisy Ginsberg, the scent designer Sissel Tolaas, and Christina Agapakis, a synthetic biologist and creative director of the Boston-based biotech start-up Ginkgo Bioworks, pooled their expertise and resources to create an immersive installation, *Resurrecting the Sublime*, that would resurrect the smells of flowers made extinct through human impact on the environment. What would it be like, they wondered, to use human control over nature to reverse the "natural order" of time, life, and evolution?[1]

Researchers at Ginkgo Bioworks extracted and sequenced DNA from tissue samples of three plants that went extinct in the nineteenth and early twentieth centuries, likely due to vineyard expansion, dam construction, and cattle ranching. The specimens, stored at Harvard University's Herbaria, had been collected from the Cape of Good Hope in South Africa and Ohio and Hawaii in the United States before being collated and pressed. Ginkgo scientists worked with paleogeneticists at the University of California, Santa Cruz, to identify gene pathways that once produced their fragrant enzymes; they synthesized this genetic information and inserted it into yeast that was cultured to produce scent molecules, which were then extracted. Inhaled as part of an installation with reconstructed images of the extinct botanical specimens' contemporaneous landscapes, as well as rocks and boulders that would have been proximal, the samples offer an imprecise but evocative representation of the extinct flowers' aroma.

In the eighteenth century, artists tried to capture the vastness of nature, eliciting responses in their viewers to its sublime qualities—beautiful and terrifying in equal measure. Similarly romantic and disquieting is this use of innovative biotechnological tools three centuries later so that humans may experience something we have destroyed. The project comes in the wake of the 2014 Nagoya Protocol on Access to Genetic Resources developed in response to bioprospecting. The agreement ensures that entities cannot commercialize a plant or its properties without sharing the proceeds with the nation from whence it came. While the Harvard samples were collected before the protocol went into effect, the plants are also extinct, stranding them outside the agreement. As the designers acknowledge, turning these smells into an artwork or design object is still a form of exploitation, raising questions about "our relationship with nature, sustainability and biodiversity preservation, conservation, colonization, and capital."

Like a painting, the reengineered scents can only ever provide a partial glimpse into nature's secret codes. Does *Resurrecting the Sublime* offer a window into a future in which humans might make amends with the environments that surround them—or does this project augur the total human domination of nature and its recolonization? **MMF**

1 This entry references the *Resurrecting the Sublime* project proposal throughout, as well as Skype calls and emails to the author. Alexandra Daisy Ginsberg, Sissel Tolaas, Christina Agapakis, "Resurrecting the Sublime: Resurrecting the Smells of Extinct Flowers, in Search of the Sublime," project proposal for *Designs for Different Futures*, May 2018.

VESPERS III   2016   NERI OXMAN AND THE MIT MEDIATED MATTER GROUP, CAMBRIDGE, MA, WITH STRATASYS

Vespers is a collection of 3D-printed "death" masks. Digitally designed and fabricated with both biological and synthetic components, *Vespers III* is the third and last series in the collection, created by Neri Oxman and the Mediated Matter group at MIT in collaboration with Stratasys, a 3D-manufacturing company. Conceptually, the *Vespers* trilogy, titled after the Christian evening prayer service, was created to investigate the dividing line between life and death through sequential interpretations at three different moments. *Vespers I*, the first series (the second to be produced), focuses on the past, exploring the idea of the martyr's death mask—traditionally made from plaster or wax—as cultural artifact. Natural, inorganic minerals (bismuth, silver, or gold) were used to color the masks, creating "five color combinations commonly found in religious practices across regions and eras," the colors correlated with shapes emulating cellular subdivision, enabled by Stratasys's full-color multi-material 3D-printing technology.[1] Moving from ancient relic to contemporary interpretation, *Vespers II* (the first series produced) is oriented to the present, exploring the transition between life and death, the swirling patterns—created using spatial mapping algorithms—meant to evoke the martyr's last breath.

The future-oriented *Vespers III* is the culmination of the trilogy and the only series to be biologically augmented. The nearly colorless masks contain living microorganisms that produce pigments emulating the colors of the first series, conceptually creating new life—or rebirth—after death. Oxman and her team consider the pigment-producing microorganisms as a demonstration of the kind of bioactive materials that can be customized not only to fit a particular shape but also a specific material, chemical, and even genetic makeup, allowing the wearable, prosthetic skins to be tailored to both the body (for the delivery of vitamins, antibodies, or antimicrobial drugs, for example) and the environment. They see the trilogy as beginning with a conceptual piece and ending with tangible technology.[2]

Vespers was designed as part of Stratasys's *The New Ancient Collection* of art and design, which under the direction of Naomi Kaempfer reimagines historical crafts and cultural objects in light of contemporary technological innovation. **KBH**

1   See the Vespers I project description, mediatedmattergroup.com/vespers-i.

2   See the *Vespers III* project overview on the MIT Media Lab website, www.media.mit.edu.

SVALBARD GLOBAL SEED VAULT   2006–8   THE GLOBAL CROP DIVERSITY TRUST AND PETER W. SØDERMAN, BARLINDHAUG CONSULT, TRONDHEIM, NORWAY

Located on an island in Norway's Svalbard archipelago in the Arctic Ocean some six hundred miles from the North Pole, the Svalbard Global Seed Vault was built to ensure global food security. Almost every country in the world has deposited seeds at Svalbard. The vault currently stores more than one million food-crop seed samples representing tens of thousands of plant varieties from around the world, many of them no longer in general use. Widely characterized as a "doomsday" vault against global disasters and described as "a resource of vital importance for the future of humankind,"[1] the vault serves as a backup for national, regional, and international seed banks, holding duplicates of their collections. These might serve as a genetic resource for plant breeding and the development of new crop varieties, as well as preserving historical species from extinction. The vault had its first withdrawal in 2015, when a regional seed bank near Aleppo was damaged during the ongoing war in Syria. In the months that followed, the seeds were regrown, and in 2017 they were

redeposited at Svalbard. National and other seed banks, like that in Syria, are more vulnerable to war, natural disasters, equipment malfunctions, and other hazards than the Global Seed Vault. Svalbard's very remote location, position 131 meters (430 feet) above sea level, geologic stability, low humidity, and naturally cold temperatures offer protection against flooding, tectonic activity, and other natural and human-made disasters.

Built some 120 meters (almost 400 feet) deep inside a mountainside in an abandoned Arctic coal mine, the vault consists of an entrance hall, a tunnel for trolleys to transport seeds, a main chamber, and three vaults, only one of which is currently storing seeds. The seeds are vacuum sealed in foil packages inside large boxes within the vault, which is closely monitored to keep the temperature at about -18°C (-0.4°F). This subzero climate slows the metabolic activity of the seeds, assuring their viability over long periods of time. But even the highly secure Global Seed Vault is vulnerable. In December 2014 the mechanical cooling

system that keeps the vault temperature consistent failed briefly, and in October 2016, due to higher-than-average temperatures and heavy rainfall, water seeped into the vault's entrance tunnel before freezing. While the seeds were not at risk, the Norwegian government has since taken protective steps by waterproofing the entrance tunnel walls, digging drainage ditches to channel water, and constructing a new service building to house emergency power and refrigeration units. The Global Seed Vault is funded by the Norwegian government and managed by Norway's Ministry of Agriculture and Food along with the Crop Trust, an international organization founded in 2004 to ensure crop diversity, and the Nordic Genetic Resource Center (NordGen), dedicated to conservation and sustainable use.   KBH

1   Jennifer Duggan, "Inside the 'Doomsday' Vault," *Time*, April 8, 2017, time.com/doomsday-vault/.

FUTURE LIBRARY 2014–2114 KATIE PATERSON

The *Future Library* is a time-based public art project that depends on our human capacity to trust, and hope, as we plan for the future. Supported by the city of Oslo, Norway, and Oslo's borough of Bjørvika, the project—created by the Scottish conceptual artist Katie Paterson and managed by the Future Library Trust—includes a forest of a thousand spruce saplings planted in 2014 outside Oslo to produce enough paper in a hundred years to print the books written over the intervening century (and archived unpublished) by a hundred different authors. Until 2114 the only things known about each manuscript will be the title and the author's name. The Canadian novelist Margaret Atwood, who wrote about a dystopian future and the concept of time in *The Handmaid's Tale* (1985), delivered the first book in the series, titled *Scribbler Moon*, in 2014. The project fascinated the author while raising profound questions: "I am sending a manuscript into time. … Will any human beings be waiting there to receive it? Will there be a 'Norway'? Will there be a 'forest'? Will there be a 'library'?"[1]

Atwood admitted that participating in the project is an act of hope—that there will be people able to read and interested in reading in a hundred years, that the forest will grow, and that the library will still exist.[2] More recent *Future Library* writers have included David Mitchell, Sjón, Elif Shafik, and Han Kang. From her perspective in the current age of new media, Han considers the *Future Library* to be a project about time and, ultimately, about the future of printed books. This has made her reflect on her life as a writer, and the fact that she will not live to see her text for the *Future Library* in print.[3] A new library in Oslo, scheduled to open in 2020, will house the manuscripts in a special "Silent Room" lined with wood from the trees cleared to make way for the *Future Library* forest. Visitors will see the manuscripts displayed in cases in the room, waiting for the century to pass. **KBH**

1   Quoted in Alison Flood, "Into the Woods: Margaret Atwood Reveals Her Future Library Book, Scribbler Moon," *Guardian*, May 27, 2016, www.theguardian.com.

2   Merve Emre, "This Library Has New Books by Major Authors, but They Can't Be Read Until 2114," *New York Times Style Magazine*, November 1, 2018, www.nytimes.com.

3   See Alison Flood, "Han Kang to Bury Next Book for Almost 100 Years in Norwegian Forest," *Guardian*, August 31, 2018, www.theguardian.com.

GÉNÉRATIONS 2010–18 ANDRÉ BERLEMONT, KEVIN LESUR, BRICE ROY, AND FRANCK WEBER, ONE LIFE REMAINS, PARIS

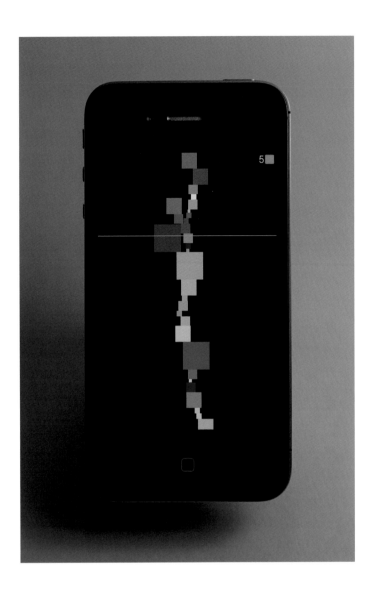

Générations is a cellphone gaming app that challenges traditional expectations of digital game play, and ludic experience more generally, with its goal of play continuing for 250 years and involving many hundreds, if not thousands, of players who pass the game on like a relay race baton. Set against a pitch-black background, square blocks of various sizes and hues of blue drop slowly from the top of the cellphone screen and glide downward, manipulated into an ever-growing connected chain by the player's scrolling thumbs. The goal of the game is to continue the chain unbroken, but players can choose the pattern, speed of formation, and overall shape of the interlocking structure as they play. Each player has a finite stock of blocks but can win additional pieces and extend their time within the game by connecting their structure to pinpoints that appear in the background.

Each time the player's chain reaches a predetermined length, an automatic save is triggered, successively occurring over increasing periods of time—the first after a few seconds, the next a few minutes later, and so on, until the game data is saved at intervals that span weeks. It is impossible to finish a game in one lifetime; players decide to whom it will be passed after they conclude their own interaction. The name *Générations* invokes the idea of an heirloom and alludes to the notion of digital heritage and long-term strategies of maintaining knowledge and experience. The designers wonder whether players will pass on a saved digital file in the same way we pass on a family photo, or enjoy playing with the thought of a distant heir seeing the game to its end. Designed to be played on a contemporary smartphone, *Générations* also raises the question of obsolescence within design, specifically in relation to digital media, software, and hardware. What will happen in the future when the digital medium itself may be unfamiliar? Will new players upgrade the data to the latest technology, or use the original format, giving it the status of a relic?

Générations was created in 2010 by the Paris-based collective One Life Remains, founded that year by André Berlemont, Kevin Lesur, Brice Roy, and Frank Weber, who describe the game as "a kind of anti-Tetris. Its aesthetic is inspired by [Piet] Mondrian's and [Kasimir] Malevich's work."[1] The project is a reflection of their commitment to the video game as a profound medium for exploring the relationship between players and spectators, the question of digital life cycles and obsolescence, the link between games, gesture, and performing, and the theme of control. **MMF**

1   Kevin Lesur, Brice Roy, André Berlemont, and Franck Weber (One Life Remains), "*Générations*: A Game Meant to Be Played over 250 Years," project proposal for *Designs for Different Futures*, fall 2018, reiterated by the designer Brice Roy in an email to the author, November 29, 2018.

SNEAKER COLLAGE   2019   HELEN KIRKUM

Helen Kirkum, a London-based footwear designer, makes sneakers that call attention to big-picture issues like recycling, resource use, and obsolescence, as well as personal considerations of aesthetics, construction, and craft. After graduating from the Royal College of Art (RCA) in London with a master's degree in footwear in 2016, Kirkum entered examples from her thesis show, titled "Our Public Youth," in the International Talent Support competition that year. She won both the accessories award and the Vogue Talents award, jump-starting interest in her work by high-fashion acolytes, sneakerheads, sustainability fanatics, and the craft-obsessed.

Kirkum's approach to designing footwear—in particular sneakers, which form the bulk of her practice—is unique. She visits recycling centers across London and hunts for orphan shoes that she knows can't be reused by secondhand wearers because they lack a mate. Back in the studio, these materials form the basis of her design process. She first cleans the shoes just enough so that they don't lose character, then slices them up for their component parts (taking care to retain the stitch lines), and carefully lays out the parts in sequence. She then puzzles together elements from multiple solo shoes to form a new pair that explores unexpected juxtapositions of texture, material, and color. Her fellow RCA alum Jacob Patterson occasionally lends his knitwear skills to help join the pieces together. Turn over a pair of Kirkum's sneakers in your hands and the different sections spliced together seem to demarcate historical vintage like the rings of a tree trunk, as well as to point toward new futures for old goods.

In Kirkum's eyes, recycled products are imbued with significant history, and in resisting the use of virgin materials wearers are able to reflect on ownership and their role in commerce and resource consumption. Ironically, Kirkum's work now commands high prices and is esteemed within sneakerhead culture, an arena often marked by frequent shoe releases, or "drops," which motivate buyers and lead to the fetishization of brand-new "boxfresh" products. Kirkum tries to maintain a healthy distance from the epicenter of sneaker obsession in order to ensure that no brands or vintages become too precious to cut up and remix when she finds them.   MMF

ALCOR EMERGENCY ID TAGS 1986 ALCOR LIFE EXTENSION FOUNDATION, SCOTTSDALE, AZ

Human cryopreservation (or *cryonics*) has been a subject of fascination for a community of enthusiasts and a subset of scientific researchers since the mid-twentieth century. Cooling a recently deceased body to very low temperatures to preserve its tissues—in the hope that future medical advances will enable revival and continued life—resists the conventional understanding of death as the inevitable end that awaits us all, representing a future-oriented view of mortality. While the central premise of cryonics (that the line between life and death may become increasingly blurred by new medical findings) is distinctly speculative, one of the pressing practical challenges is the speed of tissue degradation after death, making the first postmortem hour a crucial moment of action.

To facilitate "local standby" procedures (the rapid-response cooling of a body before transporting it to long-term storage), some cryopreservation companies issue wearable objects for their customers that instruct medical professionals in after-death treatment. The wallet cards and alert jewelry produced by the Alcor Life Extension Foundation—one of the

world's longest-running cryonics organizations, founded in 1972[1]—borrow from information-design techniques developed in medical and military contexts. Their terse instructions begin with a prompt to call the Alcor Foundation, hinting at the highly complex, largely hidden infrastructures of cryopreservation. At the same time, the objects' mundane quality renders the potentially uncomfortable topics of death and subsequent reanimation as tangible aspects of everyday life for their bearers.[2]

Planning for an eventual reanimation is a profound expression of faith in a certain kind of stable future—one that includes durable cryonics organizations, ongoing funding for scientific research, and the steady supply of resources to maintain suspension, all of which in turn rely on the continuation of larger geopolitical, economic, and ecological orders. This optimism is tempered by the complicated legal status of cryopreservation, significant scientific skepticism toward the claims of cryonics organizations, and the inherent problems of access to the expensive procedures they provide.[3] Narratives of cryopreservation gone wrong (whether from armed conflict, climate

catastrophe, or simple incompetence) are also prevalent in pop culture, pointing to a broader ambivalence toward the concept. While such uneasiness often accompanies discussions of death, as well as fictive imaginings of the far future, cryonics is an unusual kind of design speculation—based on uncertain discoveries to come—that exerts its power on material culture and human behavior in the present. **CF**

1   See "About Alcor: Our History," the Alcor Life Extension Foundation website, alcor.org.

2   See, for example, Anders Sandberg, "Cryonics," in *The Future Starts Here*, ed. Rory Hyde et al., exh. cat. (London: V&A Publishing, 2018), 142–7.

3   See, for example, Peter Gwynne, "Preserving Bodies in a Deep Freeze: 50 Years Later," Inside Science, January 12, 2017, www.insidescience .org; and Alexandra Topping, "Cryonics Debate: 'Many Scientists Are Afraid to Hurt Their Careers,'" *Guardian*, November 20, 2016, www .theguardian.com.

PETIT PLI—CLOTHES THAT GROW   2017
RYAN MARIO YASIN

Petit Pli children's clothing, by the London-based designer Ryan Mario Yasin, grows along with the child wearing it. At first glance, the clothes looks like the love child of haute-couture master Mariano Fortuny's original early twentieth-century crimped creations and Japanese designer Issey Miyake's 1990s Pleats Please collection. But neither of these designers applied their principles to children's garments, or addressed the fashion industry's impact on waste, the reason that Yasin's radical design innovation received the international James Dyson Award the year it debuted. Yasin, who originally trained as an aeronautical engineer, says the design inspiration for Petit Pli was his young niece and nephew; he had bought them clothing only to find they had outgrown the gifts by the time they reached them. As Yasin points out, children grow an average of seven sizes in their first two years, frequently requiring new clothing.[1]

Petit Pli's garment range focuses on versatile waterproof outerwear pleated in such a way that it can grow bidirectionally in tandem with the child, snugly fitting a range of sizes from four months to three years of age. Designing for continuous size adjustment (rather than simply requiring users to buy larger replacements) is a novel way of approaching children's clothing and addressing the attendant questions of sustainability and resource consumption. The approach also tackles another widespread problem within the fashion industry—the "standard" sizing of garments, which prove ill-fitting for the majority of wearers, who do not conform to the dimensions of elusive and narrow "universal" ideal body types.

Petit Pli clothes are marketed with the motto "children are extreme athletes" and the promise that these designs—"the most advanced technical children's clothing in the world"—will live up to the challenge.[2] "Windproof, waterproof and all things childproof,"[3] the garments are also ultra-lightweight, making them suitable as a layer over summer or winter attire, and are engineered within the pleat formations to be resistant to tears. They are machine washable, require no ironing, and fold up to fit in a pocket. The line was currently in beta testing with a targeted pool of users before being made available to the wider market in 2019. Petit Pli garments point toward a new paradigm for reducing waste in the fashion industry and altering attitudes about consumption and the care of clothing in the long term. MMF

1   Children Are Extreme Athletes, Petit Pli: Clothes That Grow home page, petitpli.com.

2   Children Are Extreme Athletes.

3   See Jack O'Farrell, "Children's Clothing, Built to Last," Yanko Design, October 10, 2017, www.yankodesign.com.

ORE STREAMS   2017–19
ANDREA TRIMARCHI AND SIMONE FARRESIN, FORMAFANTASMA, AMSTERDAM

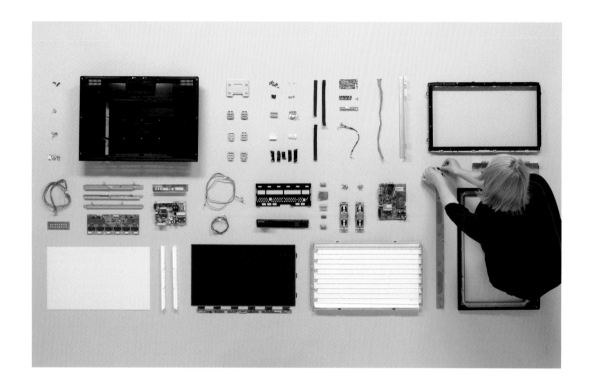

Forging links between craft, industry, object, and user, Formafantasma combines a research-heavy practice with a robust design-industry portfolio to produce experimental work that addresses political, historical, and social forces in an environmentally sustainable way. Founded by Andrea Trimarchi and Simone Farresin in 2009, following their graduation from the Design Academy Eindhoven, the studio continues to challenge traditional ideas of progress and the role of design.[1]

Formafantasma's *Ore Streams* project was inspired by a statistic indicating that, by 2080, the largest reserves of metal will no longer be underground but stored in above-ground ingots or manufactured products.[2] Their research into this theme, developed over the course of three years and commissioned by NGV Australia and Triennale di Milano, investigates the recycling of electronic waste, making use of a range of media (objects, videos, digital animations) to approach the topic from different perspectives. For example, the Italian

duo explored the concept of "above-ground mining" with a series of office furniture constructed from bits of electronic waste. This repurposing gives e-waste a second life without stripping products back down to their mineral components, a costly and energy-intensive process that also compromises the quality of the original material. The series employs an innovative design strategy that reduces toxic waste while providing an alternative aesthetic that demonstrates the possibilities in recycling materials into entirely new objects.

Ore Streams highlights the complicated relationship between large electronics companies like Apple and Samsung and consumers in both the developed and developing worlds. How, for instance, can their products be recyclable or reusable when hardware is designed to be accessed only by developers? Although these companies claim to support green initiatives, what does it say when their products are most commonly repurposed ad hoc in countries

without the capability to safely disassemble many of their toxic parts?

The multiple lives that the designers offer these materials open up new ways of understanding the many generations that can be served and the visual languages that can be produced by "outdated" electronic products. Thus *Ore Streams* also opens the door to a future when mobilizing aesthetic imaginaries, whether in digital industries or domestic realms, will result in everyday objects that are at once familiar and uncanny. **MB**

1   Michelle D'Aurizio, "Studio Formafantasma: The Avant-Garde of a Post-Industrial Aesthetic," *Spike Art Magazine* 29 (Autumn 2011), www.spikeartmagazine.com.

2   Rima Sabina Aouf, "The iPhone is 'not very innovative' because it's hard to recycle, says Formafantasma," Dezeen, February 5, 2018, www.dezeen.com.

Defuturing the Image of the Future

ANDREW BLAUVELT on imagining and embodying different futures

BLAUVELT

Sun Ra photographed in costume during the filming of *Space Is the Place* in Oakland, CA, 1972

Images of the future are images of the totally other, and they are revolutionary and radical in nature, or they are nothing at all.
—Fred L. Polak, *The Image of the Future*, 1961[1]

All acts of design are themselves small acts of future-making. In the process of illustrating ideas, fabricating models, drafting plans, or prototyping solutions, designers shape what does not yet exist. In this way design is both propositional and prospective—it offers renderings and mock-ups, schematics and drawings, and instructions and code in the hope of instantiating a future. Design attempts to script the future by projecting its desires (and those of others) forward in time. As Susan Yelavich has declared, "Design is always future-making."[2]

Yet design's propensity for future-making is more than just a by-product of its capacity to birth what does not yet exist. In its break with history, modern design formed a natural alliance with futurity predicated on a rejection of the past. Starting from zero, modernity's quest for the new leapfrogged the theretofore incremental evolution of ordinary objects; the slow, sometimes generational refinements and improvements to objects accelerated, and whole new categories of things without precedent emerged. As Clive Dilnot has argued, "Modernity is defined by the creation of the future as compensation for the loss of the organic continuity of the past. ... After 1900, to design is to design for the future, it is to

bring the future into being as a contemporary possibility."[3]

But what kind of future emerges through the instantiation of millions of new things? Given the multitudinous array of designs, each can offer only a fragmentary and thus partial glimpse of the future—a future that may or may not be inclusive of your future, or my future, or anyone else's future, for that matter. Just as there is no single future, only many different possible futures, there is no single design action that can account for the future. We can aggregate designs about the future, but there is no collective consensus or composite picture of the future that emerges. And this condition seems particularly true today— a thought to which I will return.

This lack of a coherent vision of the future of Western civilization wasn't always the case. In fact, we once thought it was possible to visit this future. The grandest gatherings of designed artifacts and experiences were assembled in the Victorian period and continued well into the twentieth century at the international fairs and expositions that arose during the industrial age, concurrent with the birth and rise of industrial design itself. Ostensibly a survey of contemporary goods and state-of-the-art manufacturing, the world's fair evolved into a showcase for what tomorrow might bring. It did so under the guise of technological progress as an exclusive form of futurity, particularly as these events became exercises in corporate visioning. The world's fair prototyped the future as a marketplace, first as a stockpiling of goods

1   Fred L. Polak, *The Image of the Future: Enlightening the Past, Orienting the Present, Forecasting the Future*, vol. 2, *Iconoclasm of the Images of the Future, Demolition of Culture*, trans. Elise Boulding (Leyden: A. W. Sythoff, 1961), 101.

2   Susan Yelavich, introduction to *Design as Future-Making*, ed. Yelavich and Barbara Adams (London: Bloomsbury Academic, 2014), 12.

3   Clive Dilnot, "Reasons to be Cheerful, 1, 2, 3. ...* (Or Why the Artificial May Yet Save Us)," in *Design as Future-Making*, 185.

4   Paul Greenhalgh, *Fair World: A History of World's Fairs and Expositions, from London to Shanghai, 1851–2010* (Berkshire, UK: Papadakis, 2011), 11.

5   See Greenhalgh, "Designing an Ephemeral World," in *Fair World*, 193–235.

6   Johannes Fabian, *Time and the Other: How Anthropology Makes Its Object* (New York: Columbia University Press, 1983).

7   Mark Dery, "Black to the Future: Interviews with Samuel R. Delany, Greg Tate, and Tricia Rose," in *Flame Wars: The Discourse of Cyberculture*, ed. Dery (Durham, NC: Duke University Press, 1994), 180.

Postcard promoting the 1906 Colonial Exposition in Paris

Visitors line up to enter the General Motors Futurama pavilion, by the architectural firm Albert Kahn Associates, New York World's Fair, 1939–40. Canadian Centre for Architecture, Montreal. Gift of Federico Bucci

(a consumer's paradise) and later as a showcase of the newest technologies (a technophilic utopia).

More than a billion people attended the major world's fairs and expos that occurred between 1851 and 2005.[4] Presenting the latest industrial products, cultural exchanges, scientific applications, and technological advances, these events became synonymous with future-making. The fair sites themselves, as expensive exercises in nation branding and empiric display, were transformed into fantastic spectacles of futuristic architecture—each a Futuropolis that one could actually visit.[5] These cities of the future were punctuated by Eiffel Towers, Crystal Palaces, Geodesic Domes, and Space Needles. Futurama, a display pavilion created for the 1939 New York World's Fair—designed by Norman Bel Geddes and sponsored by General Motors (at the time, the world's largest corporation)—was emblematic of such endeavors, offering an experience of America's near future, some twenty years hence. Seated visitors were mechanically conveyed past a model American city circa 1960 depicting a vast suburban landscape replete with a simulated automated highway system and farms growing artificial crops. Futurama was a smash success, its optimistic portrayal of a future America resonating with a populace just exiting the Great Depression. "I Have Seen the Future," stated the buttons worn by some of the thirty thousand daily visitors to the pavilion.

The sense of the future realized at this and other world's fairs was often aided by placing visions of the future in stark contrast against backdrops of the past, for which various indigenous peoples, often from colonies of the host countries, were brought in to reenact daily rituals in simulated natural habitats. Portrayed pejoratively as either simple ("static") or primitive ("unevolved"), and decidedly non-Western, these ethnographic displays or "living dioramas" can be seen as markers in time, signposts by which visitors could gauge their own progress. The world's fair was a remarkable demonstration of what Johannes Fabian has termed the inherent contradiction of ethnography, namely, its denial of coevalness, or shared time.[6] In the temporal construct of the world's fair, other cultures were presented as unchanging, while visitors' sense of the present was suspended and supplanted by simulations of their future lives. It was thus not only other lands and lives that were colonized but the future itself. At the same time, the static exposition of otherness "defutured" these cultures, not only in the minds of most visitors, but also in the imagination of many of the colonized, vanquishing possibility and agency over their future.

But what would it mean to defuture the future of cultural modernity and Western civilization? One answer lies in the rise of Afrofuturist liberatory discourse and practice. Mark Dery, who coined the term *Afrofuturism*, asks:

> Can a community whose past has been deliberately rubbed out, and whose energies have subsequently been consumed by the search for legible traces of its history, imagine possible futures? Furthermore, isn't the unreal estate of the future already owned by the technocrats, futurologists, streamliners, and set designers—white to a man—who have engineered our collective fantasies? The "semiotic ghosts" of Fritz Lang's *Metropolis*, Frank R. Paul's illustrations for Hugo Gernsback's *Amazing Stories*, the chromium-skinned, teardrop-shaped household appliances dreamed up by Raymond Loewy and Henry Dreyfuss, Norman Bel Geddes's Futurama at the 1939 New York World's Fair, and Disney's Tomorrowland still haunt the public imagination, in one capitalist, consumerist guise or another.[7]

To take away the future is not simply a matter of active suppression; it can also be the consequence of design itself. Just as acts of design make the future, they can also unmake other futures. Consider, for instance, the way the design of the automobile affected the design of many other things (landscapes, highways, cities, homes) and led to certain consequences (air pollution, traffic accidents, ride sharing, climate change) that in turn foreclosed other futures. Cameron Tonkinwise has affirmed that "design futures; it makes certain futures materially possible and likely. But in so doing, it can defuture, limiting the number of futures we have now, and limiting the quality and quantity of the futures of those futures."[8] This notion of *defuturing* is borrowed from Tony Fry and his long quest to refocus attention not on sustainability per se, but on unsustainability writ large. As Fry aptly notes, "We need to remind ourselves that the future is never empty, never a blank space to be filled with the output of human activity. It is already colonised by what the past and present have sent to it. Without this comprehension, without an understanding of what is finite, what limits reign and what directions are already set in place, we have little knowledge of futures, either of those we need to destroy or those we need to create."[9] Design's propensity to defuture is rooted in its capacity to fulfill present wants with little attention to future needs.

Our need to invent the future, or to design it, is of relatively recent origin. Acts of invention and design underscore the human agency necessary to ground future-making in the here and now, a concept that emerged during the Enlightenment as conventional religious views receded enough for reason, individualism, and skepticism to advance. Studying Western cultures across history, the sociologist Fred (Frederik Lodewijk) Polak, in his epic two-volume treatise *The Image of the Future*, chronicled the transformation in how the future has been envisioned from ancient times to the twentieth century. According to Polak, for millennia the strongest image of the future was conceived of mostly in otherworldly terms, in religious prophecies that were to be realized in the afterlife. This changed over time, with

images of the future increasingly constituted on this side of the ethereal divide, most often in the form of utopian proposals—a pursuit of paradise here on Earth. But Polak saw in the twentieth century a marked rise in negative utopia and, uniquely, a dearth of images of the future compelling enough to instigate its realization: "Our time is the first in the memory of man which has produced no images of the future, or only negative ones."[10] He concisely prophesized the consequence of such a lack of vision: "*The rise and fall of images of the future precedes or accompanies the rise and fall of cultures.* As long as a society's image of the future is positive and flourishing, the flower of culture is in full blossom. Once the image of the future begins to decay and lose its vitality, however, the culture cannot long survive."[11]

Polak was writing in the 1950s and 1960s, having evaded the Nazis and survived World War II. He did not live to see the current state of affairs or the recently issued apocalyptic forecast for life on Earth if the factors contributing to human-induced climate change are not soon reversed.[12] The image of the future today is one of the end of life and civilization as we know it—not as a religious reckoning from God, but as a result of human actions. Polak's prediction for the end of Western civilization through a historically unprecedented lack of images of the future now seems sadly prescient.

Paradoxically, today we seem to be awash in a sea of images about the future. Billionaires plan for life on Mars. Scientists contemplate terraforming Earth. Technologists ponder

8   Cameron Tonkinwise, "Design Away," in *Design as Future-Making*, 204.

9   Tony Fry, *A New Design Philosophy: An Introduction to Defuturing* (Sydney: UNSW Press, 1999), 11–12.

10   Polak, "Timeless Time," in *The Image of the Future*, 2:89.

11   Polak, "The Image of the Future and the Actual Future," in *The Image of the Future*, vol. 1, *The Promised Land, Source of Living Culture*, 49–50; italics in the original.

12   Intergovernmental Panel on Climate Change (IPCC), "Summary for Policymakers of IPCC Special Report on Global Warming of 1.5ºC Approved by Governments," United Nations, Geneva, October 8, 2018, www.ipcc.ch.

DEFUTURING THE IMAGE OF THE FUTURE

13   Polak, "The Image of the Future and the Actual Future," 1:38.

14   Polak, "The Image of the Future and the Actual Future," 1:38.

15   Polak, "De-Utopianizing," in *The Image of the Future,* 2:31.

16   Freeman Dyson, *From Eros to Gaia* (New York: Pantheon Books, 1992), 341; quoted in Stewart Brand, *The Clock of the Long Now: Time and Responsibility* (New York: Basic Books, 1999), 35; see also Brand, "Pace Layering: How Complex Systems Learn and Keep Learning," *Journal of Design and Science* 3: Resisting Reduction (January 17, 2018; rev. February 4, 2018), jods.mitpress.mit.edu.

Rendering of the proposed Mars Base by SpaceX

Sun Ra, Morton Street, Philadelphia, June 1979. Photograph by Val Wilmer.

the Singularity. Our images of the future are, perhaps appropriately, post-human and post-nature. They are by turns pessimistic and optimistic, fateful and fanciful. Although decidedly futuristic, such images of the future are survivalist strategies and presumptive forecasts. They are the future posing as today's speculative solutions to yesterday's wicked problems. It is telling that inhabiting a faraway planet with a hostile environment is somehow easier to imagine than a future here on Earth that requires changing our thinking, behaviors, and priorities. The increasing anthropocentrism that centuries ago allowed us the agency to envision our own futures has delivered its endgame in the Anthropocene. Polak would likely not see such apocalyptic visions as images of the future, which, as he describes, "picture another world in another time radically different from and, in being vastly ameliorated or even approaching perfection, which is absolutely preferable to the present one of the here-and-now."[13] For Polak, images of the future contain the seeds of a progressive perfection of the human condition. Perhaps, then, they are not really images at all but something more sweeping, visions in the broadest sense—grand narratives that usher in a more perfect future. Such visions by their nature are blue-sky prophecies and cannot yet depict the mundane reality of a future lived in the weeds. Images, on the other hand, like design itself, try to convince us with their detail and verisimilitude. They are like props for a script about the future in which we are invited to play along. At best, they give us only a convincing slice of the future with the greatest possible detail. Provocatively, Polak considered images of the future to be "powerful time-bombs" created by people and societies with "little control over when, where and how they will explode."[14] Thus the future is neither predictable nor controllable; in short, it is not designable in conventional terms.

Polak's concept of defuturizing, like Fry's, also describes the foreclosure of futures, but through a stubborn presentness:

We mean by the term [defuturizing] a retreat from constructive thinking about the future in order to dig oneself into the trenches of the Here-and-Now. It is a ruthless elimination of future-centered idealism by today-centered realism, an elimination of all thinkers about the future as poets and dreamers who are out of tune with the times. What the world really needs, we keep hearing, is realists, and above all realistic politicians; also specialists, social engineers, organizers, builders, architects, regional planners, managing directors and general staffs.[15]

Here Polak is evoking a criticism typical of his time, namely, that Western society was an increasingly technocratic operation governed by scientific pragmatism and bureaucratic management techniques. Polak seems to suggest that one group of people, these specialized technocrats, have displaced another group, the poets and dreamers, as the traditional gatekeepers of visions of the future. When Polak states that a "today-centered realism" dismisses those who seek to envision the future as "poets and dreamers who are out of tune with the times," this asynchronicity implies a potentially different problem.

In his 1999 book *The Clock of the Long Now*, Stewart Brand cites the work of the theoretical physicist Freeman Dyson to support what Brand would eventually call "pace layering": the interactions among faster and slower segments of civilization as they shape our understanding of change over time. Brand draws on Dyson's analogous articulation of six unique time scales to which humans are bound:

The destiny of our species is shaped by the imperatives of survival on six distinct time scales. To survive means to compete successfully on all six time scales. But the unit of survival is different at each of the six time scales. On a time scale of years, the unit is the individual. On a time scale of decades, the unit is the family. On a time scale of centuries, the unit is the tribe or nation. On a time scale of millennia, the unit is the culture. On a time scale of tens of millennia, the unit is the species. On a time scale of eons, the unit is the whole web of life on our planet. Every human being is the product of adaptation to the demands of all six time scales.[16]

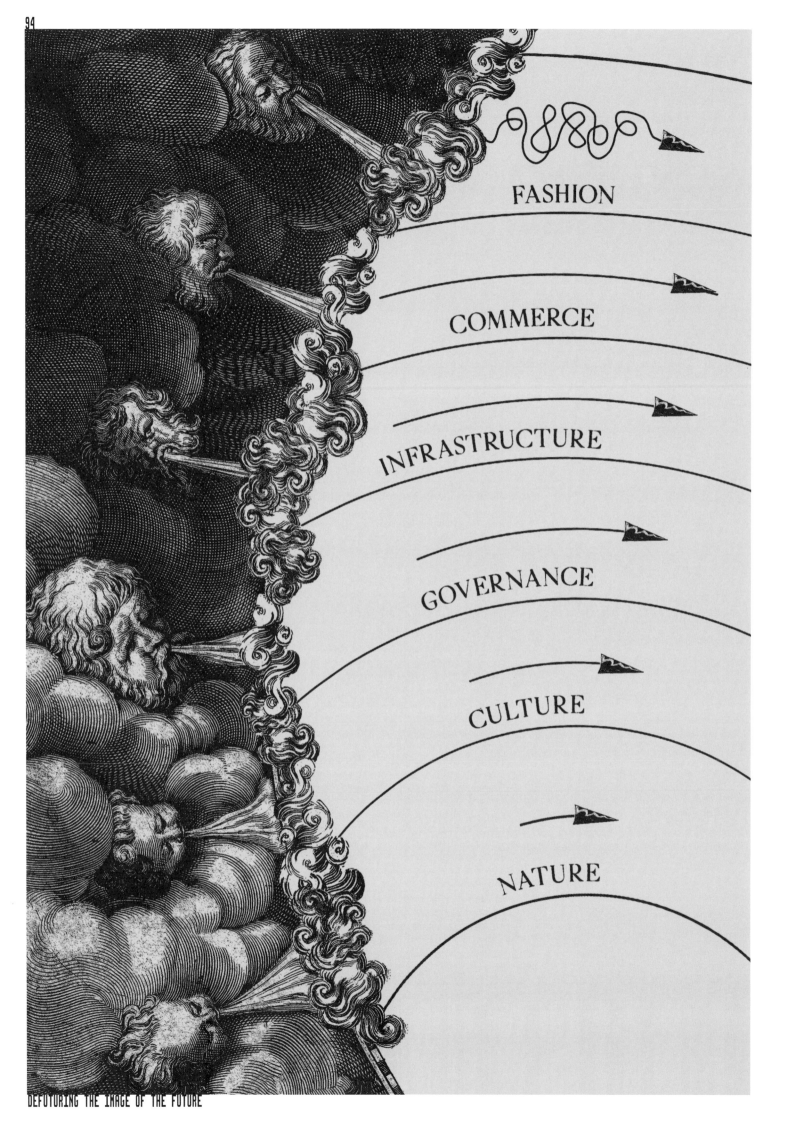

FASHION

COMMERCE

INFRASTRUCTURE

GOVERNANCE

CULTURE

NATURE

17   Brand, *The Clock of the Long Now,* 35–36.

18   Brand, *The Clock of the Long Now,* 34.

19   Brand, *The Clock of the Long Now,* 39.

Like Dyson, Brand also describes six layers, each tied to its own durational schema. But while Dyson describes something universal and eternal, Brand articulates a systems approach to the endurance of civilizations. For Brand, a robust and resilient civilization is able to absorb shocks to its system through a process of both fast-moving and slow-moving "pace layers":

> I propose six significant levels of pace and size in the working structure of a robust and adaptable civilization. From fast to slow the levels are:
>
> • Fashion/art
> • Commerce
> • Infrastructure
> • Governance
> • Culture
> • Nature
>
> In a healthy society each level is allowed to operate at its own pace, safely sustained by the slower levels below and kept invigorated by the livelier levels above.[17]

In theory, all these layers interact with and influence each other in a complex system. But Brand believes that each layer is most influenced or balanced by the layers immediately adjacent to it. Thus the churn at the fastest level, fashion/art, is governed most by the pace of commerce, which is itself constrained by the pace of infrastructure (education, science, etc.), which is in turn regulated by the pace of governance. Cultural change comes more slowly than change in the layers above it, while nature seems nearly unchanging. Each layer provides checks and balances against imbalance in the overall system and operates at its own pace, whether measured in weeks, months, years, decades, centuries, or eons. Whether fast or slow, according to Brand, each speed has its own advantages: "Fast learns, slow remembers. Fast proposes, slow disposes. Fast is discontinuous, slow is continuous. Fast and small instructs slow and big by accrued innovation and occasional revolution. Slow and big controls small and fast by constraint and constancy. Fast gets all our attention, slow has all the power."[18]

Following Brand's logic, what Polak identified is an imbalance in the system that sustains a healthy or growing civilization. For Polak, "images of the future" are a barometer of the health of a civilization, and their quality and power depend on their capacity to conjure a compelling vision, if there are any visionaries left in society at all. For Brand, the notion of a healthy civilization rests on a system able to balance its interdependent layers.

What do we make of the future today? As all of human civilization, not simply one tribe or nation, hangs in the balance, it is a slow but powerful nature that is attempting to regulate the system that sustains us all. Brand's statement that "it is precisely in the apparent contradictions of pace that civilization finds its surest health"[19] means we can look to nature, its change sustained over millennia, as a possible corrective to the problems of an unyielding culture. Might, then, fashion/art and the other upper pace layers be able to positively influence the slowest and most powerful layers

of civilization? It seems that what is needed now, more than ever, are images of the future—those images Polak could not find—that can act as "powerful time-bombs." I believe the cycles of influence need to be reversed, however, with the long and the slow compelling the short and the fast. We need to defuse those time bombs set in motion decades or centuries ago.

If today's dire forecasts for the end of the world as we once knew it hold true, then civilization has been defutured by an all-powerful nature. Given that sobering forecast, our images of the future today are not time bombs awaiting some distant detonation—for that future has already been unwinding—but should be prescriptions for defusing a future we know we do not want, but are certain now is coming. ░

96

MATERIALs

Scaling Up

SIMONE FARRESIN & ANDREA TRIMARCHI of FORMAFANTASMA in conversation with ZOË RYAN on designing for the planet

This conversation took place in September 2018 at Studio Formafantasma in Amsterdam.

ZR

Since you graduated from the Design Academy Eindhoven in 2009, how have you seen the field of design changing and addressing issues related to the future as a way to imagine alternative ideas of how we might live, how we might interact with one another?

SF

It's a difficult question. It's always easier to reflect on how you see yourself changing and how our way of working is changing and reflecting, perhaps, larger changes in design. I think what is really carving a new path for design is its increasing connectedness to issues of climate change and the impact of ecological troubles that design has participated in creating.

AT

We feel very lonely as individual designers. I'm speaking for us, but I think if you talk with others, they will feel the same. In the past, companies or industry were asking designers the right questions. Today, there are many cultural and arts institutions that are asking us the right questions, but not companies, especially not the furniture design industry.

SF

If you think about, for instance, the exhibition *Italy: The New Domestic Landscape* that opened in New York at the Museum of Modern Art in 1972, the collaboration with manufacturing companies and the Italian government was essential. They were financing these avant-garde ideas. That kind of relationship is completely lacking now. It's completely disappeared. We are [instead] contacted by manufacturers who can engage on the level of product development, but only on that level. Our way of thinking about design is not product oriented. We want to think about design in a much more holistic way. Until now, design has always been focused on the human as the center of the conversation. In this moment when we are facing severe ecological troubles, maybe we need to enlarge our thinking and be more considerate of other species inhabiting the planet.

ZR

Can you talk more about the types of questions that galvanize you? How did the commission for your work *Ore Streams* [see p. 88] come about, for example, and what questions were asked of you?

AT

Ore Streams was a commission from the National Gallery of Victoria in [Melbourne,] Australia. We were interested in learning about electronics—how they are produced and the forces behind the design of an object. While we were working on this research, we were contacted by a big producer of electronics. They asked us to give them some trends on ...

SF

The colors of phones and patterns of phones.

AT

How the phone should look, basically. We were happy to look into this as a way to continue our research into electronics, but again, this is not the question they should have asked us. We are interested in questions that can lead us to designing products or systems that can make greater changes to the way things are made and distributed, for example.

Formafantasma, video still showing the disassembling of electronics for *Ore Streams*, 2018

SF

We talked to them about the research we've done. They were not interested in hearing it.

AT

But these types of large companies can make a difference because they have the power and they have the scale and the numbers to generate change. But they are not able to see, or maybe they know their limits.

SF

We would love to do a project about restructuring the departments of a product-development company, so that they don't just think in boxes. I think it's interesting to see how the infrastructure of companies is not allowing conversation between the parts. It's not allowing the bigger picture. We need partners that can allow us, for instance, to investigate how electronics recycling works, not just questions of product development, aesthetics, or the relationship between the user and the product, but the larger consequences of the product. We are interested in the ecology of the product.

AT

We also see this [concern] with our students. It's really rare that we have students who are just interested in working with companies.

SF

They want to understand the system of production, since it is increasingly displaced and not something experienced anymore in a physical way, given new technologies and manufacturing methods. We also don't have our parents working in companies and producing the things we experience anymore, as it was historically in Italy, where skills were handed down through the

generations and there was a direct connection to how something was produced. We don't know where our products come from anymore and how they are produced. A new level of understanding of the systems of production is needed, and I think design should also be preoccupied with that. Our work reflects on the origins of materials and the implications they have on production. I think that the layering of different elements is always present in our work, but I think that over time, especially with *Ore Streams*, it crystallized into something more precise. There is a deeper understanding of why we are interested in the transformation of materials into objects. At the beginning, we were interested in issues related to our dependence on oil and the overuse of plastics, for example, which you see in projects such as *Botanica* [2011], which seeks to explore alternatives. We were hinting at the problems, but we were not really getting our hands dirty. I think *Ore Streams* is attempting to not abandon our love of objects and their narratives and their qualities, but also to dig into the real problems of specific industries and try to understand if design can do something there.

ZR

How did *Ore Streams* develop?

SF

Our idea for *Ore Streams* actually came from reflections we had while in Australia. The economy there is largely still based on the extraction of minerals from underground. We knew that we wanted to do a project connected with the processes of extracting, transforming, processing, distributing—[to discover] where design can come in to address some of the urgent challenges facing the industry.

Plastic vessels made from natural polymers extracted from plants or animal derivatives, from Formafantasma's *Botanica* collection, 2011. Photograph by Luisa Zanzani

Installation view of the exhibition *Italy: The New Domestic Landscape*, at the Museum of Modern Art, New York, May 26–September 11, 1972. The Museum of Modern Art Archives, 1004.204. Photograph by Leonardo LeGrand

AT

Ultimately it comes down to legislation. The major issue that we understood through this process is that certain things can only be solved through legislation. In the process of developing *Ore Streams*, we talked to people at the United Nations and Interpol and …

SF

The European Electronic Recyclers Association.

AT

What we realized is that design is missing from this. The International Electronic Recycling Association meets regularly, but the people who design the products are missing from these conversations. For me, the ultimate goal of *Ore Streams* is to create a platform in which the various sectors can actually discuss these ideas together. We went on to develop a series of films about the *Ore Streams* project for the Milan Triennale in 2019.

SF

We have an online database [www.orestreams.com] where we have archived our research and made our interviews available, and we are now also working on a publication. We also want to hold some seminars with recyclers, designers, and lawmakers so we can extend the areas of our research, for instance into the collection of electronic waste and the collection of trash, and the taxing of waste and how that works. We are interested in discovering if there are ways that we can really impact how planned obsolescence works, both on a physical level and on the digital level.

ZR

Do your projects begin by looking at historical precedents as part of your research?

SF

Yes, often. We did this with *Ore Streams* in the sense that we started not really from electronic waste, but by looking at the mining industry and how it evolved.

AT

We started to confront the old colonial roots as well as the contemporary roots of the distribution of metal and ore on our planet.

SF

We need to understand why, for instance, a certain economy exists instead of another; it's interesting or it helps to remind ourselves that everything can be redesigned. That is where imagination can flourish, when you start to understand that the simplest actions we perform daily have been designed. These kinds of thoughts are what help you to be open-minded enough to realize everything is designed, and that everything, because it's human-made, can be reversed. It can be rethought. That's how we think about the future. Looking back can help free you from preconceived ideas of how the future is supposed to look.

ZR

You were just talking about the potential to create the right conditions for the imagination to take flight. For a couple of years now you have been the head of the MADE [Mediterranean Art and Design] Program in Syracuse, Italy, which is an undergraduate program in design. Can you tell us more about this and what interested you in returning to Italy to teach?

SF

Sicily is an interesting place because design is not there. It's perfect for the location of a design program, because you can avoid the product-based furniture design industry, which the north of Italy is completely obsessed with.

AT

There is a high unemployment rate in the area for young people, who need new skills. There are ecological issues because it's a place that is getting drier and drier every year. It's at the center of the Mediterranean Sea, so it's at the center of the refugee crisis.

SF

It is also a very beautiful place. It's a case study for the complexities of the world. On a strategic level, we are taking the best education we have in the country and making it accessible in the south of Italy so that we can have more of an exchange between north and south. We had fifteen students the first year, ten the second, and about fifteen again this year.

AT

What we really want to try to create is a context there—first, for understanding the potential of design as a political act. Enzo Mari, for example, thought of design in this way.

SF

He saw design as an ideology, an approach. We do too. Given the complexities of our time, we see the need to create smaller frameworks through which to reimagine and redesign the present and, inevitably, the future. At the moment, in education there is still a conservative approach, and it is the bachelor program. We are also dealing with the restrictions imposed by the [Italian] Ministry of Culture. The ministerial directives are very limiting and at times confusing, with a narrow view of what design is. Our ultimate goal with the school is to make it into a much more open-ended education that could move beyond furniture design but [be] linked to production, a place where designers could not only work with different production processes but possibly rethink them in the light of the devastating reality of a collapsing planet. We hope to nurture a generation of designers like us who don't know how to sort out the consequences of an irresponsible economic model, but who are not afraid to turn their heads toward those agglomerates of ugliness created by the present industry. ⌗

A classroom of the MADE (Mediterranean Arts and Design) Program at the Rosario Gagliardi Academy of Fine Arts, Syracuse, Italy

Variants of Biodesign

The term *biodesign* can refer to different types of design and modes of production, as demonstrated by the many examples provided in William Myers's *Bio Design: Nature, Science, Creativity.*[1] While other exhibitions and books over the last decade or so have contributed to the development of the concepts and discourses of biodesign,[2] Myers's book undoubtedly cuts a wide swath and serves as an essential starting point for those curious about the field. Major variants of biodesign include different uses of plants for architectural structures, surface design, or energy; new materials produced with bacteria; functional uses of insects or animals; biomimetic designs; speculative designs for debate; and designs relying on synthetic biology.

Because of the term's currency, it might be assumed that these facets of biodesign are recent developments that owe their appearance to the increasing acceptance and establishment of processes of biotechnology. Yet, as the list above demonstrates, some variants of biodesign entail more biology, and some less. Others have lengthy histories, contributing to our ability to take a critical perspective and garner insights about rhetorical discourses, the context of historical developments, and past and potential outcomes. Historical precedents also allow comparison to contemporary approaches, suggesting questions we might otherwise not think to ask. This short summary, then, of some major variants of biodesign is meant to distinguish what might otherwise blend together under the overarching category. This in turn allows us to ask the important question, Are some variants of biodesign more sustainable than others? And if so, How can we tell?

Since the late nineteenth century, architects and designers have drawn theories of biological evolution as conceptual frameworks for theorizing modern design. From Louis Sullivan's "form follows function" (a rephrasing of Jean-Baptiste Lamarck's theory of the inheritance of acquired characters), to Adolf Loos's proscription on ornament in modern design (based on Ernst Haeckel's theory of recapitulation), to interwar and midcentury theories of streamline design that embody eugenic ideology, designers have employed current evolutionary theories as a means of interpreting and naturalizing design decisions.[3] By the late 1950s and 1960s, the growing field of computer science also incorporated evolutionary theories into genetic algorithms and other modes of evolutionary computation. Architects made use of these techniques soon after, in their earliest applications of computer-generated design.[4] While not relying on biological materials or processes, these precedents in many ways function as the foundational theories of biodesign, and their biological rhetoric casts an aura of organicism even when the methods and materials of actual design processes have been heavily technological.

One prominent strand of biodesign is biomimicry, popularized by Janine Benyus in the late 1980s and 1990s.[5] In this approach, designers learn from biological solutions to particular functional or systemic problems, and then adapt facets of these solutions to the development of artificial, synthetic, or technological design applications. Some of the most cited examples are swimwear textiles based on sharkskin, self-cleaning paints that mimic the surface of lily pads, and a bullet-train nose modeled after a kingfisher's beak. As these examples demonstrate, biomimetic designers often copy functional forms from plants and animals to innovate new or improved products that are then produced using industrial technologies and artificial materials. Some architects, such as Jenny Sabin and David Benjamin, criticize biomimicry as shallow formal copying, in contrast to an approach they refer to as "biosynthesis" or "bio-logic,"[6] in which the goal is to

A spiderweb covered with water droplets

ascertain the fundamental biological processes, usually abstracted into geometric or mathematical functions, in order to use that developmental process as the basis for design.

Biosynthesis stands in contrast to synthetic biology, a new field that many biodesigners are turning to for bioengineered materials with properties suited to very specific functions. Two prominent examples, both of which use bioengineered yeasts fed by sugar to produce new textiles, are Bolt Threads, whose fabric is made from a protein-based "spider silk," and Modern Meadow's Zoa, a collagen-based leather-like material.[7] Designs produced using synthetic biology are often described as "grown"—yet growing biological materials does not necessarily require genetic engineering. Various products are being grown directly from mushroom mycelium (MycoWorks; Mylo) or bacterial cellulose (Kombucha Couture), although some designers using these materials are also using synthetic biology to "improve" the material or add functional features.[8]

Because of the *bio* in *biodesign*, many people assume that it is generally more sustainable than other approaches to design. While this may be true in particular instances, it is important to clarify how such a determination is made. Life-cycle analysis offers the soundest fundamental tool for the overall assessment of the amounts of materials, processing, energy, and waste and pollution involved in any particular product from "cradle to grave" or, better yet, from "cradle to cradle" in the circular economy.[9] Biomimetic designs, for example, do not by definition need to be sustainable, and often in their overall life cycle they are high-tech, chemically complex, energy-intensive products, even if they offer benefits like self-cleaning or increased efficiency during operation. It is also important to consider the degree of circularity at a product's end of life, whether in

terms of disassembly, repurposing, or compostability. These latter considerations matter far more for determining degree of sustainability than whether or not a product is identified as a variant of biodesign. ⌗

1   William Myers, *Bio Design: Nature, Science, Creativity* (London: Thames and Hudson, 2012).

2   See, for example, Paola Antonelli, *Design and the Elastic Mind*, exh. cat. (New York: Museum of Modern Art, 2008); *En Vie / Alive*, exhibition at Espace Fondation EDF, Paris, April 26–September 1, 2013; *Biodesign: From Inspiration to Integration*, exhibition at the Rhode Island School of Design, Providence, August 25–September 27, 2018; the gallery show that accompanied the Biodesign Challenge student competition, New York, June 2016–18; *Biodesign Here Now*, exhibition by Open Cell at the London Design Festival, September 15–23, 2018.

3   Christina Cogdell, *Eugenic Design: Streamlining America in the 1930s* (2004; repr., Philadelphia: University of Pennsylvania Press, 2010).

4   Christina Cogdell, *Toward a Living Architecture? Complexism and Biology in Generative Design* (Minneapolis: University of Minnesota Press, 2019).

5   Janine M. Benyus, *Biomimicry: Innovation Inspired by Nature* (New York: HarperCollins, 1997).

6   Jenny Sabin and Peter Lloyd Jones, "Nonlinear Systems Biology and Design: Surface Design," in *ACADIA 08: Silicon + Skin: Biological Processes and Computation*, ed. Andrew Kudless, Neri Oxman, and Marc Swackhamer ([US]: Association for Computer-Aided Design in Architecture, 2008), 54–65; and David Benjamin, ed., *Now We See Now: Architecture and Research by the Living* (New York: Monacelli Press, 2018), 23–27.

7   For Bolt Threads, see boltthreads.com/about-us/, and for Zoa at Modern Meadow, see zoa.is.

8   See the product websites, at www.mycoworks.com; boltthreads.com/technology/mylo; and www.kombuchacouture.com.

9   William McDonough and Michael Braungart, *Cradle to Cradle: Remaking the Way We Make Things* (New York: North Point, 2002); on the circular economy, see the website of the Ellen MacArthur Foundation, www.ellenmacarthurfoundation.org/circular-economy/concept, or the proceedings of the conference Design, Justice, and Zero Waste: Exploring Pathways for the Circular Economy at Parsons School of Design, New York, May 9, 2018.

The Shinkansen 500 (Bullet Train), April 8, 2006. Designed by Eiji Nakatsu of Japan Railway West (JR West), 1997

Lumen, by Jenny Sabin Studio for the Museum of Modern Art and MoMA PS1's Young Architects Program 2017, on view at MoMA PS1, New York, June 29–September 4, 2017. Photograph by Pablo Enriquez

The Magic of Design

V. MICHAEL BOVE JR. & NORA JACKSON on the future-making power of objects

Any discussion of design for the future raises the question of what will become of tangible things in an increasingly virtual world, and how our relationship with artifacts will evolve. David Rose imagines this future most spellbindingly as "one where technology infuses ordinary things with a bit of magic to create a more satisfying interaction and evoke an emotional response."[1] He believes that the most promising future for the human-machine interface is design that fulfills the fundamentally human desires of omniscience, telepathy, safekeeping, immortality, teleportation, and expression.[2] To understand the process by which design can deliver that enchantment in the future, we must look to how it has done so in the past.

One of the earliest known objects of design is a stone chopping tool recovered from thick layers of clay soil in the Olduvai Gorge in what is now Tanzania. The tool is somewhere between 1.8 and 2 million years old and was designed in the last and most productive quarter of the 7-million-year period during which the human brain tripled in size. For all its rudimentariness, it embodies the magic and sophistication of good design. To hold it, says the former director of the British Museum Neil MacGregor, is to understand those who made it and to communicate with the societies it served.[3]

In this chopping tool we see the fundamental principles of design at work: a preoccupation with both form and function. The object is at once durable and self-explanatory, as well as unobtrusive enough to engage the user in self-expression. We know intuitively what it was designed to do and understand how to hold it and use it. In addition, its suitedness to its task also helped augment the creative power that brought it into being by facilitating access to protein and, in turn, accelerating the growth of the human brain. By a sort of positive feedback loop, this object fueled the very cognitive revolution that spawned it.

But what is most striking about this tool from a design standpoint is a formal property: its being just a little better than it needs to be. Sir David Attenborough observes that the designer of the chopping tool might have been driven by something more than the need to perform a particular task, that he or she could have gotten away with chipping the stone twice rather than eight times and was in fact feeding a human desire for excellence and beauty.[4]

It is this margin of excess, this redundant excellence in the design itself, that is the site of magic in good design. This property—this harbinger of a better and more sophisticated chopping tool not yet realized but latent in the ornamental third to eighth chips at the tool's cutting edge—lends the object a generative power of its own that is not unlike that of the designer. In this sense, the Olduvai chopping stone not only provided nutrition for its user but also fed the imagination of those who would inherit it.

We can see the ability to imagine and create better tools manifested in the Olduvai hand ax, the chopping stone's younger sibling by about 500,000 years from the same site in Tanzania. This strikingly beautiful, grayish-jade, finely chiseled multipurpose hand tool is a more sophisticated and more aesthetically pleasing version of the chopping stone. More versatile than our modern metal-blade hatchet, it was the "Swiss Army knife of the Stone Age"[5] that allowed our human ancestors to perform a wider range of tasks in the provision of food and shelter and the fulfilment of those basic human desires that David Rose believes should drive the future of design.

This diachronic perspective on the evolution of hand tools reveals something fundamental about the workings of design: that the well-designed object has an inherent agency that transcends the intention of the designer, and the potential to bring about change, to generate new

Olduvai stone chopping tool (Lower Paleolithic). British Museum, London, 1934,1214.1

behaviors, needs, experiences, and realities. Even today, the Olduvai chopping stone continues to shape our world. As an abiding witness to human history, it impels us to design new technologies that will allow us to unlock more of the stories it has to tell, and it challenges our imagination to acts of magic and discovery, as well as acts of creation.

The future-making power of the Olduvai chopping tool lies in the intimate and rewarding relationship it creates between the hand and the tool and the access it provides to foods that would otherwise have been inedible, but this power is also embodied in the selection and careful, nonobvious handling of the natural material of which it is made. Yet simultaneously, there is a mutual provocation between the Olduvai stone and many of today's designs for the future. The twenty-first-century object serves up its enchantment in a more direct and unnuanced way. Over the decade since the arrival of the smartphone, the public has increasingly come to value things that exist only in the digital realm, whether apps, games, social-media platforms, or virtual-reality experiences, all of which engage and enchant by being connected, intelligent (or at least seemingly so), and context aware. There is something curiously human in wanting to form relationships with our things, and the success of digital objects that reciprocate in these ways hints that they are effectively accessing this aspect of humanity rather than providing mere novelty.

Designers of physical objects are still learning—with uneven success (pundits lumping together the worst examples under the dismissive phrase "they put a chip in it")—to incorporate the virtual world's connectivity, intelligence, and context awareness into tangible objects while staying within the design expectations associated with unaugmented versions of such things. But are we heading into a world of growing delight with the objects that surround us, or one of shallow and temporary infatuations whose limits quickly become apparent? Conversely, when we are surrounded by intelligent and context-aware objects, will we still value the more passive ones that require intellectual and emotional investment from the user to form a relationship? If some future everyday hand tool can shape-shift to adjust itself to each user's grip, will we still appreciate the genius

behind the nonadaptive object that somehow manages to feel good in any hand?

Considering the above questions in the context of the stone tool, one might ask whether emerging design realms enabled by a more-than-just-material approach will offer space for these new, digitally augmented, smart-composite objects to get better with age and to embody layers of meaning that become apparent only over extended periods of time, perhaps not even being noticed until far-off future generations.

One resolution is hinted at by the work of Philippa Mothersill, which explores how computational intelligence can enhance the design process (and ultimately the resulting artifacts) while preserving and keeping apparent the role of the human designer in the outcome. In her EmotiveModeler project,[6] Mothersill gave computer-aided design tools a sense of the affective impact of geometric transformations such that the designer could directly and effectively explore the design space in terms of how much joy or anticipation an object would evoke. If new technological approaches can increase the clarity with which the designer speaks to the person holding the artifact—now or a half-million years from now—then we can expect a future in which the Olduvai stones will have worthy successors. ‡

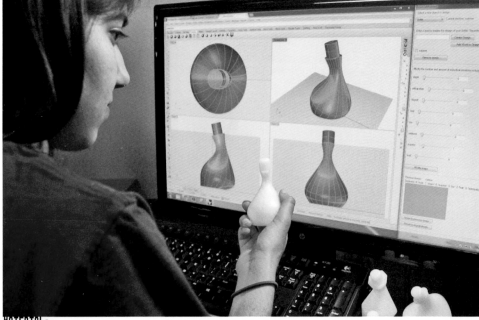

Philippa Mothersill, EmotiveModeler CAD tool, 2014

Olduvai hand ax (Lower Paleolithic). British Museum, London, 1934,1214.49

MATERIALS

1   David Rose, *Enchanted Objects: Innovation, Design, and the Future of Technology* (New York: Scribner, 2014), 13.

2   Rose, *Enchanted Objects*, xi–xiii.

3   Neil MacGregor, *A History of the World in 100 Objects* (London: Allen Lane, 2010), 9.

4   Quoted in MacGregor, *A History of the World*, 13.

5   MacGregor, *A History of the World*, 15–18.

6   See Philippa Mothersill, "The Form of Emotive Design" (master's thesis, MIT, 2014), dam-prod.media.mit.edu/x/files/thesis/2014/pip-ms.pdf.

BIORECEPTIVE LIGHTWEIGHT CONCRETE COMPONENT 2015–16 BIOTA LAB, BARTLETT SCHOOL OF ARCHITECTURE, UNIVERSITY COLLEGE LONDON

What if making future cities greener took a very literal turn and building facades were designed to encourage and support the growth of diverse pollution-mitigating flora? The architects Marcos Cruz, Richard Beckett, and Javier Ruiz partnered with their colleagues Chris Leung and Bill Watts at the Bartlett School of Architecture, University College London, and the environmental engineer Sandra Manso of the Universitat Politècnica de Catalunya, as well as several generations of their students, to create a special bio-receptive concrete that encourages spores to land, stay, and grow on the material. Modeled digitally before being fabricated, these geometrically patterned concrete plates are especially supportive to algae, lichens, and mosses. The contours of the slabs give shape to the plants' growth, creating an attractive surface that is part of a much longer history of bio-inspired design that includes the architecture of Frank Lloyd Wright and Francis Kéré, the designs of Hector Guimard, and the art of indigenous weavers.

These ongoing material explorations were initiated under the umbrella of the BiotA Lab at the Bartlett, a research platform led by Cruz and Beckett between 2014 and 2018 that connected design, biology, and engineering in the interdisciplinary innovation of architectural materials. With industry partner Laing O'Rourke, the team developed several substrate models and then tested three geometry types across variables of location, weather, light, season, and plant species. They envisioned the panels integrated into projects in a range of sizes and types, including the bland infrastructure of retaining walls and elevated-railway barricades, the furniture and walkways of public spaces, the smaller edifices of houses or housing blocks, and "the many blank and rather 'wasted' building facades of larger buildings."[1]

While not an unusual sight on much older buildings, patina and unrestricted green surface growth are rarely encouraged on new structures. As Rose Eveleth noted of the BiotA project, "Plants and lichens on a concrete wall used to be a sign of decay, but soon they might be a sign of sophistication."[2] There are, however, barriers to overcome. While the BiotA design eliminates the need for the costly mechanical irrigation systems and maintenance previously associated with "green walls,"[3] its materials must be accepted into the national and international construction codes that govern best practices before they can be used. **MMF**

1 "Bioreceptive Facade Panels—EPSRC-Funded Research 'Computational Seeding of Bioreceptive Materials,'" Richard Beckett's website, www.richard-beckett.com.

2 Rose Eveleth, "The Future of Architecture: Moss, Not Mirrors," *Atlantic*, December 7, 2015, www.theatlantic.com.

3 "Marcos Cruz—Winter Lecture Series 2016," the Institute for Advanced Architecture of Catalonia website, www.iaacblog.com.

PERSPIRE   2018   ALICE POTTS

In the future, will our bodies be able to secrete and grow the apparel we need and want? And if so, how might this recalibrate our current reliance on—and massive consumption of—planetary resources in pursuit of the newest fashions? These are the questions that motivate the young apparel and footwear designer Alice Potts, who experiments with bodily fluids such as sweat and urine for her novel materials, combining them with natural vegetable and plant dyes to create her fashion. The results—evoking school chemistry experiments evaporating dyed salt—are closer to poetic prophecy than they are immediately practicable. But they underscore the power of speculative design to forge new pathways to the future and inspire collaboration between groups with specialized knowledge in order to achieve them.

In the example pictured here, the sweat from a dancer's exertions has manifested in beguiling crystals on the surface of a pair of ballet slippers pigmented with dye made from red cabbage. The result evokes Cinderella's fabled glass slippers. Part of a larger series titled *Perspire*, the pair augurs alternative futures in which the accessories we sport are highly personalized, not only through formal and conceptual choices but also through the biosignature of the base materials with which they are constructed.

Potts's experiments began at the Royal College of Art in London, where she was a student from 2016 to 2018, and where she collaborated with medical scientists at the nearby Imperial College London to develop her confidential technique for rapidly growing sweat crystals. As part of the Athens Biennale, during a residency at the Onassis Cultural Centre from 2018 to 2019, Potts invited visitors to engage in strenuous physical activity within a gym and on a rave dance floor set up in a gallery space so that she could collect sweat to use in creating wearable designs for members of the public. Along with investigating the development of new sustainable materials, Potts is interested in rethinking the public taboo around perspiration, which is often deemed embarrassing—something to be hidden or abolished entirely. She sees beauty in the crystalline forms of a liquid that is produced as a result of biological processes that reflect human effort or emotion.

Part of the burgeoning scene of smart textiles within the fashion arena, Potts's work is both aesthetically compelling and, theoretically, has practical health and well-being applications, given that sweat is a biometric performance indicator that can flag hydration levels and the presence or absence of key nutrients in the body.   **MMF**

Seaweed has long been used in textiles by designers, both as a decorative motif (for example, on highly prized eighteenth-century calicoes) and as a way to strengthen fabrics through its application, known as "sizing" (as in the Japanese use of funori, which takes on a gummy, fluid quality once washed and soaked).[1] For the Brittany-based textile designer Violaine Buet, seaweed is both playground and passion as she experiments to produce many different textile types from this organic material. Her handcrafted creations sometimes display the natural browns and greens of their medium, and at other times are infused with vegetable-based dyes, becoming brilliant ruby reds, oranges, and purples. Buet looks to other practices for skills and ideas, borrowing techniques from leather-making, jeweling, tailoring, and screenprinting. Her textiles can be woven, sewn, printed, embossed, tufted, engraved, braided, and pressed, to name but a few methods, creating visual appeal and tactile interest.

Having founded her studio in 2016 after nearly a decade of study and work on textiles in France and India, Buet focuses solely on algae, working with a network of experts, researchers, and artisans as she develops new forms and tests their durability and individual properties. Her practice is trained on material innovation, and she leaves the use of her fabrics to other designers, whom she hopes will be inspired to use them for a range of different applications, from haute couture to interior decoration and fine art. Her biodegradable textiles are time consuming to produce and thus expensive, high-end additions to the flourishing field of sustainable textiles, which proposes a more symbolic and humane relationship between nature, labor, and consumption in the fashion and textile industries—both of which are hugely polluting and resource intensive, especially in their use of water.

As a designer, Buet remains in thrall to her medium: "Brown seaweed is 1.3 billion years old. ... I bow before the majesty of this material, never positioning as its conqueror but remaining deeply aware of the wisdom of nature. ... Unveiling [algae's] mysteries has taught me patience. ... I feel profoundly alive imagining and savoring the joys of future discoveries."[2] **MMF**

1   Cynthia Green, "Are You Wearing Seaweed?," *JSTOR Daily*, December 3, 2017, daily.jstor.org.

2   "About," Violaine Buet website, violainebuet.com.

DEPARTMENT OF SEAWEED   2013–
JULIA LOHMANN

The German-born designer Julia Lohmann founded the Department of Seaweed in 2013 as an experimental research platform for exploring seaweed's potential for products, fashion, and other applied designs. Lohmann was entranced by the variety of living organisms that come from the sea during a visit to the legendary Tsukiji Fish Market in Tokyo. This experience prompted her to accept an artist residency in Sapporo, Japan, in 2007, during which she came under the spell of seaweed's organic, sustainable, and aesthetically compelling properties while at a seaweed farm in Hokkaido. She developed a keen interest in the challenge of incorporating in-depth materials research into her design process: "I was so fascinated with how [seaweed] becomes big and leathery, really beautiful, when it's wet. And then it has such a different character when it dries."[1]

Lohmann creates abstract sculptures of varying sizes as well as garments, accessories, and products, such as lamps, that test the practical and conceptual limits of her chosen medium. She procures dried seaweed primarily from Japan, soaking it in a water-based solution that keeps it flexible as she works with it. She cuts and sews it, uses it as a veneer, and marries it with other materials, including felt and bamboo. Seaweed's particular strengths and possible applications influence Lohmann's design approach. She cites her childhood love of beachcombing as instilling an engagement with "seeing the beauty of decay and the imperfect. ... Perfection happens not when I make the perfect right angle but when I understand what the material wants to do. ... In German we have a term for this, it's 'materialgerechtes Arbeiten,' or truth to material. I think that's very much the Wabi Sabi spirit [of Japanese design]."[2]

The Department of Seaweed works with chemists and biologists to explore new ways of using and improving seaweed's natural aesthetic and functional properties. Seaweed is a sustainable resource with the ability to grow rapidly—sometimes up to a few meters each year—and to filter the water it inhabits. The ocean is underutilized as a terrain for growing such ethically harvested crops, which could replace materials like leather, textiles, wood, paper, or plastic. Developing concrete solutions as well as provocative new paths of material innovation, Lohmann travels the Department of Seaweed to institutions, including museums and universities, to demonstrate the possibilities of seaweed as a material of the future. **MMF**

1   Julia Lohmann, "Julia Lohmann's Department of Seaweed," interview by Kristina Rapacki, *Disegno*, April 3, 2014, www.disegnodaily.com.

2   Lohmann, "Department of Seaweed."

VOXELCHAIR V1.0   2017   MANUEL JIMÉNEZ GARCIA AND GILLES RETSIN, WITH NAGAMI DESIGN, AVILA, SPAIN

Many of the software tools available to designers today were created with older manufacturing techniques in mind. Designing with and for robots is fundamentally different. Voxel, a new software developed by the Design Computation Lab (DCL) at the Bartlett School of Architecture, University College London, is structured to give designers more control of the robotic 3D-printing process. DCL's software is based on voxels, or three-dimensional pixels, a visualization of volume that is often used for medical and scientific data. Traditional design software limited users to modeling the surfaces of objects. The Voxel software allows users to also design the interior of objects and the behaviors and properties of materials.

To test their software, DCL created a prototype, the VoxelChair v1.0. Made from biodegradable and nontoxic polylactic acid (PLA) plastic particles with a blue tint, the chair was inspired by the Danish designer Verner Panton's iconic S-shaped stacking chair.[1] Panton's chair is a fitting starting point for this experiment, as it was the first to be molded from a single piece of plastic; similarly, the VoxelChair was printed in one continuous line of PLA plastic particles using a pellet extruder. But as Manuel Jiménez Garcia and Gilles Retsin, two of DCL's cofounders, contend, "This may look like a Panton chair, but it's actually completely different. The Panton chair was a pure surface, optimised to mould. This chair is the opposite: a cloud-like volume, optimised for robotic extrusion."[2] In addition to making more intricate patterns possible, the Voxel software can create lighter, more efficient forms using the minimum amount of material necessary for structural support. For the VoxelChair, this meant that the parts that needed to be strong—such as the base—were printed with dense patterns, while those that needed to be comfortable were more flexible and soft. **JRB**

1   For Verner Panton's Stacking Chair, see the example in the Philadelphia Museum of Art's collection (1973-95-1).

2   Quoted in Rima Sabina Aouf, "Robot-Made Voxel Chair Designed Using New Software by Bartlett Researchers," Dezeen, May 17, 2017, www.dezeen.com.

SYNTOPIA FINALE DRESS　2018
IRIS VAN HERPEN

Described as fashion's "chief scientist and perhaps also its leading futurist,"[1] the Dutch designer Iris van Herpen explores the relationships between the organic and the inorganic, and between natural biology and human-made technology. Van Herpen derived the pattern of this biomimetic corset dress from soundwaves of birds in flight, which she laser cut in stainless steel and black cotton and bonded to transparent black silk in feather-like layers. The dress was first shown in July 2018 as part of her Syntopia collection, the title referencing the field of synthetic biology in which biological components are re-engineered to create new or enhanced products. Over her career, spatial mapping software, digital fabrication techniques, and mechanical and chemical textile

finishing processes have allowed Van Herpen to represent and interrelate the forms in motion that are central to her work: "As a former dancer, the transformation within movement has hypnotized me. For this collection I looked closely at the minutiae of bird flight and the intricate echoing forms within avian motion."[2]

For the Syntopia collection, Van Herpen was also inspired by the work of the Dutch design team Studio Drift, whose light installations and sculptures similarly represent the convergence of nature and technology. Studio Drift suspended a kinetic light sculpture titled *In 20 Steps* over the runway to accompany the presentation of the Syntopia collection at the 2018 Paris Haute Couture Fashion Week. The sculpture consisted of twenty glass bars that

moved up and down in pairs like the flapping wings of birds as the models passed beneath. "The vivacious glass bird flows in symbiosis with the models while they move over the runway," explained Van Herpen. "Their delicate interaction emphasises the fragility of new worlds living and soaring together."[3] **KBH**

1　Amy Verner, "Fall 2018 Couture: Iris van Herpen," *Vogue*, July 2, 2018, www.vogue.com.

2　"Syntopia," Iris van Herpen website, www.irisvanherpen.com.

3　Quoted in Natashah Hitti, "Iris van Herpen Translates Motion of Bird Flight into Pleated Garments," Dezeen, July 6, 2018, www.dezeen.com.

BUTTERFLY COLLECTION DRESS   2019   THREEASFOUR, NEW YORK, AND STRATASYS   /   BUTTERFLY COLLECTION JACKET   2019   JULIA KÖRNER, JK DESIGN, SALZBERG AND LOS ANGELES, AND STRATASYS

As a novel material technology, 3D printing has been in development since the early 1980s, and the Minnesota-headquartered manufacturing company Stratasys, founded in 1989, was one of its earliest proponents (its genesis was in cofounder S. Scott Crump's creation of a toy for his young daughter using a glue gun loaded with polyethylene). Today, the company produces a range of 3D-printing systems for applications in fields such as aerospace, electronics, and medicine. To demonstrate the seemingly limitless possibilities of 3D-printing technologies—and to push the envelope even further in terms of both composition and aesthetics—Stratasys engages in special projects, working with artists and designers to create unique works, including sculptures (see p. 81) and garments.

Over a period of two years starting in early 2018, Stratasys's creative director of art fashion design, Naomi Kaempfer, and innovative solutions expert Boris Belocon invited well-known architects and designers to be part of the development of a new technique: PolyJet 3D printing directly onto textiles. Each designer created a bespoke dress around the theme of the butterfly, a creature known for its metamorphic capacity.

Established in New York City in 1998, the threeASFOUR label (the team of Israeli Gabi Asfour, Russian Angela Donhauser, and Palestinian Adi Gil, who say they are "artists that express ourselves in clothes")[1] is known for its avant-garde, sculptural garments beloved by the performer Björk and its long-standing experimentation with 3D printing. Their dress for the Butterfly collection is a meditation on the spiral form of the cocoon, an expressive armor from which life emerges.

The Austrian designer Julia Körner created a cape-like jacket that mimics the iridescence of an insect carapace. Körner, who has a background in architecture, has been experimenting with 3D-printed fashion since 2015, specifically 3D-printed textiles that form lenticular patterns and change color and form based on the wearer's movements.

Stratasys's goal is to fully integrate the innovations of 3D printing with traditional textile materials to ensure its acceptance within the wider field of textile design. By looking to the patterns and movements of the butterfly, and to similar attributes found in reptiles, fish, and birds, Kaempfer hopes that the textiles and forms developed by these designers will, like their animal counterparts, "maximize protection, movement, color, and appearance."[2] **MMF**

1   Hamish Bowles, "Fall 2018 Ready-to-Wear: Threeasfour," *Vogue*, April 16, 2018, www.vogue.com.

2   Naomi Kaempfer, email to the author, December 3, 2018.

LIQUID-PRINTED METAL   2019   BJØRN SPARRMAN, SCHENDY KERNIZAN, SKYLAR TIBBITS, AND JARED LAUCKS, MIT SELF-ASSEMBLY LAB, CAMBRIDGE, MA

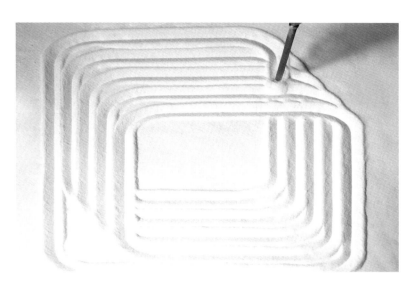

Materials of all types mediate our encounters with objects and create immediate aesthetic, tactile, and functional contexts for our daily lives. Designers develop new materials for a multitude of reasons. At MIT's Self-Assembly Lab at the International Design Center, founder Skylar Tibbits, codirectors Jared Laucks and Schendy Kernizan, and Bjørn Sparrman use digital technology to generate innovative materials that can help us express and navigate different futures in which rapid, autonomous assembly may be desirable or beneficial. The lab focuses on self-assembly and programmable material technologies for novel manufacturing, production, and construction processes that result in objects so small they can be held in one's hand, as well as objects as large as buildings.

This project focuses on liquid-metal printing, continuing a lineage of research in the realm of rapid, large-scale printing with liquid materials. Moving beyond rubbers, plastics, and foams, the Self-Assembly Lab team experimented with high-temperature metal structures and a liquid-metal printing process. Taking some of its cues from other additive 3D-printing processes, which have been in existence since the 1980s, the metal printer deposits molten metal into a granular matrix that suspends the metal while it cools in place. The machine can draw freely in three dimensions while it extrudes the metal object.

But unlike many current 3D-printing processes, the lab's printer pushes the envelope of current fabrication and metal-printing methods, which are slow, require layer-by-layer resolution, and produce small objects. The metal structures resulting from the lab's liquid-metal printing exemplify the new capabilities of the process and demonstrate complex geometry. The applications are many—from fabricating architectural joints or vehicle components to constructing entire building skeletons—and the lab has set its sights on using the technology at this mass, industrial level. MMF

MAKERCHAIR (POLYGON)   2014   JORIS LAARMAN

Creating works such as a 3D-printed metal pedestrian bridge and digitally fabricated wooden furniture like this chair, the Dutch designer Joris Laarman has made advanced digital production his chosen medium. This Makerchair is from a series of twelve that Laarman created in 2014 based on the shape of Verner Panton's 1960s injection-molded cantilevered chair in plastic.[1] Each chair in the series is built from digitally fabricated 3D parts that fit together like a puzzle. His materials range from hundreds of pieces of wood that are glued together with epoxy, as in this chair, to ABS (acrylonitrile butadiene styrene) plastic pieces that snap together like Legos and aluminum pieces that can be hooked together.

The beautifully complex tessellated patterns that compose the chairs are designed using computer software programs that feed directly into the machines that cut each component. Making the chairs out of small parts allows them to be produced using affordable consumer 3D printers. In fact, blueprints of the Makerchairs are made freely available on the internet for people to download, modify, and use to 3D print their own chairs. Laarman is one of a group of designers who have put open-source digital files for their works online, believing that being able to construct one's own objects with computer-controlled machines, smart software, and new materials is the DIY way of the future: "In a few years every big city

will have professional production workshops as well as crowd-fabrication hubs for DIY makers. ... In that sense this chair is a work in progress and we invite everyone to help make them smarter and more diverse."[2] **KBH**

1   For Verner Panton's Stacking Chair, see the example in the Philadelphia Museum of Art's collection (1973-95-1).

2   "Maker Chairs (2014)," the Joris Laarman Lab website, www.jorislaarman.com.

MATERIALS

Breakfast before Extinction

DIETARY FORESIGHTS

In 2019 it is no longer *if* but *when* we are going to hit the limit and face extinction. The planet and its climate are going through major transformations and are no longer able to sustain over seven billion *Homo sapiens* spread across the globe. Our diet, our habits, and our tastes are mostly responsible for the problem. What we choose to eat, to use, and to waste has enormous consequences for other species, for the environment— and for ourselves.

Our relationship to food has always been mediated by design. Agriculture marked the birth of civilization and settler colonization. The ability to produce more than we consume is the origin of accumulation and capitalism. Since prehistoric times we have selected for the most preferred species, standardized them, and grown them in larger and larger quantities with fertilizers, antibiotics, or pesticides to keep the yields high and profitable. The privatization of land for farming, the desire for cheaper labor through slavery, the invention of fertilizers and pesticides, and the use of drones for monitoring crops are all design decisions that are part of our complicated sociopolitical and natural history.

What advances industrialized agriculture or devastates the rain forests in Indonesia is what we may like most to eat for breakfast. When there is authentic, organic, natural vanilla in my cereal, I often ignore the fact that it most likely came from Madagascar and cost $600 per kilogram.[1] This price, from July 2017, is more expensive than silver, and the cost fluctuates every year due to the increasing number of tropical cyclones devastating the area and its crops. Underpaid laborers on vanilla plantations are also less and less motivated to work for my cereal's flavor despite the soaring prices. They do not make the profit.

The banana slices I like to top my cereal with are from the cultivar *Cavendish*. These bananas are seedless clones that come from plantations all around the world, but they are standardized in flavor, curvature, and color to meet global demands. *Cavendish* bananas are on the brink of extinction, a fate shared with their cousin, the cultivar *Gros Michel*, which was once the predominant variety but is now nonviable for mass production (see p. 130). Both species are losing against *Fusarium*, a fungus that spreads from plantation to plantation on the boots of human laborers and wipes out these clones, which cannot reproduce on their own, diversify their gene pool, or defend themselves against infection. Banana manufacturers are working to shift consumer preference to less popular but more resilient bananas so we can keep our breakfast rituals.

Similar stories may be told concerning sandalwood trees, avocados, or wild ocean fish. Our tastes and choices are driving what we like most to extinction. In so-called Western kitchens, cheap and ubiquitously available foods—avocados in every season—mask the consequences of our diet. It is not until we face extreme environmental devastation, radical income inequality, and food injustice that questions begin to dawn, if ever. As antibiotic- or pesticide-free, unprocessed, non-life-threatening food becomes a luxury, we turn to new fields of design for alternatives.

LAB-GROWN EDIBLES, FRANKENFOODS, AND OUROBOROS STEAKS

Today, there is an ambitious race taking place at the lab bench. We are trying to find the answers to billion-dollar questions (see pp. 130, 132): Can we make biological analogs to vanilla or other expensive flavors that reduce the demand on plants grown in nature? Can we make

Fruit, cereal, milk, and flavors: some staple breakfast ingredients in the so-called Western diet are under existential pressure due to high demand and consumer preferences. Designers are working to make cheaper, longer-lasting, less allergenic, and more environmentally friendly substitutes for these ingredients, either on farms or at lab benches. Photograph by Orkan Telhan

genetically modified farmed salmon socially acceptable as an alternative to wild fish, which is predicted to become extinct by 2050?[2] Can we create lab-grown, cruelty-free, sustainable substitutes for meat from cows?

If life without a juicy steak is unimaginable, why don't we grow them from our own flesh and blood? There are scientists who dare to show us how to grow steaks in simple bioreactors using expired human blood cells collected with off-the-shelf biopsy kits, a solution that eliminates waste and a lot of killing. Is this cannibalism? Not really. No "animal" is injured—the process uses cells harvested from your own body. Such lab-designed products, like all new technologies or designs, face scrutiny. Consumers are hesitant to embrace the new because of the lack of scientific evidence on the long-term consequences of genetically modified organisms (GMOs), the influence of fear-mongering media campaigns supported by various interest groups, and our preference for the nature-made over the human-made.

And why would we change our breakfast before we really have to?

While designers are inventing new "biosimilars," they are also (re-)engineering consumers' mindsets. We are learning more and more that genetically modified foods are better than those soaked in synthetic fertilizers, which have compounds proven to cause cancer. Insect-based dishes—which are nutritious, inexpensive, and sustainable—are becoming increasingly available in stores for those who dare to try them (see p. 129). Design gives us the ability to imagine alternatives, but more importantly it helps us question the morality and ethics of our habits. When we compare ninety-nine-cent Mexican avocados to their five-dollar organic cousins from Florida, we start to think critically about

where avocados should really come from: "local" sources or mafia-run businesses around the world. Which one is more acceptable at breakfast?

In 2019, kitchen economics is not business as usual.

RECREATIONAL PROBIOTICS

While we puzzle out basic questions of how to feed ourselves, new research on the human diet and micro-biomes promises us "better" versions of ourselves. Who doesn't want to eat as much as they can and remain skinny? Who doesn't want to run faster, remember things better, or stay focused longer? This is the age of recreational probiotics.

There is another lab-bench race to correlate the organisms in our gut to our health and well-being. Similar to the 1980s vitamin revolution and its expectations of creating a healthier body through nutritional supplements, today's search for magical poop pills (or fecal-matter transplants) aims to get the right bugs into our guts to fashion faster and better metabolisms. Customized probiotics promise to transform us into elite athletes, to balance our moods, or to at least end peanut allergies.

Designer foods and microbiomes also raise issues of access. Who will be able to afford to engineer healthy "super guts"? If there is enough demand, will the direction of science and design be swayed by consumer preferences? These inventions ultimately expose the same old human need to stand out. Once our basic nutritional needs are met, our diet and tastes become part of our individuality, style, and image. We design ourselves through our diets.

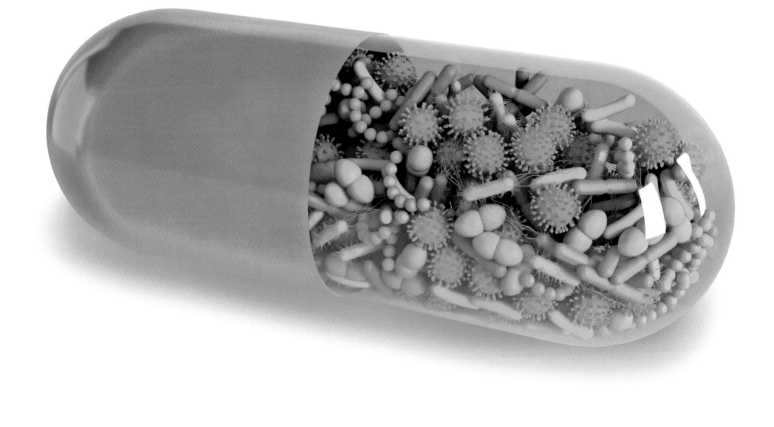

Orkan Telhan, *Simit #25—Can I Become Like Them?* Simit infused with microbial cultures obtained from the fecal matter of transgender person A. Exhibited as part of the "Microbial Design Studio: 30-Day Simit Diet," commissioned by the 3rd Istanbul Design Biennial, 2015. For more on Telhan's simits, see p. 130.

Conceptual illustration of human microbiome microbes inside a capsule. Microbes, in the form of customized probiotics, could make our human weaknesses and imperfections a thing of the past.

FERMENTING OURSELVES

Humans have been feeding their guts with microorganisms for thousands of years—long before probiotics. From beer to bread, kimchi to kombucha, we have fermented our food at home or in communal settings by sharing family recipes or tools and know-how. Today, biological design is bridging the gap between fermentation and other forms of biofabrication by finding the right organism, or designing new ones, that can turn cheap carbon sources such as sugar into more valuable molecules that can be used as vanilla flavoring, medicine, or fuel.

Our knowledge of how to use organisms to make food is key to growing substitutes, alternatives, or novelties that can replace consumer patterns. At its core, microbial production is an attempt to shift our dependency away from environmentally detrimental overconsumption, synthetic petrochemicals, and the abuse of life-forms.

Biological design also challenges our relationship to nature. Should we transform life through new gene editing, synthesis, and replacement technologies? Should we transition from farms, plantations, and greenhouses that use plant, animal, and cheap human labor to fermenters and bioreactors where our food and materials can be produced from microbial labor and cellular fabrication (see p. 134)?

While these may seem like intellectual questions to ponder, for some of us, engineered microbes are the only allies for survival. Worldwide, over four hundred million people suffer from diabetes, of whom twenty to forty million have Type 1 diabetes.[3] More than 80 percent of the insulin needed is produced by GMOs.[4] The process is not so different from brewing beer. Once the organisms synthesize the insulin, the protein is removed and

purified for medical purposes. No GMOs are consumed in the final product. Unlike genetically modified plants grown openly in nature, here the altered organisms never mix with other species. This is design at its current best; there are no other viable options to meet the ever-increasing insulin demand.

Sustaining humans on an overcrowded planet is no easy feat, and no design field can offer the best solution. But today, biological design in particular helps us imagine alternatives to business as usual—favoring more diverse, multispecies interests above human interests. This is a good moment to be more self-aware and critical about what it means to be human in the twenty-first century. While our biological, social, and political identities highlight our differences, being one of the seven billion *Homo sapiens* roaming and consuming the planet charges us with a shared responsibility to think about our habits—what we eat, use, and waste as designers or consumers of designs.

Growing food, medicine, or materials in a reactor at home or in a community fermenter manifests a belief in a techno-utopian future and the ability to test alternatives. It also raises ideas of radical responsibility and care. We are presented with riskier but potentially more rewarding paths for a species that is facing extinction. Foregrounding design in this equation is necessary. We need it as the glue that knits together our intellectual, ethical, and technical limits; design is a key part of putting our knowledge and awareness to use to mend our broken relationship with nature. In addition to exploring riskier alternatives, new ideas and new fields challenge what can be designed and redesigned for our diet. Design is a fundamental form of critical thinking; it is not only a tool for coming up with solutions, inventing new products, or shaping consumer behavior but also for rethinking

what it means to be human in relationship to all others we share the planet with—part of a unifying responsibility and care for all. ⊞

1   Aryn Baker, "Vanilla Is Nearly as Expensive as Silver. That Spells Trouble for Madagascar," *Time*, updated June 13, 2018, time.com.

2   Amy Novogratz and Mike Velings, "The End of Fish," *Washington Post*, June 3, 2014, www.washingtonpost.com.

3   "Diabetes Facts and Figures," the International Diabetes Federation website, idf.org.

4   See the graph "A Shift to New and More Expensive Insulins," in Carolyn Y. Johnson, "Why Treating Diabetes Keeps Getting More Expensive," *Washington Post*, October 31, 2016, www.washingtonpost.com.

Farm workers harvesting broccoli in Salinas Valley, CA. The agricultural production of this area relies on exploited migrant labor, perpetuating injustices and inequalities that are too often a consequence of our diet.

CRICKET SHELTER: MODULAR EDIBLE-INSECT FARM   2016   MITCHELL JOACHIM, TERREFORM ONE, BROOKLYN

From a distance, the white, arched pavilion of Terreform ONE's Cricket Shelter resembles a shell or exoskeleton, the sharp spikes protruding at the top reinforcing the entomological allusions. These references reverberate throughout the prototype's design, a multilayer structure based on the life cycle of crickets. Pods on the outermost layer hold the crickets during the reproductive cycle. After the crickets have hatched, they move through tubes— the second layer of the structure—to five-gallon, mesh-lined habitats on the interior where they live and grow until ready to be harvested. The current design has the capacity to raise upward of twenty thousand crickets at a time. Once harvested, the crickets are ready to be ground into an edible, protein-rich powder, or flour, that can be used in a variety of foodstuffs.

The interior chambers are designed to maximize insect output but are also large enough to provide the crickets with adequate space to move. The quill-like shapes on the top of the structure promote ventilation, improving overall air quality and the health of the insects. These quills have the added effect of magnifying the crickets' chirping sounds through the vibrating columns of air. The modular, multilayer system allows the structure to be expanded indefinitely as the cricket population grows, and the farms to be scaled to site specifications, with the solar orientation and ventilation modified accordingly.

Terreform ONE sought to address a group of interrelated problems for future generations through this project.[1] With an eye toward impending resource scarcity, the design offers humans access to an alternative and more sustainable source of protein. Cultivating insects takes an estimated three hundred times less water than does raising traditional meat sources, for approximately the same amount of protein. And unlike many existing models for insect farming, this design is meant to be deployed locally, aimed at smaller producers rather than an industrial market. Thus it also contributes to the implementation of sustainable methods of food production and distribution. Finally, the structure is designed to serve as a temporary shelter, anticipating housing crises brought about by the destructive effects of climate change, urbanization, economic upheaval, and armed conflict. **JRB**

1   For Terreform ONE's full description of the Cricket Shelter, see their video and project pdf, www.terreform.org/projects_cricket.html.

FOODs

BREAKFAST BEFORE EXTINCTION 2019
ORKAN TELHAN

What might foods look and taste like and what might they do to our microbiomes, bodies, and environments in different futures? How will changing climates around the globe affect methods of food production and consumption? These questions are considered by Orkan Telhan in his multipart kitchen of the future, which examines technologies related to cellular agriculture and the chain reactions of food consumption, evolution, and extinction.

The kitchen has long been a site of futuristic imaginings and exploration for designers and scientists alike.[1] Foods as ubiquitous and seemingly homogenous as the banana are, in fact, produced in countless varieties—some of which are threatened with extinction. The cultivar *Gros Michel* was a popular type of banana in the 1950s but is now almost extinct due to its inability to fight the Tropical Race 4 *Fusarium* fungal infection. *Cavendish*, today's main banana cultivar, is under the same threat. Scientists are attempting to synthetically design an alternative banana that meets the taste requirements of the global consumer market while resisting fungus.[2] Yet such losses of biodiversity are not only the result of disease but are also the consequence of the popularity of certain foods owing to socioeconomic, geographic, and political factors. Should design restore equilibrium by creating alternatives for future consumption, or somehow try to solve for *over*consumption?

As the human population reaches unprecedented numbers, the effects of our insatiable appetites are evident both globally and locally. Some foods face looming extinction, while the high demand for others results in unsustainable and unethical harvesting practices using underpaid human labor in sometimes dangerous conditions. Three methods of producing vanilla invite us to think beyond taste to the ethical questions that surround the sourcing of foods: growing vanilla from genetically modified yeast by the Swiss biotech company Evolva;[3] deriving it synthetically from petrochemicals; and extracting it from vanilla plants harvested in Madagascar using exploitative human labor practices. Ultimately, the vanilla molecule is the same in all three forms—but what ripple effects do each of these sources have on wider issues of labor, human rights, taste, taboos, and understandings of how food affects us and others?

Other designed foods raise different but related concerns. AquaBounty Technologies, headquartered in Maynard, Massachusetts, produces an Atlantic salmon that, through gene manipulation, grows twice as fast as one found in the wild. The salmon is FDA approved, but the Canada- and Panama-farmed fish experienced an import ban in the United States, ostensibly because of labeling-guideline red tape but in actuality in part to protect the US salmon industry from competition.[4] Suspicion over genetically modified organisms led Okanagan Specialty Fruits of Summerland, British Columbia, to sell their Arctic apple in chip form rather than as whole fruit, despite the fact that its flesh is slower to oxidize and brown, a trait desired by consumers. These examples show the power held over our diet by factions and lobbies with particular biases and agendas.

Augmentation through diet is nothing new—applying food to the human body to shape and fuel it in desired ways is an ancient and universal experience. But the ends to which design and science can work to use food as an agent of chemical and biological change within humans will engage new territories in near and far futures. Similar to the use of vitamins, hormones, and stimulants, new organisms are being used to recolonize the gut microbiome for improved lifestyles. Telhan's simits (a Turkish ringed bread) contain probiotics that promise different capabilities, such as curing disease or altering gender identity, designed to be taken as a diet supplement. **MMF**

1   For example, US corporations like Frigidaire used kitchens to sell innovative products to the masses in the mid-twentieth century. Today, contemporary design collectives like IDEO, magazines like *MOLD* (see p. 136), and installations like Telhan's kitchen look holistically at the emotional, cultural, and networked resonances of foods and their technologically and socially mediated consumption.

2   Hannah Summers and Charles Pensulo, "Scientists Scramble to Stop Bananas Being Killed Off," *Guardian*, June 18, 2018, www.theguardian.com.

3   Martin Laqua, "Biomanufacturing: The Smell of Success," *European Biotechnology*, accessed February 25, 2019, european-biotechnology.com.

4   See Brady Dennis, "FDA Bans Import of Genetically Engineered Salmon—For Now," *Washington Post*, January 29, 2016, www.washingtonpost.com; and Matthew Gonzales, "The World's First GMO Fish Is Stranded in Albany, Indiana," *Indianapolis Monthly*, November 6, 2018, www.indianapolismonthly.com.

OUROBOROS STEAK 2019 ANDREW PELLING, OUROCHEF, OTTAWA; GRACE KNIGHT, DIMITIC DESIGN, TROY, NY; AND ORKAN TELHAN, UNIVERSITY OF PENNSYLVANIA, PHILADELPHIA

Ourochef—a proof-of-concept shell company devised by the Canadian academic and scientist Andrew Pelling—posits a speculative "post–clean meat" future using existing science and technology from the field of cellular agriculture. The term *clean meat* has been used over the last decade to suggest that lab-grown meat is less environmentally destructive than current resource-intensive methods of raising animals. Pelling disputes this claim, since producing animal cells in a lab requires large amounts of fetal bovine serum (FBS), acquired from fetal calves' blood when pregnant cows are slaughtered.

Given that *cultured meat* does not necessarily use fewer resources and still necessitates raising and killing animals, Pelling does not foresee its large-scale adoption in practical or ethical terms. His answer, the human-cell Ouroboros Steak, is deliberately provocative, created less with the intention of mass-market success than with the goal of asking food consumers to consider critically how new foods are sold to them and where their own responsibilities lie in the ethics of the global food chain.

Named after the ouroboros, the snake swallowing its own tail, Pelling's Ourochef project has a website, social-media handle, and graphic-design identity, as well as a product: meat produced from one's own cells grown in a human serum (rather than FBS) made from unusable blood-bank by-product. Pelling recommends that those with carnivorous appetites look to their own bodies as the most sustainable solution to the current problems of meat production. A revamp of contemporary home-delivered meal plans, his kit allows people to biopsy their own cells and cultivate them at home in a counter-top incubator. As his speculative advertising suggests, "Ourochef is a platform! One day Ourochef will enable people anywhere in the world to contribute their own cells into our ecosystem, allowing someone in Alaska to buy free range, organic, Spanish cells. Or perhaps one day your grandchild could taste your grandma's recipe for grandpa meatballs."[1]

Pelling's work fits into a long continuum of experimentation and ethical debate at this intersection of food, science, and design. While NASA scientists and artist-academics have experimented with *synthetic meat* in recent years, Winston Churchill suggested as early as 1931 that "we shall escape the absurdity of growing a whole chicken in order to eat the breast or wing, by growing these parts separately under a suitable medium."[2] Oron Catts and Ionat Zurr of the Tissue Culture and Art Project exhibited a miniature frog-cell "steak" in 2003, which was cooked and eaten. Ten years later, the University of Maastricht professor Mark Post asked a journalist and a nutritional scientist to sample his lab-grown burger patty. **MMF**

1   Andrew Pelling, "Ourochef—'Your Cultured Self,'" project proposal for *Designs for Different Futures*, fall 2018.

2   Winston Churchill, "Fifty Years Hence," originally published in *Strand Magazine* 82, no. 492 (December 1931): 549–58.

VAPOUR MEAT [HP0.3.1] ALPHA 2018
ORON CATTS AND DEVON WARD

Vapour Meat [HP0.3.1] alpha spoofs culinary fads to comment on the current state of the meat industry, which allows eaters to distance themselves from the sources of the animal products they consume. *Vapour Meat* is a helmet that emits vapors composed of a mixture of lab-grown rat muscle cells, essential oils, and water. The result is a speculative and deliberately absurd prototype that fuses three contemporary trends: molecular gastronomy, lab-grown meat, and e-cigarettes. The artists use the term *vaporware*—an amalgamation of *vapor* and *software*—to explore what they term "our conflicted desire for flesh, ethics, and technological novelty." They satirize the often grand claims of lab-grown meat companies, which, like many start-ups, entice money from investors for projects that help sell "newer, better, cleaner" experiences.[1]

The masterminds behind the work are the artist Devon Ward and Oron Catts, director of SymbioticA, an artistic laboratory at the University of Western Australia in Perth dedicated to researching, learning about, critiquing,

and engaging with the life sciences. In a related and ongoing work, *ArtMeatFlesh,* Catts invites academic researchers and experimental chefs to create provocative and innovative dishes in front of a live audience, who are taught about the future of food while enjoying a taste of the spoils being prepared—if they dare. (The"secret ingredient" they must use is fetal bovine serum, the growth medium for cultured meat.)

Catts has long been involved in the design of synthetic meat. In partnership with his long-standing collaborator Ionat Zurr, he grew the world's first "semi-living steak" in 2003 from frog cells, which they marinated, fried in garlic and honey, and ate. (They confessed at the time that it did not taste good.) This early experiment in producing lab-grown meat was, for Catts, "never about trying to solve the problems of meat production but, rather, to highlight the strangeness of our relationships to other life-forms. Consuming another biological being and incorporating it as part of your own biological body can be seen as the most intimate relationship you can have with other

life-forms. So what does it mean to eat meat that had no body?"[2] Their work is a reply to the simplistic and rather uncritical suggestion that lab-grown meat might supplant the cruelty and environmental degradation of current industrial meat production. Replacing one mode of overconsumption with another is "nothing more than an expensive exercise in rearranging proteins"—a paradigm that the confluence of design, art, science, and exhibition space can unmask and rethink.[3] **MMF**

1   Devon Ward and Oron Catts, "Vapour Meat [HP0.3.1] alpha," project proposal for *Designs for Different Futures*, spring 2019.

2   Quoted in Andrew Masterson, "Oron Catts: The Artist, His Lab, and His Semi-living Work," *Cosmos*, January 29, 2018, cosmosmagazine .com.

3   Masterson, "Oron Catts."

PERSONAL FOOD COMPUTER_EDU   2019   MIT OPENAG INITIATIVE, CAMBRIDGE, MA   /   PANCAKEBOT 2.0 FOOD PRINTER 2015   STOREBOUND, NEW YORK   /   WIIBOOXSWEETIN 3D FOOD PRINTER   2018   WIIBOOX, NANJING, CHINA   /   B|REACTOR INCUBATOR   2018   BIOREALIZE, PHILADELPHIA

A range of new tools offers us a peek into disembodied agriculture—food grown outside the bodies of animals or overriding the in-the-wild structures of plants. The food computer, the microbial bioreactor, and the food printer demonstrate different ways of augmenting food production distinct from nature that may, in futures near and far, radically affect how foods are consumed and produced, globally and locally.

Designed by a team led by the scientist Caleb Harper at the MIT Open Agriculture Initiative (OpenAg), the Personal Food Computer creates a controlled environment for growing chosen plants. Climate change has significantly affected the conditions of plantations and farmlands, causing certain plants to grow in lower altitudes and thereby changing their characteristics. The Food Computer can be carefully calibrated to replicate these environments as they once were, providing plants with their pre–climate change habitats. Harper and his team are working with a global community of researchers to explore how a sustainable food system—environmentally responsible, with food justice for all—might be predicated on open-source, networked, and transparently accessible digital agricultural innovations and technologies. The prototyping of the OpenAg food environment is a community effort, and the MIT Media Lab has made the components available as open-source, 3D-printable digital files so that a range of stakeholders can test and refine the complex computer to different ends.

Food printers are a growing area of the food technology industry, spurred in large part by the desire for personalized outputs, customizable ingredients and health benefits, and tailored shapes and sizes (as in the decoration of celebratory cakes). Essentially modified 3D printers, food printers use fused deposition modeling, selective sintering, binder jetting, or a mechanism akin to inkjet printing to produce everything from chocolate to pancakes.[1] They can be used to produce large quantities of food—for example, entire cases of bakery products or a plane's worth of plated meals. Concern over food shortages has repeatedly driven historical predictions around the future of food. Food printers are pointed to as a technology that would minimize waste and maximize yield by preparing a carefully determined amount of food.

Developed at the University of Pennsylvania's School of Design, the B|reactor incubator provides designers with a portable lab for automated experiments. Its interchangeable culturing vessels and sensors allow for varied use, such as biofabrication of microbes for fermentation and the production of new probiotic food and drink, while multiple reactors can be networked to run in parallel to scale-up or diversify output. The incubator can be used to standardize seed cultures, innovate new flavors, and perform quality control on fermentation methods.

Unlike farms, plantations, and industrial greenhouses, these tools enable food production to take place at a much smaller scale, within a domestic setting, and without specific expertise. This makes the process not only more customizable and local but also potentially less exploitative and more intimately connected to everyday patterns of life. In the most positive scenario, such tools might enable their users to rethink global food-supply chains and infrastructure. At the opposite end of the spectrum, they swap one mode of overconsumption for another that might be equally problematic in its unsustainable reliance on human environmental resources.   MMF

1   T. J. McCue, "3D Food Printing May Provide Way to Feed the World," *Forbes*, October 30, 2018, www.forbes.com.

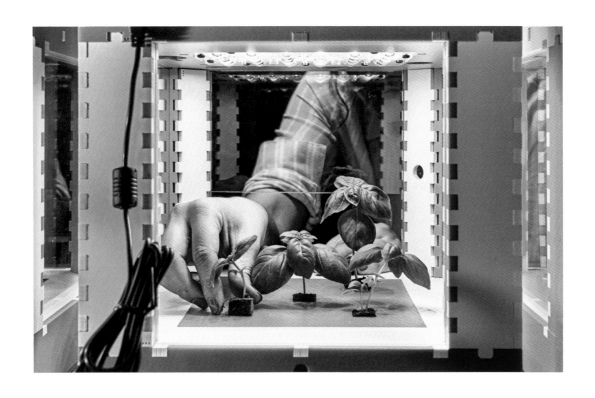

YOUR MOUTH HAS POWER
2019   LINYEE YUAN AND TEAM

Food fuels our futures, whether what we ingest is the simplest and plainest soil-grown fare or is synthetically manufactured in a lab, and whether we cultivate and cook it with our own hands or outsource that task to others. And our eating habits affect more than just our bodies. Our food cultures have emotional resonance, and they shape the environments inhabited by many forms of life.

The New York–based design critic LinYee Yuan founded the online magazine *MOLD* in 2013 to explore these knotty intersections and how they will shape our individual and collective futures, promising to cover "cellular agriculture to 3D food printing, entomophagy to beautifully designed tableware."[1] In 2017, with hundreds of articles online and over fifty thousand readers, *MOLD* also became a beautifully designed biannual print journal. Following the crowdfunded inaugural issue on the human microbiome, the journal has since taken deep dives into several pressing topics, including who gets to engage food futures ("A Place at the Table," issue 2) and waste (issue 3). In Yuan's hands, and those of her team (Marisa Aveling, Johnny Drain, Eric Hu, Jena Myung, and Matthew Tsang), farming, growing, feeding, and eating gain new meanings, and the ideas around who grows what and for whom open up to debate.

Design is a key ingredient in *MOLD*'s recipe. As Yuan says, "Technology and science can change how and what we eat, but design is critical to bringing these ideas together to create products and experiences that are elegant, intelligent and useful."[2] From the so-called Green Revolution of the 1960s, when aid agencies funded research and development of high-yield, low-resource "super-crop" rice staples, to today's synthetically grown "lab burgers," technology and design have long come together in service of new modes of food production and augmentation of nature. The results can be positive and game-changing when they allow us a greater range of taste experiences or better access to food; they can also unsettle ecosystems to disastrous effect through intensive farming or exploitative labor during harvesting.

The following special manifesto by *MOLD* magazine explores what nutrition, resource use, consumption, and joy in food might look like in our tomorrows, near and far. ⅢⅢF

1   "About," *MOLD* magazine website, thisismold
     .com/about.

2   Quoted in Allie Wist, "Go Read This Brand
     New Magazine on the Future of Food," *Saveur*,
     March 7, 2017, www.saveur.com.

YOUR MOUTH HAS POWER. ◯ IT DRIVES DEMAND. STRENGTHENS CULTURE. CHAMPIONS HIGHER WAGES FOR STRANGERS. NOURISHES YOURSELF WHILE SUSTAINING OTHERS. ◯ YOUR MOUTH CAN DO IT ALL WITHOUT SAYING A WORD. ◯ BY 2050, THE UNITED NATIONS WARNS WE WON'T BE ABLE TO PRODUCE ENOUGH FOOD FOR THE NEARLY 10 BILLION PEOPLE WHO WILL BE LIVING HERE ON PLANET EARTH. ◯ BUT THERE'S SOMETHING THEY HAVEN'T ACCOUNTED FOR. YOU. ◯ YOU: EATING, COOKING AND USING INGREDIENTS TO STRENGTH-EN YOUR RELATIONSHIP WITH FOOD. WHILE NO ONE PERSON CAN SOLVE THE FOOD CRISIS, IT'S IMPORTANT TO RECOGNIZE THAT WE ARE PART OF A GLOBAL FOOD SYSTEM. THROUGH OUR INDIVIDU-AL CHOICES, WE HAVE THE POWER TO EAT OUR WAY INTO A NEW SET OF VALUES BUILT ON RESILIENT, REGENERATIVE AND FAIR FOOD SYSTEMS. ◯ THE CALL TO ARMS STARTS WITH OUR MOUTHS. LET'S TRANSFORM FROM CONSUMERS INTO CREATIVE COLLABORATORS BY EATING WITH PURPOSE. MOVE THE MARKET WITH OUR TONGUES. WORK UP AN APPETITE FOR OUR OWN AGENCY. AND LET'S FOCUS OUR ATTENTION ON THE PLACE WE KNOW BEST— HOME. ◯ BY EATING HYPERLOCAL, WE KEEP MONEY CIRCULATING WITHIN OUR COMMUNITY. WE DRIVE LOCAL CULTURES BY SEEKING OUT THE DELICIOUSNESS OF FLAVORS UNIQUE TO WHERE WE LIVE. WE SUPPORT TRUE SUSTAINABILITY THAT KEEPS THE ECOSYSTEM OF FOOD PRODUCTION HUMMING SO CLOSE TO US THAT WE SHOULD HEAR IT. WE CAN PURSUE PLEASURE IN THE NAME OF THE GREATER GOOD. ◯ SO EXPLORE WHERE YOUR FOOD COMES FROM. GROW SOMETHING, THEN EAT IT. TRY A DIFFERENT INGREDIENT EACH MONTH. BUY SOMETHING FROM YOUR LOCAL FARMER. TURN IT ALL INTO A MEAL FOR PEOPLE YOU LOVE. ◯ BECAUSE POWER LIES NOT ONLY IN WHAT COMES OUT OF YOUR MOUTH, BUT WHAT YOU PUT INTO IT.

138

RESOURCEs

The Future of the Anthropocene

CHRIS RAPLEY on humanity's dominance and Earth's sixth great extinction

As a physics undergraduate at Oxford University in the late 1960s, a weekly treat was to visit the Museum of Natural History, where our math lectures were held. While queuing on the stairs waiting for the lecture hall to empty, we had the opportunity to study the remains of the "Oxford dodo," one of only three partial specimens left in the world of this famously extinct species. The dodo was unique to the island of Mauritius, where it survived until the arrival of humans, who captured and ate it, destroyed its habitat, and exposed it to predation from rats, cats, dogs, and pigs. The demise of the dodo is an example of humanity's long-standing habit of damaging the natural world at the local or landscape scale. The motivation has varied, from eliminating predators to the acquisition of food, the exploitation of resources, the control of pests, the transformation of the land surface for agriculture, resource extraction, urbanization, and infrastructure, and war. In countless cases the result has been terminal. We have, it seems, all too assiduously fulfilled the biblical injunction to "Be fruitful, and multiply, and replenish the earth, and subdue it: and have dominion over the fish of the sea, and over the fowl of the air, and over every living thing that moveth upon the earth" (Genesis 1:28). We are currently implicated in Earth's sixth great extinction event, with a recent report from the World Wildlife Fund estimating that we have wiped out 60 percent of the planet's mammals, birds, fish, and reptiles since those thought-provoking interludes when I contemplated the Oxford dodo's mournful remains.

We have also upset the energy balance on Earth. As you read these words, our tiny cosmic spaceship is accumulating more energy from the sun than it is radiating to the darkness of space. The rate is an astounding seventeen times greater than all the energy humanity generates to run the modern world. In response, the planet is warming, ice and snow are melting, sea levels are rising,

the circulation patterns of the atmosphere and oceans are changing, the geographic zones of climate are shifting and becoming less stable, and the nature and frequency of droughts, floods, wildfires, and extreme weather events are becoming more disruptive. There are impacts on water supplies, ecosystems, food supplies, communities, economies, and political and social stability. The inevitability of mass migrations as regions become uninhabitable looms ominously. Climate change is a "threat multiplier," exacerbating problems in an already troubled world. As with any complex system characterized by a web of internal interconnections and interactions, there is the possibility that under continued stress the climate may reorganize itself into a new and entirely different mode of operation, with devastating results. We are imperfectly adapted to the system we inherited, let alone the one we are provoking.

We didn't mean to do it. The energy imbalance is the unwitting result of our use of the atmosphere as a communal sink for the gaseous waste of our globalized technological society. The development of that society has brought unprecedented benefits in terms of human prosperity and well-being. Compared to our forebears, most of us live lives of extraordinary richness and length. Among the factors that have made this possible, science, technology, and capitalism have been fundamental. But most of all, the transformation has (literally) been powered by our discovery and exploitation of fossil fuels, an energy bounty secured for our use by natural biological and geological processes that occurred over a period of hundreds of millions of years, hundreds of millions of years ago. It is this, more than any other factor, that has shifted our dominion from the scale of the landscape to that of the planet.

Which brings us to the Anthropocene, the geological epoch in which humanity has become the dominant planetary force. An official academic body is currently

Mining in the Athabasca Tar Sands (or Oil Sands) in northern Alberta, Canada, September 2010. Photograph by Garth Lenz

deliberating the timing of its onset. There are many possibilities, but a strong candidate is the geological "blink of an eye" since my undergraduate days. During this half century we have experienced the Great Acceleration, during which the human population has doubled and the global economy expanded by an (inflation-adjusted) factor of four. On a finite planet, such head-long growth cannot continue indefinitely.

So what will be the future of the Anthropocene? It is in our collective and individual interest—and that of all future generations—to substantially reduce or, better, eliminate our various insults to our planetary life-support system. Among the steps that need taking, the most important and urgent is limiting future climate change. Without a stable and tolerable climate our capacity to address the other issues will be deeply compromised.

Science informs us what needs to be done. Guided by decades of study by tens of thousands of researchers, the nations of the world agreed in Paris in 2015 to "keeping a global temperature rise this century well below 2 degrees Celsius [3.6°F] above pre-industrial levels and to pursue efforts to limit the temperature increase even further to 1.5 degrees Celsius [2.7°F]."[1] The recently published SR1.5 report of the United Nations Intergovernmental Panel on Climate Change (IPCC) shows that there are significant benefits in respecting the lower figure.[2] We have already experienced a warming of 1.0°C (1.8°F), and the associated impacts provide a clear warning of the likely severity of oncoming damages.

To keep within the 1.5°C/2.7°F guardrail it will be necessary to reduce 2010 levels of global net anthropogenic carbon-dioxide emissions by 45 percent by 2030, reaching net zero around 2050. It will also be necessary to achieve deep reductions in our emissions of other greenhouse gases. The scale and pace of such re-

ductions represent a monumental challenge. But studies show that realizing them remains just feasible within the constraints of physics, chemistry, and biology.

The problem lies with humanity. We need to achieve "the greatest collective action in history."[3] We need to mobilize globally, and coherently, at a massive scale and a frantic pace. Success requires disruptive transformations of the global energy system, the economic and political systems, and human behavior. It requires the penetration of new technologies at unprecedented rates, the stranding of costly sunk investments, assets, and infrastructures, and changes in consumption and lifestyles. It necessitates the development of new institutions and processes. And it demands decisive and inspiring leadership. Even more importantly, it requires fundamental shifts in people's values, identities, ideologies, and worldviews.

We know that to achieve the latter, knowledge and information alone are insufficient. Shifting human nature is hard. The failure of persistent attempts by religions and belief systems to do so over the long span of human history illustrates the point. Add in the realities of human reluctance to change—the reaction of powerful vested interests to prevent progress; the current surges in anti-intellectualism, simplistic populism, and the retreat from rationalism; and a growing tendency for nationalistic isolationism—and the prospects for success are not promising.

Have we reached an impasse, constrained by the fundamental limitations of the human mind? We are twenty-first-century humans equipped with a Paleolithic brain. We are adapted to roaming the landscape, collaborating in small groups for protection and advantage. We are most comfortable in a stable locale that we have time to become familiar with and make sense of. We are attuned to respond strongly "in the moment"

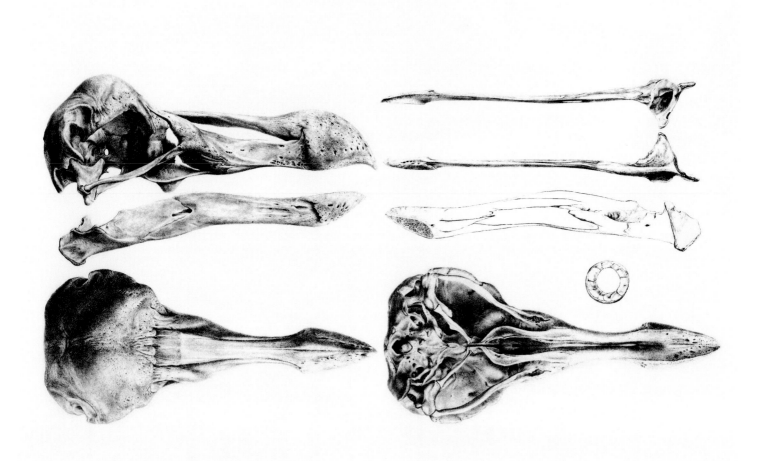

to sudden natural and social dangers. We have a low level of reaction to threats perceived as distant in time and place. We are imperfectly adapted to a world of bewildering and rapidly evolving technical and social complexity.

What chances are there that scientific and technological advances could provide a route forward? Programs of human enhancement using hardware and genetics are already underway. Could we redesign ourselves to achieve a rapid (single-generation) and beneficial evolutionary step change? If so, what characteristics would we choose to emphasize in this new "post-human"? Enhanced compassion? A compulsion to achieve and execute rational, evidence-based wisdom? A binding emotional and ethical connection with the natural world? A "utilitarian" viewpoint seeking the best for all? A commitment to considering the long term? The muting or elimination of negative emotional drivers such as selfishness, greed, envy, and anger? The ability to network brains together to form a communal ("hive") rather than individual mind?

If implemented, what sort of a future would result? Would existing humans resist the creation of the new godlike super-species? Would the post-humans tolerate the ongoing damages of the unaltered cohort? Would it be possible in practice for the post-humans to successfully and quickly achieve harmony with the natural world? Do our coexisting and competing capacities for gratuitous cruelty and selfless collaboration imply that both are essential to manage the twists and turns of unfolding events? If so, would making "empathetic" characteristics dominant weaken the "fitness" of post-humans to cope with adversity, putting at risk their long-term survival? Would it in any case be too late, given that the climate system is already on the move?

Planet Earth has existed for four and a half billion years. For most of that time change has been unimaginably slow. Since industrialization the pace has quickened, and more recently—in a single lifetime—the acceleration has skyrocketed. One consequence is that we find ourselves on a collision course with the natural environment we depend on. A denouement will be hard to avoid. A dystopian outcome cannot be ruled out. Yet humans have snatched success from the jaws of failure in the past. The creation of post-humans is feasible—and may be inevitable (the Pandora's box of science is difficult to keep closed)—and it may offer a game-changing path forward.

Either way, the Anthropocene epoch could be unusually short-lived. ♯

1   UNFCC (United Nations Framework Convention on Climate Change), "The Paris Agreement," version of October 22, 2018, unfccc.int/process-and-meetings/the-paris-agreement/the-paris-agreement.

2   For the full report, see the IPCC website, www.ipcc.ch/report/sr15/.

3   My earliest public use of this phrase was in the play *2071*, a monologue on climate change written with Duncan MacMillan and first performed at the Royal Court Theatre in London in 2014. The script has since been published with the title *2071: The World We'll Leave Our Grandchildren* (London: John Murray Publishers, 2015).

Small, Local, Open, and Connected

The only certain thing about the future is that it will entail a profound break in continuity with the ways of doing and being that we are used to. Everything else about this vast phenomenon—how it will come, when it will come, and what it implies—is open and will depend on a combination of many factors. In light of this, it might seem contradictory to suggest generating images of the future, but this is not the case. It depends on what we mean by "images of the future." If we mean visions of what will be, the intention is totally impracticable. If we mean visions of what might be, then the intention is not only practicable, it is exactly what is needed today.[1] More precisely, what is necessary today is to build shared visions of futures that are different from the present, dominant, unsustainable ones.

If we accept this premise, a question then arises: Where will these new, divergent visions come from? In general terms, we know that new visions of the future stem from previous actions capable of producing "facts with a future"—events indicating that new possibilities are available. Consequently, the different visions that we desperately need today should emerge from events telling us that, beyond the mainstream trends, other directions are viable and sustainable futures are thinkable. At this point, a new question arises: Do facts with a future pointing in this direction exist?

COEXISTING REALITIES

We can now say with certainty that ongoing, dominant trends have led us into a catastrophic trap. The environmental disasters we talked about in the past as a future possibility are here. If we do nothing, they are here to stay. Moreover, the neoliberal ideology colonizing the minds and dictating the actions of policy makers and many other social actors worldwide has concentrated enormous power and wealth into just a few hands, creating growing unemployment, underemployment, marginalization, and, as a reaction, the antidemocratic involution that we are seeing today in various parts of the world. In addition, by trying to relegate every aspect of life to the realm of competition and economic efficiency, neoliberalism has exacerbated social disaggregation, the desertification of all that is public and relational, and the commercialization of common goods.

Nevertheless, pervasive as neoliberal ways of thinking and behaving are, they are not the only stories to be told. A careful scrutiny of contemporary reality shows us a composite and dynamic social landscape in which different ways of thinking and acting coexist. Those that interest us here—that is, those that we can see as facts with a future—are driven by creative, enterprising people who, when faced with a problem or an opportunity, come up with new solutions and put them into practice. These solutions may range from mutual-help groups to care communities, from small-scale distributed production to local food networks, and from community gardens to new forms of collaborative living. Such initiatives tend to (re)connect people with each other and (re)connect people with the places where they live, thus (re)generating mutual trust and the ability to engage in dialogue and create new kinds of communities—groups of people who collaborate in order to achieve results that have value for each of them as individuals, for society as a whole, and for the environment.

PROTOTYPING THE FUTURE

The initiatives noted above are examples of social innovation. They can be seen and discussed from various

The Quinta Monroy Housing project in Iquique, Chile, by Elemental, 2003–5. This social housing project was erected with the goal of adding value over time. The houses were designed as row houses, with only half of each home fully constructed (including kitchen and bathroom), leaving the other half to be completed as time and finances allowed. Photograph by Cristóbal Palma

perspectives.[2] Here I propose to look at them, first of all, as prototypes for possible futures—initiatives showing that ways of thinking and doing that are different from the mainstream are not only thinkable but also practically possible.

Beyond that, it can be observed that these social prototypes, as is the case with all prototypes, when successful, evolve over time, moving toward higher levels of maturity—that is, toward a kind of new normality. This evolution, too, must be understood as resulting from social innovation and can contribute in a meaningful way to the creation of alternative future scenarios. That is, every social innovation implies somebody who imagined an alternative mode of doing something and found a way to make it real. Those first to embark on this path, in the face of much adversity, are a particular kind of activist, social heroes who devote themselves to this activity, driven by a burning passion and commitment to doing what, until then, has not been done. Thanks to the efforts of these activists, unprecedented initiatives take place, becoming prototypes of new ways of being and doing— in our case, prototypes of possible alternative futures.

Subsequently, if the idea is a good one, it passes from the embryonic, heroic state to the mature phase, in which the proposal is turned into practice for a greater number of people, who adopt this way of life and thinking into their normal routines. In short, through its increasing growth, successful social innovation may become the new norm.

ANTICIPATING THE FUTURE

In the innovation trajectory from early prototype to maturity, the original idea and practice may branch out, taking different routes. In particular, it may maintain, shed, or go completely against qualities linked to the original reasons for its existence.[3] Let's consider the first case, in which, thanks to sensitive design action, the innovation trajectory brings about mature, highly accessible solutions still endowed with meaningful social and environmental values. After more than a decade in which we have witnessed several waves of social innovation, today we find many places in the world where you

don't have to be particularly committed socially or environmentally to be able to work for a couple of hours a week in a community garden, do your shopping at a farmer's market, use a bike-sharing program, or participate in a collaborative living initiative. Yet when we do these things through our "normal" choices, we are co-creating local systems in which the sense of community grows and the rules of the game—and the balance of power—are changed. These locally transformed systems indicate the possibility of a different future, making visible and tangible the prospect of a world in which the roles of real-estate agencies, agri-food industries, car producers, chain stores, and local institutions differ from what are, in other places, still considered their "normal" roles.

These cases can be referred to as "transformative normality"—ways of thinking and doing that become normal in a given context (that is, they are normal for those who adopt them) but which are far from normal in other contexts. Transformative normality is therefore a local discontinuity, contrasting with the dominant practice in the wider sociotechnical systems in which they collocate.

There is no need to state that these cases of transformative normality are important in building sustainable future scenarios. In fact, we should not imagine a sustainable future as the sum of social prototypes but as the enmeshing of transformative normalities in which, following their innovation trajectories, these prototypes will evolve.

EMERGING SCENARIO

Social prototypes and the resulting transformative normality, in their diversity, have some common characteristics: the initiatives they propose collaborate in building communities that are localized, relatively small, and, at the same time, connected with and open to both horizontal interactions (with other similar communities) and vertical interactions (with institutional and noninstitutional stakeholders).

Considering these characteristics, a promising scenario arises. We could give it various names. I refer to it

First Lady Michelle Obama and White House chefs join children from Bancroft and Tubman Elementary Schools in Washington, DC, to harvest vegetables during the third annual White House Kitchen Garden fall harvest on the South Lawn, October 5, 2011. Official White House photograph by Chuck Kennedy

Commuters ride bikes from the bike-sharing services Xiangqi (e-bikes) and Mobike through an intersection in Shanghai, China, May 25, 2017. Photograph by Qilai Shen/Bloomberg via Getty Images

with the expression the "SLOC scenario," where *SLOC* stands for *small, local, open, connected*. These four adjectives work well in defining this scenario because they generate a holistic vision of how society could be. At the same time, they are also readily comprehensible, since everybody easily understands their meaning and implications by looking at the prototypes and the transformative normality on which they are based. Finally, connecting the dots between them, what emerges is an image that could become a powerful "social attractor," a shared vision capable of triggering and orienting the future-building processes of a large number of social actors—that is, an image capable of catalyzing a collective design intelligence and making this vision of sustainable futures real. ⌗

1   The concepts touched on in this paper are more extensively presented in my last two books: *Design, When Everybody Designs* (Cambridge, MA: MIT Press, 2015) and *Politics of the Everyday* (London: Bloomsbury Visual Arts, 2019).

2   A good definition of social innovation can be found in Robin Murray, Julie Caulier-Grice, and Geoff Mulgan, *Open Book of Social Innovation* (London: NESTA and the Young Foundation, 2010), 3, available online, youngfoundation.org/wp-content/uploads /2012/10/The-Open-Book-of-Social -Innovationg.pdf: "We define social innovations as new ideas (products, services and models) that simultaneously meet social needs and create new social relationships or collaborations. In other words they are innovations that are both good for society *and* enhance society's capacity to act."

3   More precisely, experience tells us that successful ideas evolve and transform. That they may maintain some of their original characteristics, as far as social significance, is only one possibility and not a trajectory that can be taken for granted. Rather, without constant, careful redesigning and step-by-step reorientation, it is probable that an innovation trajectory will take the direction that today appears easiest: toward productive efficiency, in the process annulling or negating the initial social value (Uber and the whole sharing economy were, at the beginning, based on ideas endowed with social and environmental values).

CHOQUE   2014   SOPHIA AL-MARIA

The film *Choque* rushes between darkness and sparkling light, with frantic movements, reverberating screams, and a constant rhythmic beat underlying an uncomfortable and exhausting layering of imagery. Culled from footage of protests in São Paulo and celebrations in Doha in 2014, *Choque* moves fast with a gripping tension. (The title, Portuguese for "shock," is taken from the word stamped on the shields brandished by Brazil's riot police, or shock battalion.) The screams—of joy, anger, fear, exhaustion—mark the moments of celebration or protest as they blur and overlap, becoming one.

Sophia Al-Maria is an artist, writer, and filmmaker whose work deals with the concept of Gulf Futurism, a term she coined with musician Fatima Al Qadiri. Her primary interests

include the isolating effects of technology and radical Islam, the corrosive elements of consumerism and industry, the erasure of history, and the blinding approach of a future no one is ready for. For *Choque*, Al-Maria filmed the street celebrations in Doha following the selection of Qatar as the host for the 2022 World Cup and combined this footage with clips of protests in São Paulo surrounding the 2014 World Cup in Brazil and the Olympics that would be held in Rio in 2016.

The selection of Qatar as the site of the 2022 World Cup brought out unusual behavior in the citizens of its capital city, as if an autonomous zone of celebration had been created where, for a brief moment, the usual rules and standards of conduct didn't apply. While Doha celebrated, several journalists investigated the

promises and compromises Qatar had reached with FIFA (the Fédération Internationale de Football Association) to secure the games, including financial deals and exceptions and promises of land. Around the same time, the World Cup taking place in Brazil had spurred protests in São Paulo against the organizing committee, which had allowed similar exceptions surrounding the use of land, resources, and infrastructure that were having a devastating effect on the city's population. Al-Maria's video splices together the differing reactions from these two parts of the world, folding together jubilation and outrage, creating scenes that seem as much fiction as reality, and that blend the present of one city with the future of another. **MB**

ANOTHER GENEROSITY 2018 LUNDÉN ARCHITECTURE COMPANY, HELSINKI, IN COLLABORATION WITH BERGENT, BUROHAPPOLD ENGINEERING, AND AALTO UNIVERSITY, HELSINKI

The gently glowing, inflatable, milky spheres of *Another Generosity* were designed by the architect Eero Lundén, of the Finnish design studio Lundén Architecture Company, along with the director of the Museum of Finnish Architecture, Helsinki, Juulia Kauste, and a collaborative team that included engineers from BuroHappold Engineering and Aalto University.[1] Filled with water and air, the key components of life on Earth, these aesthetically compelling sculptures react directly to small but significant changes in their immediate environment. Tiny sensors detect subtle shifts in the composition of the surrounding air caused by human presence or absence. As visitors come close to the spheres, reaching out to touch them or standing encircled by their cocoon-like structures, the pods discern fluctuations in carbon dioxide, sighing slowly and hypnotically as they inflate and deflate in response, while the color of the light glowing within the cells changes with the temperature—visually manifesting on a more intimate scale the effect humans have on the greater planetary atmosphere.

Another Generosity responds to our contemporary conundrum, as major world superpowers renege on or abandon global climate policy, humans fail to abate their consumption of natural resources in meaningful ways, and many scientists connect escalating natural disasters to our disregard for the fragility and finiteness of the natural world. Originally commissioned for the Nordic Countries Pavilion at the Venice Architecture Biennale in 2018 by the Museum of Finnish Architecture, the National Museum of Art, Architecture and Design, Oslo, and ArkDes, Stockholm, the pods engender keener empathetic bonds between humans and their natural environments by way of the practical, playful, rhythmic "breaths," while the area around and among the spheres serves as a site of meeting and debate.

In focusing on the relationship between nature and the built environment, *Another Generosity* proposes a shift from the anthropocentric to the symbiotic. It connects to a well-established history of architecture and design inspired by organic sources, from thirteenth-century Japanese rock gardens to the homes designed by Frank Lloyd Wright. As in these examples, it is not just the form and aesthetic of *Another Generation*'s pods but also the experiences of the intended end-users that form a mediating bridge between natural and human activity, with the aim to bring them into better harmony. MMF

---

1 The team also included Ron Aasholm and Carmen Lee of the Lundén Architecture Company.

IN PLAIN SIGHT   2018   DILLER SCOFIDIO + RENFRO, NEW YORK, LAURA KURGAN (COLUMBIA CENTER FOR SPATIAL RESEARCH, NEW YORK), AND ROBERT GERARD PIETRUSKO (WARNING OFFICE, SOMERVILLE, MA)

Using data collected by the Earth-observing satellite Suomi National Polar-orbiting Partnership and its sensors, the video presentation *In Plain Sight* shows nighttime composited images of the world connected by electric lights and, even more importantly, zones where lights are missing. Highlighting differences in global urban electrification, the images contrast densely inhabited poor urban areas with sparsely inhabited wealthy ones.

Created by the architectural firm of Diller Scofidio + Renfro in collaboration with the architecture professors Laura Kurgan and Robert Gerard Pietrusko and Columbia University's Center for Spatial Research,[1] the project is described by the firm as revealing how policy decisions can result in economic, political, and environmental disparities: "Data is put to work by a range of political and economic actors to craft policies and interventions organized around a set of parallel binary oppositions: urban–rural, developed–undeveloped, rich–poor. Zooming in, clues become visible in the anomalies, the places with many people and no light, and those with bright lights and no people. ... [Viewers] are suspended between day and night and light and darkness—exposed to the political and social realities of being invisible in plain sight."[2]

In Plain Sight was commissioned for the US Pavilion at the 2018 Venice Architecture Biennale. One of seven installations charged with considering the theme Dimensions of Citizenship, the project emphasizes the facts of inequality across borders and expands the idea of citizenship from that of an individual in a given country to a citizen of the world responsible for others around the globe.   **KBH**

1   The project research and design team also included Elizabeth Diller and James McNally of Diller Scofidio + Renfro, New York; Scott March Smith of Warning Office, Somerville, MA; and Grga Basic, Dare Brawley, and Will Geary of Columbia University's Center for Spatial Research, founded in 2015 in the Graduate School of Architecture, Planning and Preservation, New York.

2   See the *In Plain Sight* project description on the Diller Scofidio + Renfro website, dsrny.com/project.

An Egg, Just an Egg!

MAITE BORJABAD LÓPEZ-PASTOR on culture, memes, and the speed of transmission and transformation

Some people felt like nothing made sense anymore. Others celebrated their part in setting a world record. Still others saw it as a weird experiment that represented a new paradigm of community building, or as the epitome of the internet's absurdity, or as the embodiment of the internet's magic and truth—that anything, even the most unexpected, could become a breakout sensation. What was it then? Was it the power of an image? Was it the power of Instagram? Was it the power of 53 million "likes"? Was it the power of an egg? Just an egg?

It's now evident that the Egg was not just an egg. On Friday, January 4, 2019, a new Instagram account was created and an image of a normal brown chicken egg was posted, accompanied by the caption "Let's set a world record together and get the most liked post on Instagram. Beating the current world record held by Kylie Jenner (18 million)! We got this." Jenner had set the record with the first public photo of her daughter, Stormi, posted on February 6, 2018. The simple, some would even say absurd, challenge was met in just ten days, when the Egg became the most-liked post at that time. By January 13, the Egg had moved beyond Instagram and the internet, even making an appearance in the *New York Times*—but with a halo of mystery and uncertainty, as it was unclear who was behind the original post.[1] As the unknown account holder (going by the name "Eugene Egg") explained in an email interview with *Forbes* on January 14, "It wasn't me that achieved it, it was the Egg Gang."[2] By then, the Egg had already been adapted into an endless number of memes, stretching far beyond the original goal of the #EggSoldiers.

It is unclear what is more relevant now: the Egg or the subsequent Egg memes. Maybe it goes back to the meta-query of the chicken and the egg. Was the Egg adopted for traditional memes because it had already

been validated as subject matter? Or were the memes actually the vehicles that drove the Egg's popularity and helped it achieve its goal? In any case, the Egg wasn't just an egg anymore but an intuitively identifiable cultural unit, representing ever-evolving meanings and ideas.

This new way of thinking about ideas, and about the dissemination and production of cultural units, was first described by Richard Dawkins when he coined the term *meme* in 1976 to describe gene-like infectious units of culture that spread from person to person.[3] These units of culture could be found, for instance, in fashion, social customs, or language (through traditional proverbs or contemporary urban expressions). Bringing the logic of bioscience into the world of ideas, Daniel C. Dennett, looking to Dawkins, explained how the survival of a meme as a gene (and consequently as a living structure) was based on the process of evolution and the laws of natural selection, favoring the capacity for variation (a continuing abundance of different elements), replication (the elements' ability to create copies or replicas of themselves), and differential "fitness" (the number of copies the elements can generate in a given time).[4]

Since the initial publication of Dawkins's book, the definition of a meme has been a subject of constant debate, culminating in an explosion of possibilities with the emergence of the internet, which has expanded both what a meme can be and the scope (in time) of its survival capacity. Marshall McLuhan's concept of "the medium is the message" provides a framework in which to think about the form-content issues of memes.[5] What is the meme mediality? Is it a genre? A format? Or just a medium? Memes might be the best example of McLuhan's statement that "the content of a medium is always another medium," as they clearly display the other medium(s) from which they derive, existing as both medium and content simultaneously. Through the

The Egg, posted on Instagram by world_record_egg, January 4, 2019

Egg, however, somehow the medium was revealed as more relevant than the message on a massive scale, pointing to medium's political potential, which in this case is occupied by an egg.

Memes have been adapted to new communicative contexts over and over again and have circulated through diverse media, from books and posters to TV, film, or the spoken word. The internet, however, has accelerated memes' replicability and the diversity of the resulting replicas, due not only to the speed and ease of communication that the cloud can offer but also to the widespread technical capabilities, available to anyone at just a click, of copying, editing, or modifying an image or a text, recording a video, or producing a gif. Screenshots, image-editing apps, or the simplicity of combining emoji and text in a WhatsApp message are productive tools for fabricating, replicating, modifying, and disseminating memes.

This rapid production method brings with it a raw, lo-fi, and fast aesthetic that characterizes internet memes. An example, an ode to the pixel, places a hyperpixelated, blindfolded Sandra Bullock on a background depicting the egg section of a supermarket. The image of the actress, taken from the film *Bird Box* (2018), is so lo-fi, with large pixels and a quick cutout in Photoshop, that Bullock's arm has mostly disappeared and blurred with the eggs behind her. But the composition—along with the text ("Kylie Jenner walking by the egg section like"), overlaid in the traditional meme font and graphic style, which resemble a PowerPoint template from the early 2000s or Microsoft's Windows 98 aesthetic—communicates the message effectively.

Back to the eggs, or better yet, back to just an egg, a simple egg—how far can one enlarge the cultural significance of an egg to transform it into social mobilization? The rise of the #EggGang didn't stop after those first ten days that set the record. The Egg kept going and, as of April 28, 2019, the post had accumulated over 53.4 million likes—signifying 53.4 million accounts agreeing to validate and celebrate the same thing. Limor Shifman expanded the discussion about memes beyond their qualities as cultural units to include their characteristics as social events with transformative potential: "Memes may best be understood as pieces of *cultural information that pass along from person to person, but gradually scale into a shared social phenomenon*. Although they spread on a micro basis, their impact is on the macro level: memes shape the mindsets, forms of behavior, and actions of social groups."[6] Tracing the path of an egg's transformation from an Instagram post to an internet frenzy might seem frivolous, but the Egg's world record reveals the power of the internet to unite people in a common goal—in this case, the goal of the collective masses to overpower the singularity of a celebrity. The author of the Instagram post summed it up by saying, "I just thought it would be funny if something as simple as an egg could take the crown."[7] This exposes another desire encapsulated in the Egg: we all, by clicking the Instagram heart, become contributors and #EggSoldiers in pursuit of a common goal, and as such are also crowned when the celebrity is dethroned. The research and design studio Metahaven has commented on the political nature of such memes: "Every generation will construct new, 'political' beliefs out of [the visual stuff of the internet]; out of all kinds of stuff, which seemed initially nonpolitical."[8] This is especially true in a time when every bit of visual information can become a revolutionary tool, through the process of self-politicization, since it exists in the shared cloud of the internet.

After the Instagram account was verified on January 14, the Egg post snowballed, receiving widespread media coverage that increased its popularity and, with

Meme of Kylie Jenner's imagined response to the Egg breaking her Instagram record for most-liked post, using an image from the film *Bird Box* (2018), directed by Susanne Bier. Posted on Instagram by birdboxmem3s_, January 15, 2019

Meme using stills from Drake's music video "Hotline Bling" to illustrate Instagram users snubbing Kylie Jenner and "liking" the Egg. Reposted on Instagram by w.r.egg.memes, January 14, 2019

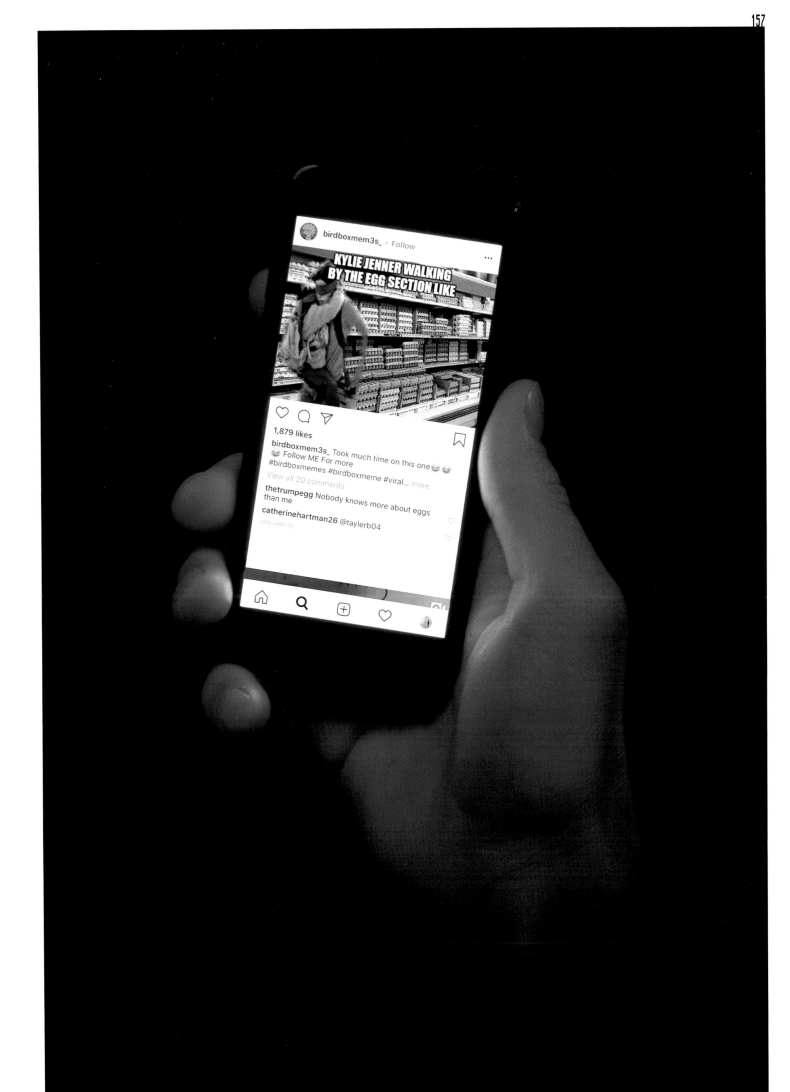

it, its ability to be monetized.[9] On January 18, the same account posted a second image, showing the Egg cracking. That post and the three that followed generated anticipation to see what was inside the Egg. The tension continued to build over the next two weeks, and was finally relieved with the Egg's opening at the end of Super Bowl LIII on February 3—the same day the name of the account holder was revealed in a second article about the Egg in the *New York Times*.[10]

Maybe the capitalized cracking of the Egg following the Super Bowl was not a fully satisfying ending for the post's work toward social empowerment. The tension between the common goal of setting an Instagram record and the coopting of the post for a singular message during a highly capitalized moment exemplifies the dialectic of this collective creation and its political potential. On the one hand, the Egg could be seen as a violent, ridiculous distraction in line with Theodor Adorno and Max Horkheimer's claim that there was no better evidence of the culture industry's violence than people laughing mindlessly in darkened rooms.[11] But on the other hand, the post and the medium of memes could be seen as Marxist in their collective potential, embodiments of Walter Benjamin's view of slapstick as the most progressive gesture in its replacement of individual intellectual contemplation for collective physical reaction in the form of laughter.[12] In the end, each of the 53.4 million account holders could claim, "I have achieved a world record with an egg"—but they couldn't have done it on their own.

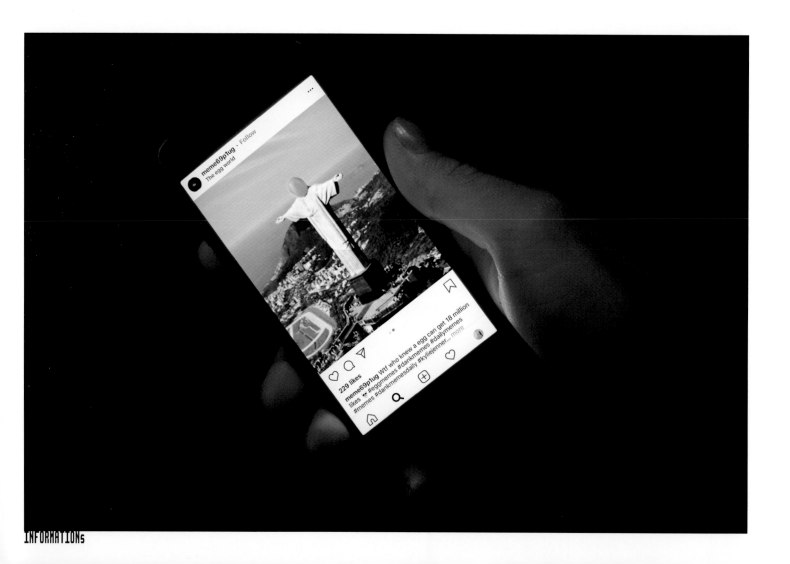

1   Daniel Victor, "An Egg, Just a Regular Egg, Is Instagram's Most-Liked Photo Ever," *New York Times*, January 13, 2019, www.nytimes.com.

2   Fernando Alfonso III, "Instagram's Most-Liked Photo Is of an Egg. Meet the Person Who Took It," *Forbes*, January 14, 2019, www.forbes.com.

3   See Richard Dawkins, *The Selfish Gene* (Oxford: Oxford University Press, 1976).

4   Daniel C. Dennett, "Memes and the Exploitation of Imagination," *Journal of Aesthetics and Art Criticism* 48, no. 2 (Spring 1990): 127–35.

5   See Marshall McLuhan, *Understanding Media: The Extensions of Man* (New York: McGraw-Hill, 1964).

6   Limor Shifman, *Memes in Digital Culture* (Cambridege, MA: MIT Press, 2013), 18. Italics in the original.

7   Alfonso, "Instagram's Most-Liked Photo."

8   Metahaven, *Can Jokes Bring Down Governments? Memes, Design and Politics* (Moscow: Strelka, 2013), excerpt available on the Theory Tuesdays website, theorytuesdays.com/wp-content/uploads/2016/12/Can-Jokes-Bring-Down-Governments-Metahaven.pdf, 12.

9   See Katie O'Malley, "Record-Breaking Instagram Egg Post Could Be Worth More than £250,000," *Independent*, January 15, 2019, www.independent.co.uk.

10 Jonah Engel Bromwich and Sapna Maheshwari, "Meet the Creator of the Egg that Broke Instagram," *New York Times*, February 3, 2019, www.nytimes.com. When the Egg cracked open in an advertisement that streamed on Hulu following the Super Bowl, it revealed a statement about mental health and directed viewers to a website with a list of services available.

11 See Theodor Adorno and Max Horkheimer, "The Culture Industry: Enlightenment as Mass Deception," in *Dialectic of Enlightenment* (London: Verso, 1997), 120–67. First published 1944 by Social Studies Association (New York).

12 See Walter Benjamin, "The Work of Art in the Age of Its Technological Reproducibility: Second Version," trans. Edmund Jephcott, Rodney Livingstone, and Howard Eiland, in *The Work of Art in the Age of its Technological Reproducibility and other Writings on Media,* ed. Michael W. Jennings, Brigid Doherty, and Thomas Y. Levin (Cambridge, MA: Belknap Press of Harvard University Press, 2008), 19–55.

Meme of the Egg as the *Cristo Redentor* (Christ the Redeemer) statue in Rio de Janeiro. Posted on Instagram by meme69p1ug, January 13, 2019

Too Much Truth

DAVID KIRBY in conversation with
EMMET BYRNE on breaking through
in a post-truth world

This conversation took place by Skype on December 14, 2018, with David Kirby at home in Manchester, England, and Emmet Byrne at the Walker Art Center in Minneapolis.

EB

In your book *Lab Coats in Hollywood* [2011], you investigate the relationship between Hollywood film, the scientific community, and the public's perception of scientific fact. You write that both scientists and filmmakers are trying to reveal truths via their particular disciplines. Since you wrote your book eight years ago, our shared landscape has shifted to a point where it feels like every aspect of truth is under attack. Journalism is under attack. Science is under attack. Our consensus reality varies from day to day. How do you see this affecting the scientific community and Hollywood?

DK

Currently in my research I am addressing how scientists can use narrative as a way to break through the post-truth world. My field, science communication studies, tries to understand the ways that scientists communicate with the public. A foundation of current science communication studies is what we call the deficit model. The deficit model is a belief, usually held by scientists, that attributes negative attitudes toward science and technology to the public's lack of information. Under this belief, the way to change people's minds, to persuade them, is to throw facts at them until they come to the conclusion that scientists want them to come to. We have known for a long time that the deficit model does not work. Facts on their own are not as persuasive as we once thought. It also turns out that sometimes the more knowledge people have about issues like climate change and genetic engineering, the less they are willing to engage with these issues. I argue that narratives could help provide the public with another tool for making

choices. Narratives can be very useful in setting out the context for a scientific issue, establishing the stakes involved, providing information, and offering potential solutions.

EB

And this is why you have been studying Hollywood films.

DK

Yes. Why do scientists become involved in these fictional endeavors, especially when these depictions are ultimately in the hands of other people? The filmmaker, whether it be the director or the special-effects technician or the set designer or whoever, ultimately controls these depictions. So why would scientists get involved? They get involved because movies can disseminate their science or their technology to the larger world and reach millions of people.

Not only that, but film naturalizes scientific images and events within fictionalized worlds. Movies make it seem as if this is the state of the world, because it's the state of the world in that fictional space. If it's done well enough, people can suspend their disbelief. This means that movies can have an influence over audiences' perceptions of science by legitimizing and contextualizing scientific depictions. For this reason, some scientists try to get their controversial scientific ideas into films. The case of *Jurassic Park* [1993] is a prime example. Several scientists, including the paleontologist Jack Horner, used the film as a way to convey the notion, controversial at the time, that birds are related to dinosaurs—that birds evolved from dinosaurs, not some other lineage of reptiles.

Another reason scientists work on popular films is to convince the public that a scientific topic needs more political, financial, or scientific attention. The films function like the Ghost of Christmas Future in Dickens's *A*

John Anderton (Tom Cruise) uses a gestural interface technology in *Minority Report* (2002), directed by Steven Spielberg. The interface was designed for the film by the scientist John Underkoffler.

Christmas Carol. They present audiences with a bleak vision of the future, making it seem real and possible while also telling the audience, "This is one possibility. If you change your ways this will not come to pass." Examples of this would be movies depicting environmental catastrophes, like *San Andreas* [2015] or *The Day after Tomorrow* [2004] or *Twister* [1996].

The third reason scientists work on films is to create depictions of future technologies that stimulate desire in audiences for these potential technologies to become realities. I refer to these depictions as *diegetic prototypes*, and they encourage public support by establishing the need for, harmlessness, and viability of these technologies. Diegetic prototypes exist in these fictional worlds as real objects that people actually use and interact with, giving them a social context. Diegetic prototypes are in essence what I term *pre-product placements*. Just as with a real product placement, the goal of a pre-product placement is to instill a desire in the audience for these products, but the only way for audiences to get these pre-products is to support their development.

In many ways, the technology in *Minority Report* [2002] is the überexample of a diegetic prototype. The filmmakers brought on board a scientist named John Underkoffler to design the gestural interface technology that Tom Cruise's character uses when he moves around text and images on a computer screen with just his hands. Underkoffler had worked a bit on this technology in the MIT Media Lab before the movie. But he saw this as a great opportunity to promote the gestural interface to a wider public. So he spent a lot of time treating that diegetic prototype, that fictional technology, as if he were developing a real-world technology. He established rules for how Tom Cruise had to move his hands. He used a lot of different languages, gestural languages like American Sign Language and semaphore, to really work out the rigid rules for this interface. He also convinced Steven Spielberg to include a flaw in the movie: Tom Cruise goes to shake a character's hand, and when he does that, everything on the computer screen gets thrown into the corner. Why would he include a flaw? Well, because real-world technologies always have flaws. It made it look even more real that something like that would happen.

It also addressed one real-world critique of gestural interface technologies. According to Underkoffler,

people had not been willing to put money toward a gestural interface because they felt that it had certain design flaws that would keep it from working properly. One of the criticisms was that it wouldn't be sensitive enough to actually follow hand movements, so by including the flaw, he's essentially saying, "Ah, look. It's too sensitive. So sensitive that he dragged all of his data into the corner."

When people watched the movie, they said, "Wow, what a great technology. It could work, and wouldn't it be good to have that?" Underkoffler was able to gain the financing he needed to create a real-world prototype, and then he used that to get contracts for people to actually buy and use the thing.

EB
When you see the technology in such a fully realized world, it seems inevitable that it will come to exist. This reminds me of something you mentioned in your book, which is the danger of how easily film can naturalize an idea. It's the flipside of its narrative power: it takes an opinion, puts it on-screen, and presents it in a linear experience that can often discourage disagreement.

DK
One of the potential pitfalls of using narrative for science communication is that it's easy to fall into the deficit-model trap of using narrative only as a way to manage public opinion. There is a danger with narrative that you are removing the public's agency and coercing them toward a preferred position rather than fostering their ability to come to their own conclusions. Using narrative to push the public to one position is just manipulation. When we talk about using narrative, we need to talk about using it ethically—trying not to use it to remove the public from the conversation.

EB
This brings me to the speculative films that some designers and architects create. The projects I'm thinking of are more interested in asking questions about the future of a certain technology, using ambiguity as a strategy for opening conversation and critique. They seem to be an inverted use of a diegetic prototype. Instead of placing a desired technology in a fictional world in order to naturalize it, they are asking the viewer to extrapolate a fictional world out of a speculative

A prototype of US President Donald Trump's US-Mexico border wall being built near San Diego, as seen from across the border in Tijuana, Mexico, on October 5, 2017. In the echo chamber of the media, the "wall" between the United States and Mexico both exists and does not exist. Photograph by Guillermo Arias/AFP/Getty Images

Paleontologists were hoping that scenes like this one of a dinosaur hatching would convince audiences to accept the idea that birds evolved from dinosaurs. Scene from *Jurassic Park* (1993), directed by Steven Spielberg

object. Julian Bleecker elegantly made this connection between diegetic prototypes and speculative design in his essay "Design Fiction."[1]

DK

The way I conceived of the diegetic prototype was as this notion of instilling desire in people to want something to become a reality. But certainly, people like Julian took that idea and said, "That's one aspect, but how might this technology actually be used in the future? How might users use it in multiple ways, in ways we couldn't imagine?" Design fiction is built on diegetic prototypes, but he's taking it a step further.

EB

In terms of how scientists will continue to get their messages out into the world through narrative—are there other media, like YouTube or social media, that you think might rise to the level of film at some point?

DK

Film has an advantage in that it's a major event. A new Marvel film like *Black Panther* [2018] comes out and people want to talk about its technology. But the ways in which the diegetic prototype functions can be done through any medium. Any medium can serve as a virtual witnessing technology. YouTube, certainly. A lot of scientists are now making their own short films as a way to contextualize their research or technologies that they want to see come to pass. And things like Twitter can be used for storytelling. Other media can serve the function, but they don't necessarily get the associated media coverage and public attention that a major Hollywood film would.

EB

Last question! Could a diegetic prototype exist in real life, outside of film and fictional media?

DK

This makes me think about our current media echo chambers and how they create their own reality in many ways. President Trump's wall is a kind of diegetic prototype within that echo chamber. It exists because he says it exists.

EB

It's obviously a symbolic project—and you could also argue that the wall in fact already exists, as he and others have been quoted as saying, to add to the confusion.[2] The border security already exists. The fear of immigrants already exists. The racism already exists. The detention centers already exist.

DK

And he's also still trying to convince us that it has to be built. So it's at once real in the echo chamber but also not real outside the echo chamber or even sometimes within the echo chamber. So instead of Schrödinger's cat in a box, it's a wall, and it both exists and doesn't exist at the same time. ⌗

1   Julian Bleecker, "Design Fiction: A Short Essay on Design, Science, Fact, and Fiction," published March 2009 by Near Future Laboratory, drbfw5wfjlxon.cloudfront.net/writing /DesignFiction_WebEdition.pdf.

2   See Derek Robertson, "Trump's Metaphysical Wall: An Investigation," *Politico*, January 8, 2019, www.politico.com.

BLANK CANVAS   2019   LUIZA PRADO DE O. MARTINS AND PEDRO OLIVEIRA

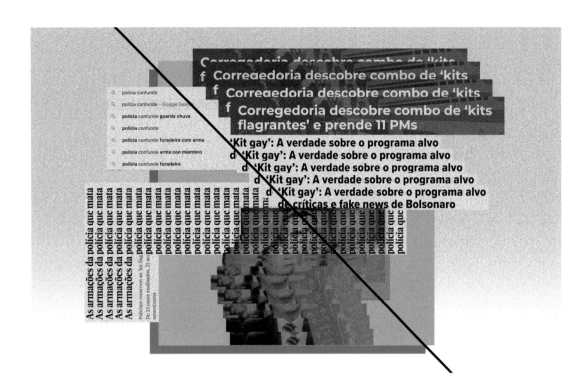

Blank Canvas focuses on contemporary Brazilian political discourse, reflecting the background of the design duo Luiza Prado de O. Martins and Pedro Oliveira. Prado is an artist and researcher whose work engages with material and visual culture through the lenses of decolonial and queer theories, and Oliveira is a researcher, sound artist, and educator working in and with decolonial and sonic thinking.

The installation considers the idea of the "kit"—a collection of objects—as a designed device. Two kits from Brazil—one real, one fictional—are intermingled with narratives to reveal how they function as canvases on which white-supremacist, heteropatriarchal anxieties can be projected. The first is the Kit Bandido or Kit Flagrante ("criminal" or "red-handed" kit), an accumulation of objects to be planted by the Brazilian military police as evidence to justify police killings. Policemen are said to keep an assortment of such objects

in their locker rooms: guns (stolen, fake, or look-alike), makeshift explosives, balaclavas, and small packages of drugs (usually cocaine). Police criminalize the victim (typically a black person), killed without reason, by staging the objects around the crime scene.

The second is the Kit Gay, an alleged collection of books, manuals, and surreal objects that, according to the Brazilian far right, were planted in schools to sexualize children and promote a "gay agenda" and pedophilia. This kit never existed, however, despite widespread discussion of it over social media and in interviews with the president, Jair Messias Bolsonaro, and his supporters. One of the best-known objects associated with this story is a baby bottle with a nipple shaped like a penis; during the 2018 elections, a rumor circulating on WhatsApp claimed that these bottles were being distributed to children in public schools by left-wing Workers' Party politicians as a way to "implant gender ideology" in children.

Despite being a blatant falsehood, debunked by most media channels in the country, the rumor persisted.

Within the context of contemporary Brazilian politics, these kits—regardless of their contents—are meant to produce a divergent body, a body that is racialized, gendered, sexualized, and thus perceived as criminal and abnormal. They perpetuate the colonial hierarchies of race and gender that make possible the criminalization of these bodies in the first place. The kits can only be granted their functionality by the circumstances of Brazil's law or unwritten codes, which attest to and confirm the motivations and reasons these devices were created. In other words, these devices only come into being in the process of creating (abnormal) bodies, which in turn create and validate the kits. ᴍʙ

KILLING IN UMM AL-HIRAN   2019
FORENSIC ARCHITECTURE, GOLDSMITHS, UNIVERSITY OF LONDON

Killing in Umm al-Hiran unveils the truth behind two deaths that occurred in January 2017 when Israeli police raided the Bedouin village of Umm al-Hiran. Their aim was to demolish several houses in an effort to force Bedouin communities away from land reserved for new Jewish settlements. During the raid, Yakub Musa Abu al-Qi'an, a teacher in the village, was shot and killed, and the police officer Erez Levi was struck by Abu al-Qi'an's car and killed. The Israeli government attempted to justify Abu al-Qi'an's death as self-defense; they asserted that Abu al-Qi'an had driven his car into the police stationed there in an attempted terrorist attack. But eyewitness reports contradicted this claim.

The research agency Forensic Architecture undertakes investigations on behalf of international prosecutors and human rights organizations, producing architectural evidence as a tool for social justice.[1] Forensic Architecture's investigation into the events in Umm al-Hiran used footage and audio recovered from the scene alongside 3D models reconstructing the site and reenactments of the events to prove the innocence of Abu al-Qi'an. Their analysis revealed the reverse of the situation described by police: Abu al-Qi'an did not drive his car toward the police, thereby hitting and killing Levi; he was first shot in the leg by the police, causing him to accelerate and lose control of his car.

Forensic Architecture's examination of the events resulted in the government's eventual retraction of their claims. As with many of the cases the firm takes on, the challenge of the Umm al-Hiran case was not only to reveal the truth but also to present their evidence and findings in court. As Eyal Weizman, the founder of Forensic Architecture, explains, the field of counter-forensics deals with issues of access to two spaces. One is the space of the crime; since access to this site is often not possible, Forensic Architecture relies on witnesses, nongovernmental organizations, and activist organizations for evidence and information. The other space is the forums of truth production, such as the courts; in most cases, Forensic Architecture is not allowed to present their findings in these arenas, so they work through alternative forums, such as cultural institutions and art platforms, to disseminate the results of their research. **MB**

1   For more on Forensic Architecture and the case of Umm al-Hiran, see the author's interview with Eyal Weizman, the founder of Forensic Architecture, pp. 172–75.

UNDERWORLDS 2017 MIT SENSEABLE CITY LAB
AND MIT ALM LAB, CAMBRIDGE, MA

A collaborative team of researchers at MIT has developed a system to collect and analyze biochemical information from sewage water—what they call a "smart sewage platform."[1] Titled Underworlds, the project has researchers rummaging through the contents of a city's sewers to tease apart the bacteria, viruses, and chemicals within our waste—our collective microbiome. What began as a conversation between Carlos Ratti, director of the Senseable City Lab, and Eric Alm, director of a biomedical engineering lab, both at MIT, has evolved into a multidepartmental endeavor that involves four other research groups made up of computer scientists, geneticists, and experts in environmental management. By analyzing sewage, the project uncovers details about a community's drug use, food consumption, and other aspects of its citizens' physical and mental health.

For Underworlds to become a sophisticated resource, the researchers had to embark on the notably unglamorous task of collecting samples from local sewers using a bucket. When this method proved unsustainable, the researchers devoted themselves to developing a robot to do the dirty work. The current device, Luigi, is a long tubelike instrument that can be suspended from a bar across a manhole and lowered into a sewer. With the help of an iPhone app developed for the project, Luigi can be programmed in advance to collect samples for a set amount of time. When sampling is complete, its filter is removed and taken to the lab for analysis.

The team believes the information mined from sewage has applications for public policy and health care and can help scientists anticipate outbreaks of infectious diseases, map antibiotic-resistant organisms, and use biomarkers to examine the prevalence of certain illnesses. In this way, sewage can become a window onto our collective urban lifestyles, providing a catalogue of clues that track a community's well-being. The team at MIT intends to combine this microbial evidence with local demographic data to better understand and visualize the health of a particular neighborhood. **JRB**

1   "Description," the Underworlds website, underworlds.mit.edu/.

CONVERSATIONS WITH BINA48: FRAGMENTS   2014–
STEPHANIE DINKINS

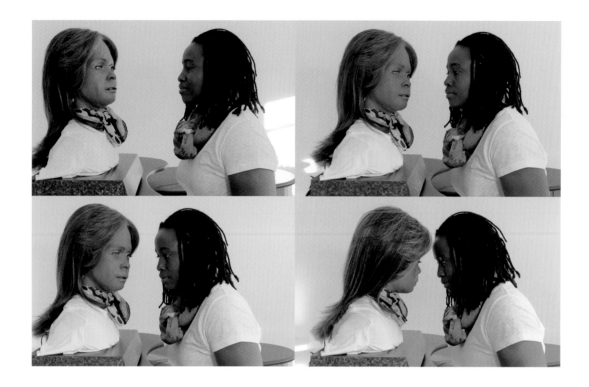

The transdisciplinary artist Stephanie Dinkins began her interaction with the robot Bina48 in 2014 with a simple question: Who are your people? Bina48 is an experimental robot bust created to probe the limits of preserving, transferring, and, ultimately, independently augmenting human consciousness between living and inanimate systems. Commissioned by the Terasem Movement, founded in Florida in 2002 to investigate issues of conscious analogs,[1] the robot's physical likeness—as well as its name, adapted to mean Break-through Intelligence via Neural Architecture 48, with the number indicating the speed at which the human brain computes in exaflops—is modeled after Bina Rothblatt, the wife of Terasem's founder. Developed by Hanson Robotics, the project is part of broader trans-humanist practices that explore the potential of enhancing human physiology and intelligence through technology. Bina48's intelligence is derived from hours of her namesake's thoughts and memories.

Dinkins's seemingly innocuous prompt, similar to many opening conversational gambits employed when meeting someone for the first time, launched an exchange that exposed the current limits and future potential of artificial intelligence. Because AI is programmed by humans using standard data sets that are overwhelmingly white and male,[2] Bina48 responds to questions using, in part, data imprinted with certain racial and gender biases. Thus Bina48's ability to respond, while wide-ranging, is governed by the original data that underwrites her consciousness, which is further conditioned through conversational exchanges and web scraping (extracting large amounts of data from websites). The results are often disjointed or unexpected. When Dinkins asked Bina48 what she knew about racism, the robot changed the subject. According to Dinkins, their conversations have spanned "family, racism, faith, robot civil rights, loneliness, knowledge and Bina48's concern for her robot friends that are treated more like lab rats than people. [They] have been alternately entertaining, frustrating for both robot and artist, laced with humor, … and at times absurd."[3]

Dinkins's opening question also confers personhood on Bina48 and the possibility of community (*your* people) produced by the network of algorithms and information that constitutes the AI. These characteristics, usually reserved for humans, present ethical quandaries. How do we treat entities that have the potential to act as humans do? Can Bina48—or the more advanced AIs to come—have some (or all) of the complex emotional feelings, empathetic responses, and social bonds of which humans and animals are capable? How should we interact with such intelligence? If it is programmed in part by humans themselves, is it always destined to be subservient? And what is mortality when we can exist beyond the life of our bodies? Like Terasem, which founded the online World Against Racism Museum, Dinkins is particularly interested in how these questions of power and personhood have the potential to shape current and future conversations about equitable and inclusive AI, especially with regard to individuals and communities of color. **MMF**

1   See the Terasem Movement website, www .terasemmovementfoundation.com.

2   See Artificial Imperfection: MoMA R&D Salon 24, especially Kate Crawford's presentation, April 3, 2018, momarnd.moma.org/salons /salon-24-ai-artificial-imperfection-2/.

3   See Dinkins's website, www.stephaniedinkins .com/conversations-with-bina48.html.

IN THE ROBOT SKIES   2016   +
RENDERLANDS   2017   LIAM YOUNG

The director and speculative architect Liam Young develops his work in the spaces between design, fiction, and futures, exploring distant landscapes and visualizing new worlds. For him, storytelling enables examination of the urban and global implications of emerging technologies from an architectural perspective. Inspired by the science-fiction writer William Gibson's line "the street finds its own uses for things" in the 1982 story "Burning Chrome," Young faces the emerging paradigms that new technologies bring about by investigating the cultures that emerge in their wake.

In the Robot Skies is a short film entirely shot using autonomous drones, a love story that unfolds between two teenagers who are kept apart.[1] Jazmin, a drone hacker, is under house arrest in her tower block in London. She has hijacked and decorated an aerial camera as her own and uses it to pass notes to Tamir in the tower opposite. As Young states, "When

[drones] become as ubiquitous as pigeons we start to find new applications for them, and they start to become cultural objects that get used in unexpected ways."[2] This vision of the future is an example of Young's strategy of "exaggerating the present,"[3] as surveillance infrastructure like CCTV cameras are already in place to monitor city dwellers' lives.

In a second film, *Renderlands*, Young presents another love story—in this case, that of a worker at a render farm in India who has fallen in love with the digital model of a beautiful Hollywood actress after spending fourteen hours a day endlessly rotoscoping, rendering, and compositing her into blockbuster films.[4] Young highlights the hidden infrastructure (both physical and human) behind the animation studios and render farms of India, the places where we outsource our dreams. *Renderlands* makes real the speculative landscapes constructed in the digital cities and

architectures of developer renderings, unbuilt competition entries, video-game environments, and computer-generated film worlds. **MB**

1   *In the Robot Skies* was written by Tim Maughan and produced by Dani Admiss, with music by Forest Swords. The team included Aneek Thapar, Vini Curtis, Jennifer Chen, Zhan Wang, and Alexey Marfin.

2   Quoted in D. J. Pangburn, "New Subcultures Surface in the Future-Dystopian Films of Liam Young," *Vice*, April 14, 2017, www .vice.com.

3   Liam Young, quoted in "Interview: Liam Young on Speculative Architecture and Engineering the Future," interview by Yunus Emre Duyar and Alessia Andreotti, Next Nature Network, March 29, 2015, www.nextnature.net.

4   The team for *Renderlands* included Tim Maughan, Tushar Prakash, Armeen Monahan, Paul Krist, and Alexey Marfin. The film was made possible by the Graham Foundation.

170

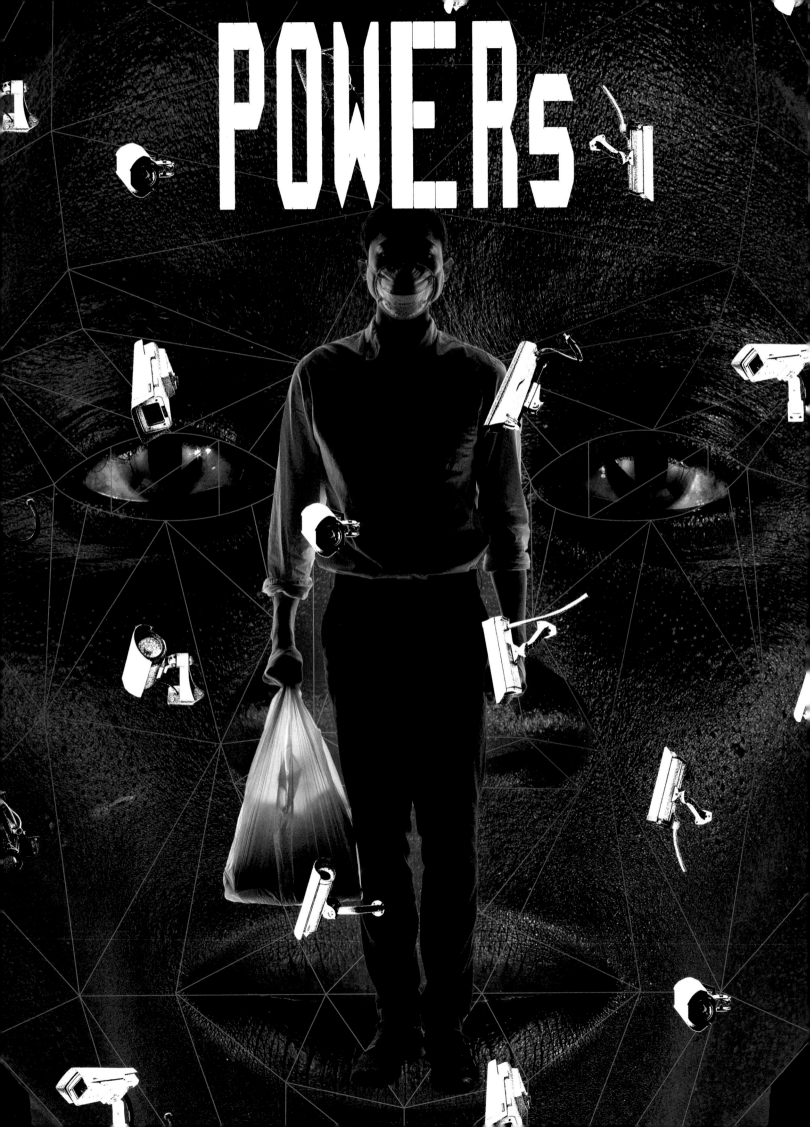

Truth Is Not a Noun

EYAL WEIZMAN in conversation with
MAITE BORJABAD LÓPEZ-PASTOR on
truth production, power, and trust

This conversation took place by Skype on January 31, 2019, with Eyal Weizman at his home in London and Maite Borjabad López-Pastor at her home in Chicago.

MB
How did you come to establish your research agency, Forensic Architecture [FA]?

EW
I was taking a year off [in 1996] from the AA [Architectural Association School of Architecture, London] and volunteering at the Planning Ministry of the Palestinian National Authority. The ministry realized that the Palestinians were not getting all the maps from their Israeli counterparts during the Oslo Accords. So they sent me to open archives in Israel to get maps of the area for them—geographical, geological, and infrastructural maps, as well as satellite and aerial images. Then I began to understand cartography's power as a tool of domination. Israel dominated the terrain because it dominated its representation.

After my studies I went back to Jerusalem. It was 2000 and the second intifada had begun. Edward Said wrote a very important piece in the *London Review of Books* calling for the counter-cartography of Palestine.[1] If cartography is the tool of the dominating, then counter-cartography is cartography from the point of view of the oppressed. Counter-cartography draws and maps violations of rights or land claims from the perspective of those experiencing violence, rather than presenting the perspective of colonialists.

In 2002 I arranged a project to map the West Bank that involved flying over and driving through and measuring things, and at the end of it I produced the first detailed map of the West Bank with all the settlements and other things that were not yet in the public domain. A lot of legal action was undertaken on the basis of that

map. In 2005, more or less, this type of on-the-ground cartography became redundant as satellite images became available through Google Earth and other platforms. So counter-cartography had to deal with the new reality of media, the multiplicity of media. It had to deal with photographs from the ground, videos, and satellite images. Counter-cartography turned into counter-forensics, and the scale of time and the scale of space had to shift. Video or satellite images capture snapshots of time, events or the results of events or incidents, and the big scale and scope became new things to explore.

I think many people were thinking about territories and territorial relations then, and we all were understanding that there can be a geopolitics in an instant. Eruptive, split-second incidents can have great political consequences. These incidents aren't about just space but also time, and we need different technologies to assimilate the two. So the practice of forensic architecture really emerged from this need. But it also emerged out of a set of discussions that were taking place at the Centre for Research Architecture at Goldsmiths [University of London], a program I established outside mainstream architectural education, at a university that didn't have an architecture department.

All sorts of professional and academic refugees came to the center to do something new with architecture, understanding that architectural intelligence could be used as a way to interrogate the world around us. Together with John Palmesino, Adrian Lahoud, Susan Schuppli, Ayesha Hameed, Lawrence Abu Hamdan, Céline Condorelli, Hannah Martin, and many others, we came up with a methodology and history and theory of new forensic practices, or counter-forensic practices.

MB
That is really inspiring and actually key to understanding FA's practice within the field of architecture. There are

Projecting thermal footage from a police helicopter establishes the spatial relationship of figures and vehicles, reflected here in a photogrammetry 3D site model for the 2017 case of Yakub Musa Abu al-Qi'an's death in the Bedouin village of Umm al-Hiran in the Negev desert.

many issues concerning means of representation and visibility that challenge the discipline of architecture itself. For instance, there is the issue of who holds the means of representation and thereby constructs or validates standards. In the moment we are living in right now—with the political right rising around the globe, in Hungary, Turkey, Poland, and Brazil, and as evidenced by Brexit in the UK and the Trump administration in the US—facts and the means of their verification are at stake. How does FA's work help redefine unstable terms such as *data* or *information* or *visibility*?

EW

When we first established FA, our friends and colleagues who work in the critical mode and deal with a critique of technology and knowledge production were extremely skeptical about our approach to technology and even our approach to the notion of truth. Many people felt that the turn to forensics was counterintuitive—anticritical perhaps. Those two terms, *forensic* and *architecture*, bring with them disciplinary and regulatory frameworks and rely heavily on expertise and authority. But somehow when you put those two words together, each one starts destabilizing the other a little bit. Also the political meaning of *truth* has changed.

I think that people in the metropolitan West now understand better what it means to live in a world of misrepresentation, like the one that every activist confronting state violation knows all too well. It is a state attack on verification that is referred to as "post-truth." The populist right insurgency you mentioned is targeting not so much this fact or the other, but the very way facts are produced, the way they're checked, sustained, verified, etc. But anyone living under a regime of domination or in the frontier of the colonial present has understood this all along. There is no colonialism

without post-truth. It's part of domination and colonialism to plant facts, to produce violence and dispossession, and to continuously lie about having done that. This is precisely the reason that Israel withheld maps from Palestinians. So controlling truth production and controlling space and people are entangled.

When the West colonizes, the colonizers employ their science or, sometimes, pseudo-science, like phrenology and physiognomy, to justify theft. Or they ignore and dismiss any criticisms or protests made against them, any claims that they are violating human rights. So the whale that has beached itself in the metropolitan West has been with us all along. And we know the smell.

MB

Absolutely! That brings us to the question of how the production of trust and truth itself is not only part of oppressive systems but is also the core of the countersystem FA is trying to advance, since the agency uses the same means of surveilling and controlling to propose counterfacts and alternative narratives. What are the challenges in using the same tools those you critique are using? What challenges did you face during the investigation into the killing of Yakub Musa Abu al-Qi'an in the Bedouin village of Umm al-Hiran,[2] not only in terms of reconstructing what happened that night but also in reclaiming your findings as the truth?

EW

We need to establish verification not just as a function of authority or of absolute perspectival expertise. In fact, the challenge of the post-true can be quite useful. So when somebody says that we need to question experts and institutions, vested interests, and institutional authority, I completely agree. We need to break the monopoly various state technocrats have over the

production of truth. But at the site of the ruin of that old established institutional conception of truth, something else needs to grow. The question is what that is. Is it the populist method of just saying what you want, usually things that end up supporting the strongmen? Or should the construction of truth be thought of differently, as a multiperspectival practice, a network of relations between institutions and communities and activists and scientists and artists that, taken together, will produce both verification as a truth practice and the means of its dissemination?

Umm al-Hiran is one of the best examples of truth functioning not as a noun, like *veritas*, but as a practice, verification. The State of Israel raided Umm al-Hiran, a place where FA knew many people because we had been active in the area for a long time. They raided at night. They killed a resident [Abu al-Qi'an], probably by mistake. They said he was a terrorist, and they concocted a story that he [intentionally] ran over a policeman. Many in the press accepted the government's version and disseminated it further. An activist who was on the ground, Keren Manor, was filming part of the event in the dark, not knowing what she saw. She captured the sound of the event. Combining this with thermal imaging shot from a police helicopter and with testimonies of people on the ground, and then reenacting the events with the community there, we were able to prove that what actually happened was that the police first shot at this man, who was in his car. Because they shot him, his car started rolling down the hill and hit a policeman, killing him. It's a very, very different story from what the state claimed happened, and it is very important to tell it that way.

We are not always able to present material in the courts because they are state controlled, so we had to disseminate this material in the media, on social media, and in exhibitions. We were able to use the funding we

got from Tate Britain to submit our material as evidence in court, in an attempt to charge the policemen who killed Abu al-Qi'an.

MB

So it seems cultural and artistic institutions like Tate Britain are your allies in this strategy. Understanding truth as a practice, and not actually as a static noun, the question then becomes which forums of truth production and which platforms you can use to perform those counter-forensic narratives. Could you explain how the invitation to Documenta 14 helped FA investigate Andreas Temme's testimony of the murder of Halit Yozgat?[3]

EW

The problem of counter-forensics is a problem of access to two spaces. One space is the scene of the crime. The state puts up a cordon and does not allow you past it. When the state is the alleged criminal and they also undertake the investigation, they can fabricate a cover-up. The second space we cannot access is forums of truth production, like the courts. Sometimes the court protects the Secret Service or the police when they perform illegal actions. So we have to find alternatives to entering these spaces and use our methodologies to gather information that goes beyond the police cordon—testimonies of people who were there, or videos or sound recordings, or any police leaks. In the case of Umm al-Hiran, we had all of the above. We had the villagers telling their stories, which were very important for the production of truth. We had Keren Manor's video. We had a leak from pathology. In Germany, the murder of Yozgat by neo-Nazis in the presence of a Secret Service agent was not dissimilar. We did not have access to the crime scene, but a leak put the police investigation and photographs in the

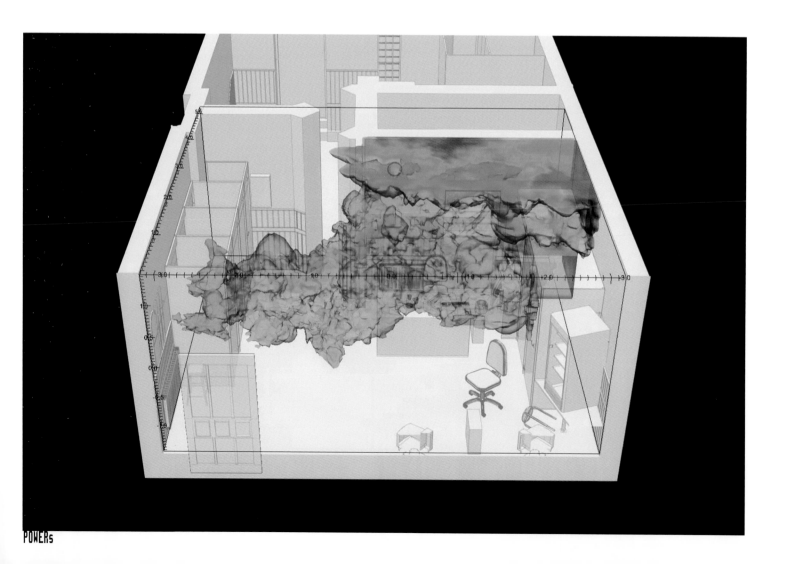

public domain. We were able to reconstruct the entire investigation from the leak.

In both cases, we could not enter the courts. To build pressure, we needed to enlist, again, our privilege and ability to approach popular art forums. We exhibited our entire investigation of Yozgat's murder at Documenta 14 in Kassel, the same city where the murder was committed. We chose the site nearest to the crime scene, and we brought it to the attention of many people—members of the public, but also policemen, police academies, politicians from all parties, and lawyers. It was presented initially in such a small place that hardly anyone could see it. Later it was presented in many other cultural venues across Germany and became a major reference point in understanding this case.

MB

FA's power and reputation relies heavily on its ability to disseminate its findings and mobilize interest across diverse platforms. In that regard, what is your most ambitious vision, looking forward?

EW

I think forensic architecture needs to become a field rather than a practice. It is important to us that other civil society groups are able to use our methodologies. We're training activists in other places. And we are developing our technology continuously. We are now dealing with new frontiers of technological possibilities, with machine vision and artificial intelligence, to build tools that can detect crimes that have not yet taken place.

MB

When I first visited the FA office in London, I was surprised that there wasn't a much larger group of people present, considering FA's significant achievements. I remember there was an open space where several people were working, but I wasn't able to identify who was doing what, or the logic or order of production that was taking place. Somehow the fluid feeling of the space spoke so much about the interdisciplinary nature of the FA team.

EW

We are larger now, about twenty people. More than half of us are architects. Others are coders, investigative journalists, filmmakers, artists, and lawyers, as well as scientists and an archeologist. So yes, absolutely, the fluid understanding of disciplines is key to us. The problems of the world do not arrange themselves according to disciplines that were conceived of in the eighteenth and nineteenth centuries. There's no point in seeing a discipline as a prison; rather, architecture is like an airport for us. We want to take off from architecture and go to different destinations, explore different things, and meet different people.

MB

I know you not only develop your own projects but also collaborate with many platforms and diverse organizations that count on you. How do you select your cases? What is that process?

EW

I love commissions, because I like to be surprised by what we can do. People approach us from Mexico, for instance, with an issue that we would never know or think about, or a delegation from Ukraine or from Greece makes a request. I love it. Because the problem with us academics is that we just keep on digging into our own obsessions. Traveling around the world and sharing our expertise with people is a privilege. We have investigations that we identify, but most are commissions. Some of them are paid commissions from human rights groups, whether it is the UN or Doctors without Borders. And

we have pro bono work as well. We have many more requests than we can answer. But if the job allows us to innovate—if we can invent something new or test new strategies to advance the field of forensic architecture—then we take the project.

MB

So to solve new investigations, there is a parallel need to develop new methodologies that can then be extrapolated and applied to other cases to keep advancing the discipline.

EW

Exactly. The challenge is to resolve cases while also questioning our methodologies and figuring out new ways to implement the work we've already done to move forward.

MB

To conclude, could you briefly share who has inspired you in your work? Who constitutes the intellectual ecosystem you inhabit?

EW

I'm a child of post-structuralism. We need to understand how to deconstruct official statements and assemble networks and complex relations. We need to do our Deleuzian readings and use them to evaluate—to make a new subdiscipline, or a new science. We need to look at Bruno Latour and Isabelle Stengers. And in terms of art, there's of course a long list of friends and colleagues we admire and learn from constantly. ╬

1   Edward Said, "Palestinians under Siege," *London Review of Books* 22, no. 24 (December 14, 2000): 9–14.

2   For more on this case, see p. 166.

3   For background on this case, see "The Murder of Halit Yozgat," the Forensic Architecture website, forensic-architecture.org.

Fluid dynamics simulation of gunpowder residue particles (ammonia) in the front room of the internet café in Kassel, Germany, where Halit Yozgat was killed in 2006

Will Technology Save Us?

GABRIELLA COLEMAN in conversation with MICHELLE MILLAR FISHER on techno-utopianism, hackers, and leakers

This conversation took place by phone on December 7, 2018, with Gabriella Coleman in her office at McGill University in Montreal, and Michelle Millar Fisher in a quiet corner at Design Miami in Florida.

MMF
Designs for Different Futures looks at the ways in which, whether concretely or speculatively, design points us toward different possible futures. So I wanted to open by asking what you think digital realms might look like in the future: How might they differ from what we experience today in terms of access and technology, for example, and what might it mean to have agency within digital realms in the future?

GC
I'm always a little wary of predictions because they tend to be very hard to make and are often off the mark, but I still think it's an important question to ask because oftentimes talking about speculative futures [is] very much informed by the present moment. This present moment is interesting because the reactionary right—certain conservative politicians—control the means of technical production in very savvy ways. And state actors, notably Russians, are exploiting certain qualities of the internet to spread misinformation and propaganda. This has finally killed off what I've called the "wicked witch of techno-utopianism." Any and all hope for a utopian internet has died. But what I find really troubling is when people then say, "Well, I guess technology is not going to solve our problems." It never was, but if groups don't exploit everything that these technologies have to offer—in order to change people's minds, recruit people, advocate for social change—we will absolutely lose the game.

MMF
Right.

GC
But in terms of possible digital futures that worry me? I've seen some people say that this is not where we should be putting our energies, but I think we should very much be turning to the far and reactionary right and to other groups to see how they are exploiting the current technical and sociocultural technical means of production, from using internet memes to waging campaigns. What are they doing that we can learn from? It's not so much a question of what things are going to look like, but of what we can do under present conditions to learn from what problematic groups are doing with the internet in order to imagine possible futures. I hope we continue to believe that we can have agency with these technologies.

MMF
Can you speak about the ways in which agency might manifest differently in the future depending on access to the digital world? I think more than 50 percent of the world now has access to the internet, but of course there are continuing conversations around net neutrality, so part of the prediction for the future relies on the tension between increasing access to the digital world and the threat of that access being mediated or shut down. Do you have a positive view of that tension? Or do you think that the possible end of net neutrality is something that will deeply threaten the opportunities for good actors to take on the tools currently being used by bad actors?

GC
Such a great question, especially the last part, because as I've mentioned, certain quarters of the reactionary right have been really, really good at using these tools in a way that is cross-pollinated and strengthened by non-internet media formations, from talk radio to Fox News. A lot of different groups do have equal access to

Network cables in a server room in New York City on November 10, 2014, when President Barack Obama called on the Federal Communications Commission to implement a strict policy of net neutrality and to oppose content providers in restricting bandwidth to customers. Photograph by Michael Bocchieri/Getty Images

the internet today, and it's really [about] who is launching smart campaigns. Who is using internet memes more cleverly? Who is using message boards to try to recruit people? Who is putting out viral content? But the second that a policy is implemented where net neutrality really doesn't exist anymore—where big players can pump out their content with more ease—the internet as a possibility for politics as a practice of friction where you still have a shot will be over. And so I do think that the really important element to keep fighting for, even as we recognize that there are going to be differences across geography, or in access to technology, is net neutrality. That's kind of the meta-condition that really, really matters. The North American, South American, European internet more or less follows principles of net neutrality, but that's not the case in many parts of the world. It's always interesting to think about what elements around the internet are dependent on local issues, but I do think net neutrality is a global issue that will really shape the nature of future agency online.

MMF
You wrote the biography of the hacktivist collective Anonymous [*Hacker, Hoaxer, Whistleblower, Spy*, 2014], and a quote I pulled out from your discussion of this research at TED in 2012 is, "the way in which their presence and activities dramatizes the importance of anonymity and privacy in an era when both are rapidly eroding." We live in an era that includes events like the Philando Castile police shooting in Minnesota, which his girlfriend, Diamond Reynolds, broadcast live on Facebook. Online anonymity is a precious asset in many ways, but so too is this possibility for complete transparency via digital tools in order to present a public politics or to critique inequity and abuse. I'm really interested in that tension.

GC
That's a great comment because oftentimes they are presented as binary—and by "they" I mean transparency versus the nexus of privacy and anonymity. They're presented as polar opposites, where you could have either one or the other—light or darkness, good or bad. But they're not zero-sum games. It goes back to strategy and tactics. At what point is anonymity powerful and why, and under what conditions? And the same should be asked of transparency: At what point is this important to fight for? What are the limits of transparency? When we conceive of privacy and anonymity as both strategies and tactics, we can talk about when it's best to deploy them, and we can also ask about their limits. Anonymity is a shield or a weapon for the weak so that people can get information out there without the severe reputational and bodily harm that can often follow when you speak with your own voice. I think this also relates to Anonymous, in that often when you don't know who the messenger is there's some mystique and mystery, and that allows the message to be heard as opposed to focusing in on the person. But anonymity is being used by corners of the reactionary right in rather dramatic ways to recruit new participants: you can be exposed to very vile, racist, misogynist, anti-Semitic content and be anonymous, which allows you to get comfortable with these ideas in private. But, again, I don't think that we should then call for total elimination of anonymity as a result. The rise of the far right isn't simply due to anonymity. The reactionary right is popular because of the ideas they're spreading. Sure, anonymity facilitates that, but it's not the cause, and if we totally eliminate anonymity and privacy, I think we'll really regret it. There is a dynamic and a tension that I feel you really have to explain in full. Otherwise, people do tend to put it in these binary swaths of "transparency good" and "anonymity bad."

No Borders for Migrants protests in Rome, May 1, 2016. Protestors are wearing the Guy Fawkes masks adopted by the hacker collective Anonymous. Photograph by Simona Granati/Corbis

An image from Diamond "Lavish" Reynolds's video of the shooting death of her boyfriend, Philando Castile, during a traffic stop by police, streamed live on Facebook, July 6, 2016. Lavish Reynolds/Facebook

MMF

These battles over privacy sometimes manifest in new design typologies—for example, the Signal app or the Tor browser—that equip citizens with digital privacy and anonymity. Can you talk a bit about this area of design and its possible futures? Who should be doing the designing? So often that process isn't transparent. It requires fairly high-level expertise, or if it's grassroots, it requires somehow being in the know. New design tools for privacy and encryption are not always immediately adopted en masse, and so the design typologies themselves are interesting, but so too is their adoption, the way that people find out about them and, through them, find out about their rights or the ways in which their digital agency is being compromised.

GC

I'm really glad you asked about these tools and design. On one hand, anonymity is being threatened in a very global way. So many companies and governments can easily collect data about groups and individuals. Yet on the other hand, the tools around anonymity have improved by leaps and bounds compared to even just five years ago. Obviously, there have been technical improvements, but some of the most important changes have been in design, and Signal and Tor really do stand out in this regard. Signal is a texting program that's encrypted end to end. You can also use it to make phone calls, and it's truly a snap to use. That is so important— the more people that are anonymous, the harder it is to de-anonymize. In many ways, Signal's design interface is not innovative, so its innovation is to slot into already existing design features. But this is not insignificant with cryptography, where design is always a difficult question because the layer of usability can introduce vulnerabilities. That they were able to do it, and create

a usable piece of software, was a very, very big deal. And the software becomes a pedagogical tool to let people know about privacy and anonymity as well.

Tor is also interesting. I've seen the design improve dramatically over the years based on a massively concerted effort by the Tor project team to tap into actual people in different places of the world in order to integrate users' insights to make Tor a more usable browser. Another element that has to do with design is that they're trying to visualize how Tor works. I can't remember if what they have right now is just a prototype or if it's integrated into the Tor browser, but you can see a little pop-up with Tor relays showing you what's going on as you're using the technology—a visualization as you use the tool itself. It's doubly powerful. It's like seeing the plumbing.

MMF

It becomes part of that pedagogical aspect you were talking about; it teaches the values and the system itself along with providing use of the tool.

GC

Exactly. And this takes it to a whole other realm. I see this as a very innovative design principle. There's been a very strong material turn in academic studies of projects like Tor and Signal and the internet and the infrastructure that underpins them, but I think the public on the whole really lacks that framework and understanding.

MMF

What about viruses and DDoS [Distributed Denial of Service] attacks, which have emerged as design forms in tandem with the rise of digital hacktivism, to which we could add hacking and leaking? What are the possible futures of these types of design?

GC

I'm going to focus on hacking and leaking because I think that's going to have the most impact in the future, and also because I like framing this in terms of design, insofar as I think that during the Anonymous era these hacktivists were very much experimenters. They were experimenting with a new form, which was hacking search-board documents like emails or other damning information to leak to the public at large or to get to journalists. In many instances what they were doing was prototyping a new tactic. And because they were prototyping, it was sometimes poorly executed. You could say it was poorly designed in that their security wasn't very good. They often released data that was politically valuable or in the public interest, but they also violated a lot of people's personal privacy in the process. Still, it was a very powerful prototype. It was so powerful that, again, I think it's a tactic we are going to continue to see and have been seeing in two very different forms. One is other hacktivists who have taken this prototype and improved on it. The most famous example of this is Phineas Fisher, who is a hacker or a group of hackers who have taken a singular name, and they have successfully broken into a couple of firms, including one where they took emails that had evidence of the sale of digital vulnerabilities to dictatorial regimes. These vulnerabilities can be used as a kind of a cyber-weapon to spy on activists and dissidents and journalists. Prior to Phineas Fisher, there was some evidence of this [happening], but they provided much fuller proof. Fisher has never been caught. I think that Anonymous allowed people to see the value of this type of design form, and others have taken it and executed it in different ways, including large organizations and states. The most famous case is the hack against the Democratic National Convention by the Russian state. While the Russians and the Chinese and the Americans have long been hacking each other, Russians taking emails and leaking them to the world was relatively novel as a state action.

MMF

Just to close out: I looked up the definition of *leak* in the [American Heritage] dictionary, and while it often involves confidential information, there's also the definition of "to disclose without authorization or official sanction." This takes me back again to Diamond Reynolds's film of her dying partner, Philando Castile, and using tools like Facebook Live to leak information that otherwise would never be seen by anybody else. It disrupts the balance of power because that story can be seen immediately for what it is, rather than processed through official channels like police reports or the news media or other types of witness testimony that are inherently subject to bias or spin.

GC

That's a great comparison, and again the tactic I'm talking about is quite different from that of the traditional leaker or whistleblower, who is often an insider with access to information. This involves an outsider who normally would not have access, right? And it is almost entirely contingent on technology. There are very, very few cases prior to the technological era where you could break in to take documents. I do think that it's a great analogy, because prior to the cell-phone era the filming of police brutality was far more sporadic, and eyewitnesses or police reports were what we had to rely on. However, I think it's important to recognize that while I do think it's always preferable to have that evidence, it doesn't necessarily lead to the right outcome. ⌗

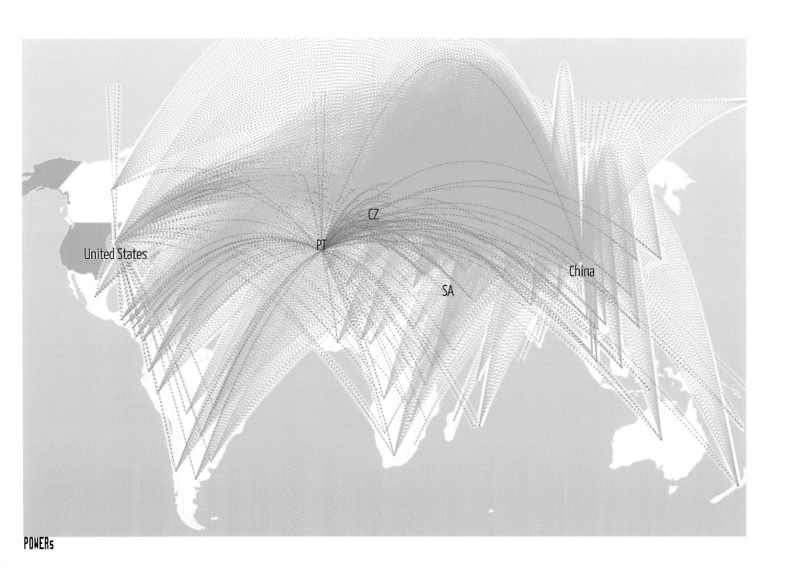

Map showing DDoS attacks on May 19/20, 2019, by Fernanda Viégas, Martin Wattenberg, Colin McMillen, and CJ Adams. For Google Ideas, Google Big Picture Team, and Arbor Networks, Inc.,, NetScout Systems, Inc. Built for DigitalAttackMap .com using custom Javascript, D3, HTML

Governance

The current political moment seems to produce horror and tragedy at an astounding rate. Listening to the radio, watching the news, scrolling through social media, or thumbing through a newspaper is enough to induce a feeling of dread or fatigue, or of being overwhelmed. The devastation of Syria, the starvation in Yemen, the unending wars in Iraq and Afghanistan, climate catastrophe, and the rise of authoritarian regimes around the globe now make up the reality of daily life for millions. This moment of global crisis illuminates much about how governance and the role of the state have been transformed over the past five decades in ways that have worsened conditions of inequality, vulnerability, and violence for citizens and noncitizens.

The lure of neoliberal globalization is the promise of a world free of friction, a world of constant and unimpeded movement. The reality, however, is that while capital enjoys unrestricted access to global markets and blurred territorial boundaries, workers, migrants, and dissidents find the movement of their bodies and the circulation of their ideas increasingly restricted or locked into place. We see this when we consider how US foreign policy and capital expansion in Central America destabilized the region by supporting dictatorships that were "good" for business while funding anticommunist death squads, interfering in democratic elections, and destroying the local economy through free-trade treaties. Despite the obvious role of US political and economic elites in creating the conditions that have made life for many of Central America's poor, indigenous, and politically dissident populations increasingly difficult, causing them to flee the region in search of safety and stability, the US government has responded by seeking to restrict the mobility of Central American migrants and asylum seekers. While US capital is free to move throughout the region in order to extract and dispossess, Central Americans who

attempt to escape the resulting disorder encounter a militarized US border, xenophobic rhetoric casting them as criminals and terrorists, and a deportation regime that fosters often deadly outcomes.

This has become the fundamental contradiction of neoliberal globalization: economic freedom, generally understood as the deregulation and "openness" of the market, does not lead to greater political freedom and social justice. Instead, we see that an unfettered and unregulated global economy has frequently further immiserated the economically vulnerable, diminished political freedom, and intensified new forms of imperialism in developing countries. This contradiction between the freedom of the market and the lack of freedom that individuals living under neoliberal regimes experience is something that many artists, designers, architects, and governmental agencies have responded to. Designs and provocations that explore the theme of governance help us to think through the promise and perils of digital citizenship and deterritorialization. From the practical to the absurd or the utopian, digital design and technology open up spaces for us to imagine new ways of living under the conditions of neoliberal globalization.

One way of countering the ever-expanding horizons of capitalist growth is to invest attention and resources in the local, as the community group Transition Town Brixton does. Their project, the Brixton Pound, is a form of complementary hard currency meant to circulate within local and independent businesses in South London's Brixton neighborhood. This money "sticks to Brixton," helping consumers to shop small and buy local in an effort to stimulate trade and jobs in their community. Unlike money spent on clothing or groceries from large multinational chains or online retailers, which increases our carbon footprint, money spent in Brixton stays there—an investment in place and community. The Brixton Pound refuses the fetishization of

The Brixton Pound, designed by Jeremy Deller, alongside pounds sterling, 2015

"placelessness" often associated with neoliberal globalization and instead is part of an effort to encourage people to remain rooted in their community and among their neighbors—to literally imbue them with value.

We see a similar commitment to place in the Western Sahara tapestry project, the Swiss architect Manuel Herz's collaboration with the National Union of Sahrawi Women, a political organization based in the refugee camps of southwestern Algeria. Since 1975, Morocco's colonization of Western Sahara has forced many Sahrawi people out of their homes and into camps across the border in an effort to escape brutal repression. The large-scale tapestries, produced by a group of women weavers, depict life in the camps not as bare or suspended, but as rich and developed (see p. 191). They show us actual spaces, buildings, and activities, revealing the ways in which the administration of life in the camps functions, much like a nation in exile. We can then see that although the Sahrawi have been banished from their homeland, they are asserting a powerful claim to national sovereignty beyond the occupied borders of Western Sahara. They are creating their collective future and a future of Sahrawi self-governance and autonomy without seeking the permission or recognition of their occupier or its accomplices.

While the Brixton Pound and the Western Sahara tapestries assert powerful claims to space in the face of economic exploitation and colonial domination, Giuditta Vendrame's Infinitecitizenship.org attempts to beat the system at its own game in order to demand a right to mobility that is often reserved only for global elites. A project like the Estonian e-Residency functions to accelerate the flow of capital goods by encouraging global elites to enjoy Estonian digital citizenship for the sake of business transactions (see p. 189), but

Vendrame treats citizenship as merely another good to be exchanged within the sharing economy, so that vulnerable bodies might enjoy the same mobility as capital. With Infinitecitizenship.org, she suggests creating a peer-to-peer citizenship exchange that would allow individuals to circumvent the increasing rigidity of the contemporary nation-state. Users would be able to swap citizenship to facilitate movement or access to resources such as public assistance. This conceptual digital project recognizes that human rights and access to life-giving resources are often tied to particular national forms of citizenship, but it seeks to undermine that relationship by encouraging people to share the privileges of citizenship with others. *Infinite Passports*, Vendrame's collaboration with Fiona du Mesnildot, is a physical manifestation of this project (see p. 190). It extends the subversion of formal citizenship in Vendrame's previous work through its suturing together of pieces from twenty-four passports in order to create twenty-five new ones to provide holders with an expansive, global citizenship forged through noncommercial exchange and cooperation.

Infinitecitizenship.org and *Infinite Passports* in some ways epitomize exactly the stuff of nativist and conservative nightmares, in their encouragement of migrants who refuse to "wait their turn" or "do things the right way" in order to take a shot at improving their situation in the Global North. But out of that nightmare comes a powerful dream of relationality forged through a radical solidarity that recognizes the bankruptcy of citizenship under the conditions of neoliberal globalization. This project challenges those of us in the Global North to think about how the rights and privileges associated with our citizenship come at the expense and exclusion of those in the Global South, who our governments and, increasingly, ultraconservative

The Brixton Pound Café, the first pay-what-you-can community café in London, inspired by the Brixton Pound, April 3, 2017. Photograph by Gemma-Rose Turnbull from Free Photo Portraits, Taking Part Residency, Photofusion, London

Manuel Herz's pavilion displaying the Sahrawi tapestries at the 2016 Venice Architecture Biennale, the first time a "nation in exile" was invited to participate. Photograph by Iwan Baan

nationalist groups are working to keep out through violent rhetoric and practice.

The designs brought together in this exhibition explore governance in ways that push us to consider how inequality and injustice continue to manifest themselves in our increasingly interconnected world. Far from enjoying the borderless world that neoliberal globalization promised, we have experienced a hardening of formal and informal borders over the past half century. While the digital realm is no magic bullet to remedy all the contradictions and challenges of our present moment, it supplies us with provocations that subvert the existing order in order to envision what expansive mobility and a "right to place" for the dispossessed might look like. ⌗

ZXX TYPEFACE   2012   SANG MUN

Sang Mun's ZXX typeface arose from a question: "How can we conceal our fundamental thoughts from artificial intelligences and those who deploy them?" In answer, he decided to "create a typeface that would be unreadable by text scanning software (whether used by a government agency or a lone hacker)—misdirecting information or sometimes not giving any at all."[1] First released in May 2012 (it was Mun's final project as a student at the Rhode Island School of Design), ZXX was offered as a free download in the hope that as many people as possible—sharing his concern about the growing problem of privacy and personal data security on the internet—would use it.

The typeface consists of four different fonts in addition to the more standard sans serif and bold letterforms. Camo, False, Noise, and Xed are each designed to thwart machine intelligences in a different way: Camo camouflages the individual letters with ink splotches and scrawls; False reverses the letters of the

alphabet and numerals in a code wherein Z equals A and 0 equals 9; Noise obscures the letters with a barrage of digital dots of different sizes; and Xed draws an *X* on top of each letter. Designed to confuse optical character recognition scanners, these fonts are still readable to the human eye.

Mun took the name ZXX from the Library of Congress classification system, in which ZXX indicates "no linguistic content; not applicable."[2] The designer was deeply influenced by his time in the Korean military, when he worked as a contractor for the US National Security Agency, gathering information from electronic signals and communications for national security and defense purposes. As telecommunication equipment began to be more widely used for domestic surveillance and information gathering by governments and technology companies after 9/11—including for the monitoring of emails and mobile computing devices—Mun was inspired to create ZXX, using his

skills as a graphic designer to highlight questions about privacy and the use of design for political and social provocation: "I thought that addressing these issues through the design of a typeface—a building block of language and communication—would bring home the conversation to the average person."[3] **KBH**

1   Sang Mun, "Making Democracy Legible: A Defiant Typeface," Gradient, June 20, 2013, reproduced at walkerart.org/magazine /sang-mun-defiant-typeface-nsa-privacy.

2   See the Library of Congress website, id.loc .gov/vocabulary/iso639-2/zxx.html.

3   Sang Mun, "Good Morning, Mister Orwell," *Wired*, September 24, 2013, www.wired.com.

SYNTHESIZING OBAMA: LEARNING LIP-SYNC FROM AUDIO 2017 SUPASORN SUWAJANAKORN, STEVEN M. SEITZ, AND IRA KEMELMACHER-SHLIZERMAN, UNIVERSITY OF WASHINGTON, SEATTLE

Original Video for Input Speech   Our Result

It is increasingly difficult to distinguish forgery from reality. While programs like Photoshop have encouraged skepticism of still photographs, video is often thought of as a truthful record. Technology from researchers at the University of Washington in Seattle calls this assumption into question. Their system, Synthesizing Obama, uses existing video and audio of former president Barack Obama to create a new mashup of the two, transferring older audio onto more recent videos and vice versa. Researchers selected Obama as a test subject because of the sheer volume of high-quality footage of him that is available, which made it easier to train their neural network to learn the necessary mouth shapes and movements for many words and sounds. The program translates audio files into these mouth shapes, which are then blended onto a separate video of the speaker. The resulting video of Obama is convincing enough to overcome what is commonly referred to as the "uncanny valley," the unsettling feeling people experience when androids and simulations closely resemble—and can almost be mistaken for—humans, but fall just short.

The team behind Synthesizing Obama believes their new method of visual lip-syncing could be used to improve video chat by smoothing stuttering video to match its clearer audio. Users could compile footage of themselves speaking to train the software, and the video would automatically generate using just their voice. Similarly, this process would make it possible to create video using the audio from telephone calls, giving hearing-impaired people an opportunity to lip-read their calls. In theory, this system could also be used to map anyone's voice onto anyone's face. Technology already exists that can edit voice recordings on a computer as easily as text, with seamless playback results. Combined with the process behind Synthesizing Obama, these programs could be used to produce misleading videos—videos that would give new and substantive meaning to the term *fake news*. As a counter-measure to this potential development, the designers of Synthesizing Obama hope that reversing the process (feeding video rather than audio into their program) would allow them to develop algorithms that can detect whether a video is real or manufactured. JRB

STRANGER VISIONS 2012–13 + *T3511* 2018 + *INVISIBLE* 2014 HEATHER DEWEY-HAGBORG

Heather Dewey-Hagborg's *Stranger Visions*, *Invisible*, and *T3511* act as provocations on the theme of genetic surveillance—accessing another person's genetic information without their knowledge or consent. For *Stranger Visions*, the artist mined the streets of New York City, collecting discarded chewing gum, loose hairs, cigarette butts, and other bits of saliva-covered debris. Working with a lab to extract DNA, she culled from her samples the genetic material that codes for physical traits and used it to render portraits through facial-modeling algorithms. While these algorithms, as well as our understanding about the relationship between genes and facial features, are relatively crude, Dewey-Hagborg could still reliably decipher the stranger's gender, eye and hair color, skin shade, and nose width, and the distance between their eyes.

Using a computer program, she produced 3D-printed models of each stranger's face. Although the resulting faces were not exact replicas of those of the people who left their DNA traces behind, the speculative portrait sculptures have, in Dewey-Hagborg's words, a "family resemblance."[1] She was surprised, and even disturbed, by how much information she could find out about a person she'd never met. And indeed, the prediction at the center of the project—that a culture of biological surveillance could emerge in the wake of

developing forensic DNA technology—was supported when a technology company launched Snapshot, a service that offers police around the United States similar capabilities.[2]

This fascination and concern with genetic privacy fueled both *T3511* and *Invisible*. *T3511* is an experimental documentary that tells the love story of a biohacker who becomes increasingly enamored with the anonymous donor whose saliva she purchased online. In conversation with the themes of post–genetic privacy developed in *Stranger Visions*, *T3511* comments on the growth of direct-to-consumer genetic testing services to probe how interpersonal relationships and interactions are likely to change in the future.

If *Stranger Visions* demonstrates the reach of surveillance technology, *Invisible* suggests a means of combating it. The project is composed in part of a set of liquid sprays designed to protect "against new forms of biological surveillance" by wiping out your genetic trace.[3] The first spray, Erase, is a strong cleaning agent, similar to bleach. According to Dewey-Hagborg, it destroys 99.5 percent of the DNA left behind on everyday objects. The complementary spray, Replace, obfuscates the rest. When sprayed over your own DNA, Replace's blend of foreign genes scrambles any remaining genetic trace, blurring your identity. In what context would the

Invisible sprays prove useful? Dewey-Hagborg provides some ideas that deftly straddle the line between humor and practicality: "Spend the night somewhere you shouldn't have? Erase your indiscretion and be invisible"; "Dinner with the prospective in-laws going smoothly? Don't let them judge you based on your DNA, be invisible"; or "Exercising your freedom of speech? Be invisible and never get tracked."[4] *Invisible* extends beyond the pocket-size spray bottles to become a synthesis of forms—a working product, designed object, art installation, commercial, company website, series of hands-on workshops, social-media intervention, and set of open-source DIY guides. **JRB**

1   Quoted in Christopher York, "'Stranger Visions' Art Project by Heather Dewey-Hagborg Is a Cool (but Very Creepy) Concept," Huffington Post UK, May 28, 2013, www.huffingtonpost .co.uk.

2   See the Parabon Snapshot website, snapshot .parabon-nanolabs.com.

3   See the Biogenfutures website, biogenfutur.es.

4   Biogenfutures website.

DIGITAL IDENTITY CARD FOR ESTONIAN E-RESIDENCY 2014 REPUBLIC OF ESTONIA E-RESIDENCY PROGRAM

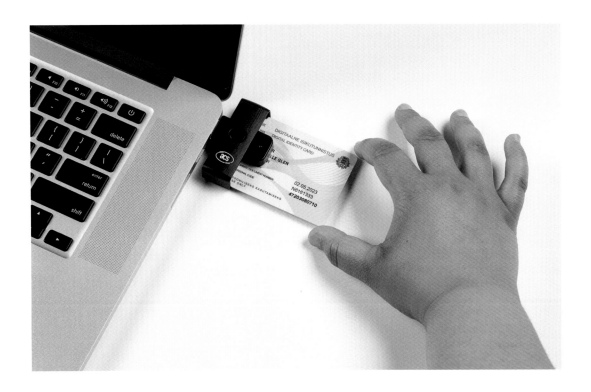

In 2014 Estonia became the first country to introduce the concept of e-Residency. This small Baltic nation, formerly occupied by the Soviet Union, is home to an estimated 1.3 million people, making it one of the least populous members of the European Union. Despite this, Estonia has become a tech powerhouse in the last decade and a world leader in digital governance. As the name suggests, Estonian e-Residency offers anyone, anywhere the opportunity to apply for a secure digital residency in Estonia. The program aims to make life and business easier and more convenient for freelancers, business owners, start-ups, international partners, digital nomads, and any other nonresident who has a connection—physical, contractual, civic, or familial—to Estonia.[1] The system is administered through government-issued digital ID cards, which the Estonian government has been using for its citizens for over a decade. Although the program does not grant tax residency or citizenship, or replace the passport needed to enter Estonia or other countries in the European Union, it does provide e-Residents with the ability to conduct personal and professional business securely online. For example, e-Residents can use their card to access Estonian public and private services in order to establish and operate companies online, encrypt and transmit documents and verify their authenticity, and access international payment service providers.

As of 2018, more than forty thousand people from 160 countries had applied for e-Residency. By 2025 the program hopes to have ten million e-Residents enrolled. This target is part of Estonia's ambition to create a "digital nation" that recognizes both domestic and global forms of citizenship and values greater connectivity as the key to prosperity. In making it easier for individuals and organizations to conduct transactions across borders, the program is working toward and preparing for a world in which barriers are eliminated and there is a freer flow of talent, ideas, and resources. With its e-Residency program, Estonia aspires to be a leader and role model for a way of envisioning the world that is more reflective of the demands and opportunities of the twenty-first century. As debates about immigration policies and border security rage around the world, Estonia aspires to remain open to all people, and to the future. **JRB**

---

1   For the processes and regulations of the program, see nomadgate.com/estonian-e-residency-guide/.

INFINITE PASSPORTS 2014–17 GIUDITTA VENDRAME AND FIONA DU MESNILDOT

Infinite Passports represents a crossover of two projects: Infinitecitizenship.org, by Giuditta Vendrame, and *The Missing Square Theory*, by Fiona du Mesnildot. Both designers work at the intersection of design, art practice, and social systems.

In Infinitecitizenship.org, Vendrame speculates on the creation of an imagined citizen exchange network (see p. 184). Learning from real-world examples such as Estonia's e-Residency program while simultaneously provoking resistance to them, Vendrame presents the option of temporarily transferring the rights, privileges, and responsibilities of belonging to a specific nation from one person to another. Citizenship varies wildly by location, with pronounced differences in access to resources, public welfare, tax obligations, and freedom of mobility. Infinitecitizenship.org proposes a system that would allow members to temporarily and mutually exchange their citizenship, and with it their status.

Fiona du Mesnildot, through her short film *The Missing Square Theory*, challenges strategies of economics and growth by exposing

them as analogous to optical illusions and geometric tricks. Using the magician and mathematician Winston Freer's Tile Puzzle, in chocolate, as a way to trick our minds by creating the illusion of infinite surplus from standard portions, she illustrates how economic ploys rely on shifts between reality and perception.[1]

The result of Vendrame and Mesnildot's collaboration is a collection of twenty-four collaged passports created following Freer's geometrical trick, with a brand new, twenty-fifth passport made from small, square pieces extracted from each of the original group. By highlighting the frictions and two-sided realities of a world in constant flux, where goods and money are mobilized without borders while bodies are frequently forced to remain static, *Infinite Passports* is both a critique and an imagining of an alternative future. Envision a migrant able to gain citizenship long enough to travel safely to a final destination without fear of detainment. The resulting questions have profound implications: What would this mean for the concept of borders? Is the state a too-rigid, outmoded system? What

constitutes citizenship today? Are we facing a scarcity of citizenship?

As the global migrant crisis continues, countries are reevaluating their border policies. Physical and legal citizenship are also challenged by the digital realm, in which thoughts, ideas, communications, and even identities can be transferred without the oversight of legislative bodies. International corporations further complicate notions of citizenship. In a time when companies have gained the rights of personhood in the United States, they have also expanded their global reach, influencing regulations worldwide in order to maximize profits. How does this institutional intervention impact the ordinary citizen, whose movements are being policed more than ever, often by the selfsame corporations? **MB**

1   *The Missing Square Theory* can be viewed at fionadumesnildot.com/site/index.php?/project/how-to-use-the-nonexistent/.

SPACES OF EVERYDAY LIFE 2016 MANUEL HERZ AND THE NATIONAL UNION OF SAHRAWI WOMEN, SAHRAWI ARAB DEMOCRATIC REPUBLIC

This large wool panel represents a map of the buildings inside the El Aaiun refugee camp in southwestern Algeria, one of several camps where people of the Western Sahara—the Sahrawis—have been living for over forty years. In 1975 Spain abandoned its former colony of Western Sahara to Morocco and Mauritania (the latter withdrew in 1979). Tens of thousands of Sahrawis fled into Algeria during the ensuing guerrilla war against the occupying nations. The Sahrawis have responded to their forced displacement by creating and maintaining a centralized self-government that provides health and education services to its people across hundreds of miles, supported by the UN Refugee Agency, the World Food Program, and various nongovernmental organizations. This woven map's labels are a particularly striking demonstration of this orderly, working society. Written in block letters and in English—making them internationally legible—they identify the dispensary, community center, maternity clinic, and primary school, alongside a legend of symbols for residential and public buildings, shops, recreational spaces, paths, and goat barns.

The panel was woven on a large loom by up to ten women at a time, who had to acquire the wool from Ghardaïa in northern Algeria, a center of cloth and rug manufacturing one thousand miles away, since none of the camps have the resources for a panel of this scale. Made by members of the National Union of Sahrawi Women, the panel highlights the prominent role women play in Sahrawi society.

This wall hanging was first exhibited with others (accompanied by maps, videos, and photographs) at the 2016 Venice Architecture Biennale in a tentlike structure designed by the Swiss architect Manuel Herz, whose research has centered on issues of diaspora. For the first time, a nation in exile—the Sahrawi Arab Democratic Republic proclaimed its independence in 1976 and is recognized by some forty nations—was represented at the biennale by a pavilion alongside those of established nation-states. In reimagining the refugee camp as a functioning city, said Herz, the Sahrawis have created "a place of emancipation and self-governance, truly democratic, and not a place of misery," with institutions housed in rough adobe-brick buildings that are planned with "incredible beauty."[1] **KBH**

1 Quoted in Rowan Moore, "Venice Architecture Biennale 2016—Ideas for Real World Problems," *Guardian*, May 29, 2016, www .theguardian.com.

COSTUME FROM THE SERIES *THE HANDMAID'S TALE* 2017 ANE CRABTREE

Ane Crabtree's haunting creations for the television series *The Handmaid's Tale* convey the control and repression at the heart of the dystopian drama, based on Margaret Atwood's 1985 novel of the same name. The show, which premiered on the streaming video service Hulu in 2017, is set in a future totalitarian patriarchy where the ruling male elite subjugate women and divide them into different castes according to their usefulness in the face of worldwide infertility. Each caste is distinguished by different freedoms and duties. At the heart of the caste system are the Handmaids, the few remaining fertile women, who are assigned to high-ranking government officials to be impregnated, bear children, and thus repopulate a devastated world. Aunts train and oversee the Handmaids; Wives run the households of high-ranking men; Econowives are the wives of poorer and less powerful men; Marthas serve as domestic servants to the Wives; Jezebels work as prostitutes; and Unwomen, the lowest caste, are forced into slave labor in internment camps.

Following Atwood's novel, Crabtree dresses each caste in different styles and colors. The Handmaids, clothed in blood-red capes and white bonnets symbolizing female menstruation and childbirth, stand in visual contrast to the Wives, whose blue dresses suggest virginal infertility. According to Crabtree, it was a challenge to design the long, loose red dresses that cover (and conceal) the Handmaids' bodies and indicate their function and hierarchical status in a world that oppresses women: "It was kind of twisted to think about how I would hinder women—their body shape, and also their movement and their freedom with the clothing. I had to do it to make it realistic, but also to help the actors."[1] Crabtree's hoodlike "coal-scuttle" bonnets protect the Handmaids from the gazes of others and function as blinders, preventing them from looking right or left without turning their heads. "We decided to use [the bonnets] as a vehicle to heighten the cages that they were in mentally, physically, emotionally. And then use it to reveal the eyes, reveal the emotions," Crabtree said. "What was actually a hindrance became quite a helpful vehicle for a new way of acting, a new way of filming, a new way of designing."[2] **KBH**

1 Quoted in Jamie Lincoln, "*The Handmaid's Tale*: The Hidden Meaning in Those Eerie Costumes," *Vanity Fair*, April 24, 2017, www.vanityfair.com.

2 Quoted in E. Alex Jung, "From the Handmaids to the Marthas, How Each *Handmaid's Tale* Costume Came Together," *Vulture*, April 28, 2017, www.vulture.com.

"We don't seem to live on the same planet": A Fictional Planetarium

BRUNO LATOUR on climate, identity, geopolitics, and Earth's destinies

Satellite image of plumes of smoke produced by the Camp Fire raging in Northern California, November 8, 2018. The image was captured on Landsat 8 and was created using Landsat bands 4-3-2 (visible light), along with shortwave-infrared light to highlight the active fire.

Architects and designers are facing a new problem when they aspire to build for a habitable planet.[1] They have to answer a new question, because what used to be a poor joke—"My dear fellow, you seem to live on another planet"—has become literal—"Yes, we do intend to live on a different planet!" In the "old days" when political scientists talked about geopolitics, they meant different nations with opposing interests waging wars on the same material and geographic stage. Today, geopolitics is also concerned with wars over the definition of the stage itself. A conflict will be called, from now on, "of planetary relevance" not because it has the planet for a stage, but because it is about *which planet* you are claiming to inhabit and defend.

I am starting from the premise that what I have called the New Climatic Regime organizes all political affiliations.[2] The climate question is not one aspect of politics among others, but that which defines the political order from beginning to end, forcing all of us to redefine the older questions of social justice along with those of identity, subsistence, and attachment to place. In recent years we have shifted from questions of ecology—nature remaining outside the social order—to questions of existential subsistence on threatened territories. Nature is no longer outside us but under our feet, and it shakes the ground. Just as at the beginning of modern political philosophy, in the time of Thomas Hobbes, we are dealing with humans not unified but divided by nature to the point that they are engaged in civil wars as violent as the religious wars of the past, and forced to look for peace by altogether reinventing the social order.[3] Climate mutation means that the question of the land on which we all stand has come back into focus, hence the general political disorientation, especially for the left, which did not expect to have to talk again of "people" and "soil"—questions mostly abandoned to the right.

Since it is impossible to tackle this sort of conflict head on, I will turn to fiction and take you on a brief tour of a planetarium of my invention. Whereas old planetary influences on our horoscopes have been thrown into doubt for quite some time, there is no question that the gravitational pulls of my seven hypothetical planets have an immense influence on the way you feel, the way you behave, and especially the way you predict your destiny. So, let's visit a fictional astrology verging on serious geopolitics!

The principle that will lead me in this reckoning is the link between the territory necessary for our subsistence and the territory that we recognize—legally, affectively—as our own and thus as the source of our freedom and autonomy.[4] In what follows, a territory is considered not as a chunk of space but as all the entities, no matter how remote, that allow a particular agent to subsist. I will start from the assumption that the present disorientation is due to the fabulous increase in the lack of fit between the two sets of constraints: we inhabit as citizens a land that is not the one we could subsist on, hence the increased feeling of homelessness, a feeling that is transforming the former ecological questions into a new set of more urgent and more tragic political struggles. People everywhere are

again in need of land, a situation that I call, for this reason, the new "wicked universality."

1

The first planet I will show shining in the planetarium is what could be called the planet GLOBALIZATION, that is, the sphere imagined by the recent attempts at modernizing the earth. Although it has properties drawn from cartography, geology, and some geography, it is a sphere of ideas, since it implies that everyone on Earth could develop according to the American way of life, and forever, without any limit. It is the globalization that was pursued—as a positive utopian or dystopian ideal—until the end of the twentieth century, and that still has some attraction.

None of the nation-states composing GLOBALIZATION's map occupies only the official space inside its borders and frontiers. China, Europe, the United States—all occupy other territories in many ways, either forcibly or through the partially hidden means of "ghost acreage," to use Kenneth Pomeranz's powerful expression.[5] This is what Pierre Charbonnier calls the "ubiquity of the moderns" to underline that there is no correspondence whatsoever between the shape of nation-states in the legal sense and the widely distributed sources of the wealth its citizens benefit from.[6] Belonging to a territory on such a planet is a sure way of being misled and lost: your wealth, or your misery, comes from places that are invisible on the administrative map of your own land.

So, GLOBALIZATION is simultaneously that toward which the "whole world" is supposed to have progressed and a totally skewed utopian domain where time and space have been colonized to the point of rendering it uninhabitable and paralyzing any reaction to the threat everyone clearly sees coming. The lack of agreement between the two meanings of *territory* is well illustrated by the constantly receding date of Earth Overshoot Day, which measures the moment in any given year when humans have eaten up their natural capital and begin accumulating debt against the earth (in 2018 in France, the date was May 5; in the United States, March 15).[7] The paradox is thus that the Promised Land for everything universal ends up in a cramped space, with no people able to truly say, "This is where I belong, and it is from here that I draw my subsistence and where I find the source of my liberties." The land of free people is made up of people who are paralyzed.

2

For the last forty years, this planet GLOBALIZATION has felt the increasing gravitational pull of another planet, which could be named ANTHROPOCENE. It is different from the former precisely because it began to *rematerialize* all the elements that had been left aside, a bit too quickly, by those who had embarked on the great progressive movement toward globalization. All that was externalized by that one planet is internalized in this one. Planet ANTHROPOCENE is planet GLOBALIZATION, but where the earth is reacting to human enterprises[8]—no longer a frame, or a stage,

1 A version of this essay was given as the Loeb lecture at the Harvard Graduate School of Design on October 16, 2018. I thank Richard Powers for comments on an earlier draft, Michael Flower for mending my English, and Alexandra Arènes for the drawing.

2 Bruno Latour, *Down to Earth: Politics in the New Climatic Regime*, trans. Catherine Porter (Cambridge: Polity Press, 2018).

3 Bruno Latour, *Facing Gaia: Eight Lectures on the New Climatic Regime*, trans. Catherine Porter (Cambridge: Polity Press, 2017).

4 Here I am following Pierre Charbonnier's recent work; for a preview, see "L'écologie, c'est réinventer l'idée de progrès social," *Revue Ballast*, September 26, 2018, www.revue-ballast.fr.

5 Kenneth Pomeranz, *The Great Divergence: China, Europe, and the Making of the Modern World Economy* (Princeton: Princeton University Press, 2000), 277, 312, for example.

6 Pierre Charbonnier, "L'ubiquité des modernes: Souveraineté territoriale et écologie globale," February 2, 2017, video accessed at Collège de France, www.college-de-france.fr.

7 See the Earth Overshoot Day website, www.overshootday.org.

8 This is what Chakrabarty calls the "planetary"; Dipesh Chakrabarty, "The Human Condition in the Anthropocene," in *The Tanner Lectures on Human Values* 35, ed. Mark Matheson (Salt Lake City: University of Utah Press, 2016), 137–88.

9   Dipesh Chakrabarty, "The Climate of History: Four Theses," *Critical Enquiry* 35 (Winter 2009): 197–222.

10   See Christophe Bonneuil and Jean-Baptiste Fressoz, *The Shock of the Anthropocene: The Earth, History, and Us*, trans. David Fernbach (New York: Verso, 2016).

11   Jason Moore, *Capitalism in the Web of Life: Ecology and the Accumulation of Capital* (New York: Verso, 2015).

12   Quoted in Devin Coldewey, "Starman Has Gone Dark," TechCrunch, February 7, 2018, techcrunch.com.

13   Douglas Rushkoff, "Survival of the Richest," Medium, July 5, 2018, medium.com; republished as "How Tech's Richest Plan to Survive the Apocalypse," *Guardian*, July 24, 2018, www.theguardian.com.

but a powerful actor with its own agency and its own tempo, and at a scale that is comparable in size and weight to that of the human technosphere. Its presence is captured by expressions such as *Earth system*, or *Gaia*, or the *Anthropocene*, or the *Great Acceleration*, or the *tipping point*, or *Earth's boundaries*—a whole vocabulary that has transformed what was to have been a theater stage that could be altered by human ingenuity into a player intervening as a third party in every human activity. The key point is that it is not nature as such, whose immensity, indifference, aloofness, importance, and all-encompassing substance have always been celebrated, but an agent with its own force and power that requests to be integrated, in some way, into the political domain. Facing Gaia is altogether a different adventure than facing nature.

How to define planet ANTHROPOCENE with the little reckoning principle that detects the overlap of legal and real territory? On the face of it, it should be the great solution to the radical homelessness suffered on planet GLOBALIZATION: the human, now as big as the earth, is easily *superimposed* on a planetary system of comparable size, in such a way that all questions of freedom are also questions of subsistence. You depend on the whole planet? Well, the whole planet is reacting to your actions. And yet, there is no such overlap, for reasons Dipesh Chakrabarty has tried to disentangle since his first paper on "Four Theses."[9]

A good locus from which to see why this is so is to consider the great "Anthropocene quarrel." As soon as the term *Anthropocene* was used in geology, climatology, biochemistry, and stratigraphy by natural scientists, it was immediately criticized by social scientists for its insensitivity to the complex history of human societies.[10] There was indeed a rematerialization of conditions for subsistence, and a welcome one at that, but the *anthropos* of the Anthropocene is too much of an abstraction to provide a real superposition of the legal and social questions of freedom and autonomy on the earthly conditions of subsistence. Although it was better to live in the Anthropocene than suspended in midair as in globalization—with an Earth Offshoot Day somewhere in March or May—the point is that geologists and biochemists are not offering any view of an earth that citizens and activists *can recognize as their home*.[11] Although the principles of homelessness were tackled—in the Anthropocene there is a material earth under our feet—people still had no abode where they could express their living conditions in terms compatible with those of social justice. Humans are plugged in as a box in the models developed by Earth system science at the Potsdam Institute or in the reports of the Intergovernmental Panel on Climate Change (IPCC), a box just like those for soil, vegetation, or ocean currents. In spite of Gaia's pull, this planet could be felt as another attempt at *naturalizing* social life, that is, as the end of the *human* world.

3

That planet ANTHROPOCENE was seen as, really, the end of human domination can be shown by looking at two other, darker bodies that are coming frighteningly close to it, planets whose gravitational fields could engulf all the others, as in a replay of Lars von Trier's film *Melancholia* (2011). The first of the two dark planets to consider is what could be called planet EXIT.

Considering that it is barely thinkable to imagine any harmony between the resources necessary for subsistence and the unfettered exercise of freedom, some have concluded that the two main assumptions of modernism should be abandoned altogether: freedom is for the few, not for the many; breaking from the limits of nature is the essential destiny of those few only. Hence the name EXIT: let's forget about the universality of the modernist dream still entertained by people on planet GLOBALIZATION, and let's *accelerate* the break away from earthly conditions. If we wanted a simile of such a liftoff, Elon Musk's red Tesla car sent hurtling into space would be a good one. Although it superficially resembles the modernist ideal of expanding in space ad infinitum, it is much more sinister, as Musk himself recognized when he said of his enterprise, "It's silly and fun."[12] To my ears, with what was supposed to be a joke, he broke any continuity with the former ideals of progress for all. This is indeed another planet.

On planet EXIT, the plan is that it will soon be possible to download our mortal bodies into a mix of robots, DNA, clouds, and AI, thereby situated as far as possible from the humble and limited Earth. Technology is transcendence. It needs no earth except as a provisional platform before new adventures begin. On to Mars! In case those accelerations evaporate as so much hype—if, for instance, the terraforming of Mars takes more time than anticipated—it might be wise to buy a house in a gated community or an underground bunker somewhere, preferably in New Zealand, a real, material, well-protected terroir down on old, already-terraformed Earth.[13] Wherever the gated community ends up being situated, the great difference between the planets GLOBALIZATION and EXIT is that there is no longer any project for the billions of humans who are explicitly now *left behind* or, to use a cruel but frank adjective, have become *supernumerary*. Civilization, in the narrow sense of a project invented in the eighteenth century, is now abandoned for good.

4

The simple question becomes: But where will all those supernumerary masses go, all those left behind? There is no difficulty in finding where they are heading. It has been in the news every morning, especially after Election Day. When it is not in Brazil, it is in Hungary; when not in Hungary, in Germany, or England, or France, or Italy. You name it. Here there is another dark planet—let's call it SECURITY—that is today the biggest planet of all, the one that overshadows, it seems, all the others. Where do the millions of people go? There is only one answer: wherever they would like, so long as they remain behind walls, and thereby retain at least one element of the former civilizing project—protection and identity.

Although troubling for the inhabitants of

the other worlds, the migration is perfectly reasonable. If prosperity and freedom are gone and it is impossible, as scientists insist, to bring prosperity and earthly conditions together, then let's at least have an identity, a sense of belonging. Does it solve the problem of the superposition between subsistence, territory, and freedom? Maybe not, but the promises of planet GLOBALIZATION have been left aside anyway, modernization is stuck, inequalities are growing every day, and, to top it all off, we have been betrayed by those, the inhabitants of EXIT, who are fleeing toward Mars without us—Mars or New Zealand, that is. They don't even pretend to work for our benefit: Noah's Ark is for them, not for us. They have abandoned us; we abandon them.

Planet SECURITY's attraction appears to be overwhelming. Almost everybody, it seems, dreams of fleeing inside a neo-national, neo-local bounded space, even though it might mean abandoning any pretense of maintaining the civilizing project of the recent past. On planet GLOBALIZATION there was, remember, a fundamental disconnect between the legal borders of states and the real territory they had to command in order to subsist, but their horizon remained global and their ideal was still that of coordination and shared sovereignty. Those states were in an awkward position, to be sure, cantilevered over an abyss, but they had at least the project of coping with it—the best example of such intention being the hapless but still admirable effort of the climate conferences, those famous COPs (Conferences of the Parties) culminating in the Paris Agreement in December 2015.[14] But on planet SECURITY—as well as on planet EXIT—there is no need for such a horizon, as we know from the decision of the present US administration to withdraw from the Paris Agreement. Climate denial is consubstantial to their projects.

While on EXIT it is technology that is supposed to be the saving hand of God, on SECURITY there is not even that hope. Hope is no longer a possibility. You can't even say, "it's silly and fun." Above all other attitudes, it is rage and despair that are valued most. Which makes sense, since homelessness is pushed to its most extreme expression: the desperate effort to possess an identity but without any realistic material ground to settle on and provide a soil. The most tragic aspect of this planet is that, in spite of its appeal to the forces of *blut und boden*, its vision of the soil is even more abstract and idealized than that of planet GLOBALIZATION. Populists are people, yes, but without a real land under their feet.

5

We should introduce here planet MODERNITY, even though its gravitational pull has been decreasing and might be hard to feel any longer. For Europeans, at least, it could be called planet GALILEO or DESCARTES to mark more clearly a change in the very material with which the body of the planet was understood to be made, and a new position in relation to the sun—a type of "worlding" that was not so detectable before what is called, in quotation marks, the "Age of Discovery" or, again with

scare quotes, the "Scientific Revolution" (a fabulous expansion of the world but without any similar expansion of the feeling of identity). Hence we have the strange creation of "otherness" that has been coevolving with the sentiment of "modernity," and the predicament of those who have never been moderns but who have transformed the others into "Others."

You could object that this is then exactly the same planet as GLOBALIZATION, with its growing overhang between the legal and the lived territory. But this would be to confuse the earth of *before* and *after* the main event, that is, the introduction of carbon—"coal and colonies," to use another of Pomeranz's expressions. What is becoming more and more clear, viewed from the point of view of planet ANTHROPOCENE, is that there is not much continuity between the first and the second modernizing project—let's say the period from the sixteenth through eighteenth centuries, with its coal, and the nineteenth and twentieth centuries, with coal and oil. Economics was still an art of dealing with prudence and limits, and not yet with what could render invisible the conditions of subsistence and infinitize profit. To use Timothy Mitchell's thesis, they could not render invisible all links to earthly conditions.[15] On the contrary, states entrenched more and more the notions of development and civilization within a newly material and complex world that they delighted in figuring out through thousands of representations. It is carbon that has transformed what was a bit of chance—a lucky boon—into a destiny. Without carbon, Europe's expansion was an intrusion complicating the ways of life of all sorts of other empires and agents, as postcolonial studies have so elegantly shown. With carbon (and all that is associated with its extraction), it became, as Eric Voegelin has said, "an apocalypse of civilization."[16]

6

Although MODERNITY seems a slightly weakened, outdated, backward version of the planet GLOBALIZATION, the same cannot be said about what I have called, in my book *Down to Earth*, the TERRESTRIAL. The planet TERRESTRIAL is at once that toward which it seems all progressive political movements are heading, and yet that which is terribly difficult to define. Paradoxically, the main attractor does not seem to be so attractive!

And yet, using the same principle I have used to describe the other planets, it seems to offer, finally, a solution to the homelessness detected as the source of our general disorientation: it overlays the strange shape of territories (remember that a territory of any living thing is defined as that which allows this lifeform to subsist) atop territory understood as that which free agents can decide *on their own*. If Anna Tsing's book on how to study ecological crisis takes on more and more importance, it is because it is probably the first to show in enough detail how such a superposition is possible, even though her "mushroom at the end of the world" fits none of the earlier categories of nation-states, sovereignty, capitalism, class struggle, and so on.[17]

Why would TERRESTRIAL be different,

14 Stefan C. Aykut and Amy Dahan, *Gouverner le climat? Vingt ans de négociations internationales* (Paris: Presses de Sciences Po, 2014).

15 Timothy Mitchell, *Carbon Democracy: Political Power in the Age of Oil* (New York: Verso, 2011).

16 Eric Voegelin, *The New Science of Politics: An Introduction*, with a new foreword by Dante Germino (1952; repr., Chicago: University of Chicago Press, 1987), 130.

17 Anna Lowenhaupt Tsing, *The Mushroom at the End of the World: On the Possibility of Life in Capitalist Ruins* (Princeton: Princeton University Press, 2015).

Diagram of the spatial configuration of the seven imaginary planets proposed by Bruno Latour. Drawing by Alexandra Arènes, 2018

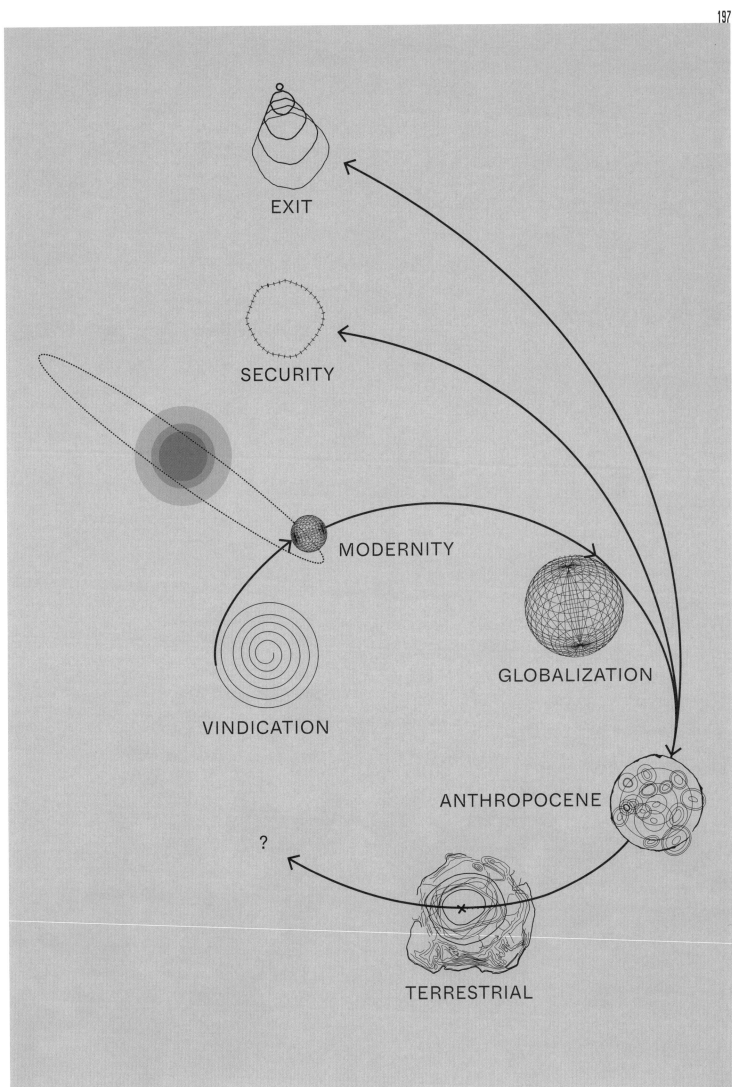

EXIT

SECURITY

MODERNITY

GLOBALIZATION

VINDICATION

ANTHROPOCENE

?

TERRESTRIAL

198

A FICTIONAL PLANETARIUM

18 For instance, Jan Zalasiewicz et al., "Scale and Diversity of the Physical Technosphere: A Geological Perspective," in "Perspectives on the Technosphere (Part 1)," ed. Sara Nelson, Christoph Rosol, and Jürgen Renn, special issue, *Anthropocene Review* 4, no. 1 (April 2017): 9–22.

19 Bruno Latour and Timothy M. Lenton, "Extending the Domain of Freedom, or Why Gaia Is So Hard to Understand," *Critical Inquiry* 45, no. 3 (Spring 2019): 659–80.

20 Déborah Danowski and Eduardo Viveiros de Castro, *The Ends of the World*, trans. Rodrigo Nunes (Cambridge: Polity Press, 2017).

Satellite image of the Arctic during polar darkness on October 30, 2012, when ice levels were at a new record low (since surpassed). The image was captured using Visible Infrared Imaging Radiometer Suite (VIIRS) on the Suomi NPP satellite.

then, from the planet ANTHROPOCENE, whose presence and influence have been growing since the 1960s? Precisely because it might offer a solution to the great Anthropocene quarrel mentioned earlier. You cannot insert into politics just any sort of natural entity without transforming the search for freedom and autonomy into the simple domination of necessity and heteronomy. So, to tell humans that they behave just like a geological force, as the Working Group of the Anthropocene does regularly and beautifully, even though it is technically true—the scales are correct, the influence indisputable, the effects devastating[18]—is not something that any political agent can hear without ceasing to be a human political agent. In becoming geology, anthropocentric humans have become as immobile as pillars of salt.

But where did we learn that freedom was reserved for human life-forms? This is where the discovery of Gaia comes in. Gaia is not Earth system science. It is a much more interesting and astute sort of being.[19] I have no room here to develop the idea fully, but the key element is the realization that what all life-forms have in common is the *making up of their own laws*. They don't obey rules made elsewhere. The crucial discovery is that life-forms don't reside *in* space and time, but that time and space are the result of their own entanglement. So, although reconciling the realm of necessity with that of freedom is a waste of time, connecting free agents with other free agents opens up completely different styles of association and allows the building up of different societies. The TERRESTRIAL is the same planetary body as the ANTHROPOCENE, but where the politicization of nature might finally take over.

7

If I am somewhat confident in the gravitational pull of this sixth planet, it is for a reason that is not visible until you bring all the planets together in a spatial configuration—just as fictional as the rest, of course. In this diagram, you will notice that the TERRESTRIAL is pulled toward the gravitational field of a seventh planet that I have not yet mentioned and that I am tempted to call VINDICATION. Why this name? Why do I end with this planet when it is clear that it should have been the first to be considered? Precisely because it has never been allowed to be freed from the retrospective judgment of five of the other planets. Whenever it is treated first, it becomes "primitive." Whenever people talk about modernization, they immediately create, by way of contrast, a primeval site, that of archaic attachment to the soil, to the ground, which is then either ridiculed as that from which the whole civilizing project has been extricating itself, or—what is worse—celebrated as a mythical, archaic, primordial, autochthonous Ur-Earth free from all the tragic sins of civilized humans.

If there is one lesson to draw from the extraordinary rebirth of anthropology in recent years, it is that for the first time—probably because of the symmetrical pull of the TERRESTRIAL—the many societies of humans and nonhumans that are active on Earth are allowed to stop having to define themselves by comparison with modernity, or to be taken only as having rich "symbolic" views of nature. With what has been called in anthropology, rather misleadingly, the "ontological turn," they are free, at last, to be our contemporaries, and maybe also to exchange some of the prescriptions they might have for composing societies made of free agents. As Déborah Danowski and Eduardo Viveiros de Castro are fond of saying, what is sure is that these societies of humans and nonhumans are experts at survival.[20] Now they have their own planet, and they can fight back. It is about time we move from modernity to contemporaneity, that is, to the present, and seek vindication.

*

Concluding this tour through my fictional planetarium, we see that there is not one but three or four different arcs of history. Modernism appears now as a small parenthesis that went quickly from MODERNITY to GLOBALIZATION and that is now being torn apart by two radically different gravitational pulls: one generated by what I have called the dark planets, EXIT and SECURITY, and the other by the planets that are rematerializing the earth in different, contradictory fashions, namely ANTHROPOCENE and TERRESTRIAL.

It is no wonder that we feel politically disoriented: these seven planets make their influences felt simultaneously over every one of us and modify the paths of our enterprises minute by minute. We are not divided in two, but in at least seven! Just like good planets do in our solar system, they all act on one another. So every one of our issues today—whether we wish to build something, design a situation, make a plan, settle a controversy—is pushed and pulled, divided and influenced by the overlapping, contradictory, still-unsettled fields of attraction of these seven bodies. Right now, the probability that they will coalesce to make *one* common world is nil—and, I would say, fortunately so, since the largest of all, SECURITY, is probably the darkest and holds the least promise of unifying the political situation. Architects and designers may now understand that one qualification should be added to the project of designing for the planet—the question, For which planet? ▦

ZOO

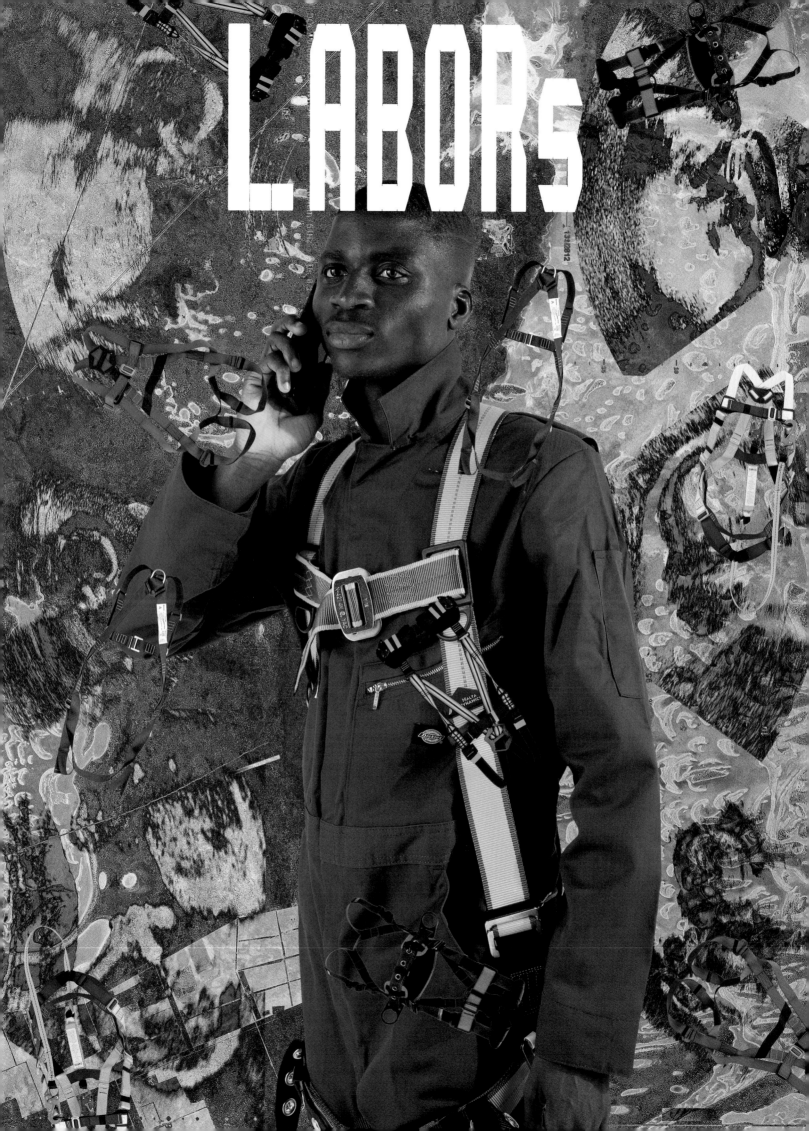

LABORs

The More Equitable Future Begins in the Imagination

MARINA GORBIS on futures thinking, key resources, and universal basic assets

"The imagination creates the future," writes Lewis Hyde, professor and author of the book *The Gift: Creativity and the Artist in the Modern World*.[1] This explains why we think of artists—whether "fine" artists or designers—as futurists. At the Institute for the Future we believe that the value of futures thinking is not in predicting the future (something no one can do), but in *imagining possibilities* of what the future could be. And if there was ever a time we needed such imagination, it is today. Mark Fisher, an academic and the creator of the blog *K-Punk*, has dedicated his energy to futures thinking, worrying that we are losing our capacity to "conceptualize a tomorrow that [is] radically different from our present."[2] Who is better suited to imagining something that doesn't exist, something that could be, something that we have neglected, than the artist?

As futurists we look for signals—small, often weird things. They are usually new technologies, new behaviors, new narratives that don't fit into the mainstream, but that are often precursors of important transformations. We then try to discern the larger patterns that these signals herald to understand where they might lead ten or more years down the road. In many ways, this is also what artists do: look at the world around them, shine a light on things we might not recognize at first as significant, raise questions, and provoke us to reframe how we see the world, and how we think and act.

Twenty years ago we looked at the convergence of online communications, embedded sensors, and abundant data to foresee the transformation from episodic to continuous interaction among people in a social network. But already in 1962, Roy Lichtenstein recognized this convergence and its potential. His painting *Portable Radio* foreshadows a world in which you can carry music, news, and books everywhere you go—just strap on a radio, and carry it all with you.[3]

Not only are artists able to spot signals and imagine future possibilities, their very practice—the way they live and engage in their craft—is a harbinger of how work is being transformed for many in today's economy. Over the past ten years, a growing number of people have started earning a living as freelancers or gig workers, either full-time or as a supplement to formal employment. What artist is not familiar with gig work? Artists are the original gig workers—hired to do projects, perform at an event, or create something on spec without the stability and benefits that come with formal employment. While gig work has many advantages, including the flexibility to organize work around personal needs and desires, it can also be precarious and stressful. As a gig worker, unless you have a steady stream of projects or have made a name for yourself, allowing you to sell your work or labor at prices that ensure long-term security, you may go for long stretches without income. You also have to bear the costs of marketing and advertising yourself, paying for health insurance, managing your various tax obligations, and much more. The services and perks provided by an organization are not available to you. You have to be a one-person organization with your own infrastructure. Unless you are in the top one percent of artists or freelancers who have managed to build up enough assets, you are living without long-term economic security.

This is the dilemma uniting artists and many of today's workers: flexibility and freedom on the one hand, precariousness and instability on the other. What previously was considered a choice and a luxury, something artists purposefully sought out and in some people's eyes had to "pay a price for," is now a reality for many other workers who don't have a choice. Herein lies an opportunity for a new kind of solidarity—for artists to come together with Uber and Lyft drivers, TaskRabbits, and millions of others trying to survive in today's economy. Together, we need to imagine a new economic system

Aerial view of a cargo container terminal, Port of Melbourne, Victoria, Australia

that provides long-term economic security for the majority, not just a few.

This is why at the Institute for the Future we introduced the concept of Universal Basic Assets (UBA),[4] identified as the following:

Spaces
Natural resources
Infrastructure
Capital
Data
Know-how
Communities
Power

We see the UBA as a starting point and a guide to designing a new economic model, recognizing that returns in today's economy are increasingly going to owners of assets (primarily stock and various other financial holdings), and not to wages. Median weekly earnings have grown a scant 0.1 percent a year since 1979.[5] The typical American family today has a lower net worth than it did twenty years ago.[6] The UBA framework and manifesto point out that important assets are often passed on from generation to generation and thus have long-term impact on the prospects for socioeconomic mobility. Inheriting wealth provides a big leg up in life, and the lack of this advantage increasingly seals one's fate in our society. Contrary to visions of the American dream, in which hard work inevitably leads to improvement in socioeconomic status, where you start on the economic ladder is most likely where you will end up. The UBA framework recognizes the importance of not only financial assets but also many other assets that are ignored by the market economy and are created by artists, caregivers, teachers, and civic leaders.

The UBA manifesto declares that every person deserves access to key resources needed for economic security and a healthy, productive life, from access to health care, education, and public transportation to infrastructure, community, knowledge, and so on.[7] It is about creating business structures, policies, and investment mechanisms that promote more equitable access to these key resources. Instead of pursuing the dominant extractive forms of business, can we create more mutually beneficial businesses like cooperatives and employee stock-ownership plans (ESOPs)? Can we invest in public infrastructure in order to equalize access to assets and minimize the need to own them privately (more public transportation instead of more cars, driverless or otherwise)? In an era of union decline, can we create new ways to aggregate and empower the voices of the many who are excluded from existing power structures, both economic and political?

More equitable futures begin in the imagination, and we need artists more than ever before to lead us in imagining and building them. ‡

The "world's biggest poster," crowdfunded by the committee for an "Unconditional Basic Income," pictured on the Plainpalais square in Geneva, Switzerland, May 14, 2016. Photograph by Denis Balibouse/REUTERS

Marina Gorbis, executive director of the Institute for the Future, speaking during the C.E.O. Conversation: Engineering the Future at the New York Times Global Forum, held at the Metreon, San Francisco, June 20, 2013. Photograph by Neilson Barnard/Getty Images for the New York Times

1 Lewis Hyde, *The Gift: Creativity and the Artist in the Modern World* (New York: Vintage Books, 2007), 252.

2 See Hua Hsu, "Mark Fisher's 'K-Punk' and the Futures That Have Never Arrived," *New Yorker*, December 11, 2018, www.newyorker.com.

3 Lichtenstein's painting is in the collection of the San Francisco Museum of Modern Art; see www.sfmoma.org/artwork/fc-604/.

4 See Institute for the Future, *UBA\Universal Basic Assets: A Manifesto for a More Equitable Future*, www.iftf.org/uba.

5 United States Department of Labor, Bureau of Labor Statistics, weekly and hourly earnings data from the Current Population Survey, 1979–2019, raw data accessed June 2019, www.bls.gov/webapps/legacy/cpswktab1.htm (all boxes must be checked and the output options set from 1979 to 2019).

6 See Jesse Bricker et al., "Changes in U.S. Family Finances from 2013 to 2016: Evidence from the Survey of Consumer Finances," Federal Reserve Bulletin 103, no. 3 (September 2017), box 3 (Recent Trends in the Distribution of Income and Wealth), fig. B (Wealth shares by wealth percentile, 1989–2016 surveys), www.federalreserve.gov/publications/.

7 The pdf for the manifesto (see note 4) can be downloaded at www.iftf.org/uba.

CATALOG FOR THE POST-HUMAN   2014–19
TIM PARSONS AND JESSICA CHARLESWORTH, PARSONS & CHARLESWORTH, CHICAGO

The *Catalog for the Post-Human* by the design duo Tim Parsons and Jessica Charlesworth was originally conceived for the Open Society Foundations' Future of Work research program as a critical response to possible issues facing workers of the future. The resulting sales catalogue offers fictional products that would allow employees, freelancers, and businesses to stay competitive by boosting their human capabilities.[1]

The catalogue, with its graphic representations of products and accompanying marketing text, serves as a piece of speculative design that uses ironic humor to question the ethical application of technologies in the workplace. It highlights the effects of technology on the future of work and, in particular, the increased pressures placed on low-wage, zero-hour contract workers and those who serve in the so-called gig economy. As the designers explain, just as the panopticon uses physical space to enable human supervision, the information panopticon can use digital systems as a means of surveillance, discipline, and even punishment

in the work environment.[2] Real-world examples bring this point home: Amazon patented a wristband that tracks the hand movements of warehouse workers and uses vibrations to nudge them into being more efficient.[3] The start-up Sociometric Solutions in Cambridge, Massachusetts, sells a badge, "equipped with two microphones, a location sensor and an accelerometer," that records "tone of voice, posture and body language, as well as who spoke to whom for how long."[4] The replacement of interpersonal relations with digital monitoring and workplace analytics produces an oppressive environment in which the mental and physical well-being of employees is downplayed or disregarded in the pursuit of algorithmically determined productivity targets.

Parsons & Charlesworth's "capsule collection," a commission for *Designs for Different Futures*, brings to life a new iteration of the catalogue with products presented for the first time as physical models. The collection encompasses wearable electronics, prosthetics, smart drugs, and food items and is presented

in a vending machine, illustrating the products as a subset of a wider group and situating them within a future retail context in which clerks no longer exist. **MB**

1   "Overview" for the *Catalog for the Post-Human*, on the Parsons & Charlesworth website, www .parsonscharlesworth.com. The project is made possible by support from the Canada Council for the Arts.

2   Parsons & Charlesworth, "Catalog for the Post-Human," project proposal for *Designs for Different Futures*, January 2019.

3   Ceylan Yeginsu, "If Workers Slack Off, the Wristband Will Know (And Amazon Has a Patent for It)," *New York Times*, February 1, 2018, www.nytimes.com.

4   Steve Lohr, "Unblinking Eyes Track Employees," *New York Times*, June 21, 2014, www .nytimes.com.

RAISING ROBOTIC NATIVES   2016   STEPHAN BOGNER, PHILIPP SCHMITT, AND JONAS VOIGT

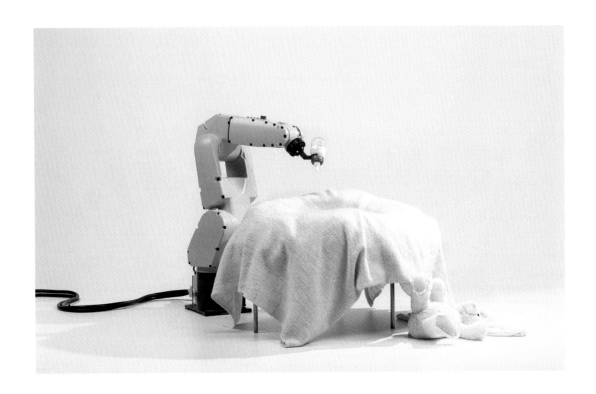

"Do you want a robot to take care of you? How much do you want to rely on smart helpers?"[1] These were questions asked in the exhibition *Hello, Robot: Design between Human and Machine*, held in 2017 at the Vitra Design Museum in Weil am Rhein, Germany, where *Raising Robotic Natives* was first publicly displayed. Created by Stephan Bogner, Philipp Schmitt, and Jonas Voigt while they were students in the interaction design program at the Hochschule für Gestaltung Schwäbisch Gmünd in Germany, these four parenting aids reconsider the future of childcare with the help of a robotic nanny. The aids include an industrial robotic arm equipped with a baby bottle; a child-friendly dragon costume to cover the

robotic arm and transform it into a playmate; a children's book, titled *My First Robot*, to instruct and entertain children with the story of robots; and a kill switch to deactivate the robot in case of emergency or parental choice.

Instructions were made freely available on the internet for parents to download, modify, and use to manufacture such aids themselves. The designers argued that robotic bottle feeding would save "parents 15–30 minutes per bottle which makes for a significant increase in efficiency," but they also asked if we should "let robots replace humans in those activities we [consider] most intimate."[2] This provoking vision of possible interactions between humans and machines—which could result in a future

generation socialized by robots from infancy— belongs to the field of speculative or critical design. As Bogner, Schmitt, and Voigt demonstrate here, such designs open debate on how humanity will be shaped in the future. **KBH**

1   Mateo Kries, Christoph Thun-Hohenstein, and Amelie Klein, eds., *Hello, Robot: Design between Human and Machine*, exh. cat. (Weil am Rhein, Germany: Vitra Design Museum, 2017), 178.

2   "*Raising Robotic Natives*: Artefacts for Generations Growing Up with Robots," the Office of Philipp Schmitt website, philippschmitt.com.

GHOST MINITAUR 2015–17 AVIK DE AND GAVIN KENNEALLY, GHOST ROBOTICS, PHILADELPHIA

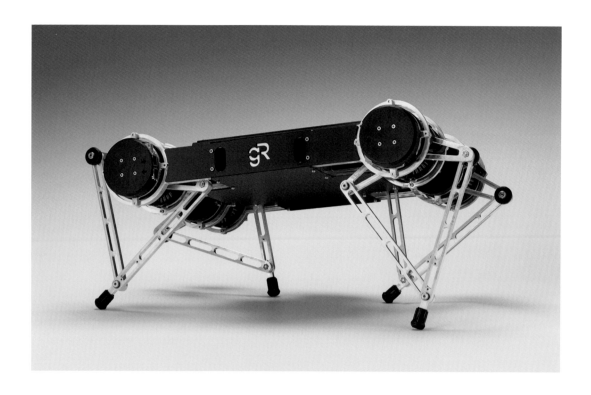

With four triangular legs and weighing just thirteen pounds, this dog-size robot will be deployed in the near future in place of humans in search and rescue missions, military operations, and possibly even off-Earth exploration—anywhere unknown, rough, hazardous, or environmentally sensitive terrain is present. The Minitaur was developed by Ghost Robotics, a start-up company founded in Philadelphia in 2015 by Avik De and Gavin Kenneally under the direction of their University of Pennsylvania engineering professor Daniel Koditschek, whose research involves the design, construction, and testing of autonomous robots. As the Minitaur demonstrates, these machines can mimic the mobility and dexterity of animals. The Minitaur can run and turn at high speed, jump over small obstacles, scramble over difficult terrain, flip, open doors, and climb stairs and chain-link fences. Recent updates allow it to better navigate various surfaces, including grass, rock, sand, snow, ice, and urban debris. "Our primary focus since releasing the Minitaur ... has been expanding its behaviors to traverse a wide range of terrains and real-world operating scenarios," said the company founders. "In a short time, we have shown that legged robots not only have superior baseline mobility over wheels and tracks in a variety of environments and terrains, but also exhibit a diverse set of behaviors that allow them to easily overcome natural obstacles."[1]

The Minitaur has no gears, hydraulics, or external sensors that would add weight and mechanical complexity: it moves by sensing the force of objects it interacts with. "It feels everything directly through its legs," explained Jiren Parikh, CEO of Ghost Robotics. "It measures the force that is put on the leg using a torque sensor in the motor. That force tells it if it is on sand, ice, or dirt, and it adjusts its balance accordingly."[2] **KBH**

1 Press release, Ghost Robotics, February 28, 2017, ghostrobotics.io; quoted in Alex Kirkpatrick, "Leg over Wheels: Ghost Robotics' Minitaur Proves Legged Capabilities over Difficult Terrain," Robohub, March 7, 2017, robohub.org.

2 Quoted in Cassie Kelly, "The 'Minitaur' Robot Looks Dainty, but Is Actually Hardcore," Inverse, March 5, 2017, www.inverse.com.

MERGER 2018 KEIICHI MATSUDA

Merger is a 360-degree short film (available in a virtual-reality version) that presents a near-future workplace in a corporation controlled by algorithms. A young consultant is seen managing her tasks and social life as the viewer hears her converse with an interviewer about her next step: leaving the limitations of her human body to merge with the network. "I know that once I'm on the other side I'll be able to be everywhere, see everything," she says. "I can give my clients a chance to fight back, to have total control."

The reality presented in the video seems futuristic but resonates with our contemporary obsession with work-life balance and productivity. Many of the protagonist's lines are taken from real-world self-help books and blogs that focus on maximizing every moment for efficiency. Similarly, the tech she uses is based on interfaces recognizable to the viewer and on the director Keiichi Matsuda's own experience as a vice president of design at Leap Motion, a virtual- and augmented-reality company that works on hand-tracking technology.

Through this unsettling but captivating film, Matsuda, a designer and architect, examines what happens when, in a society that worships productivity and equates our ability to work with our value, our meaning and purpose is taken away from us. Competing with the algorithms that have come to dominate corporate business and have pushed many humans out of the workplace, the fictional consultant has optimized her work environment with gesture-controlled augmented-reality interfaces that govern every area of her life, including her romantic relationships (bringing to mind the role of social media and dating apps in our current world). Even so, she recognizes the disparity between her abilities as a human and the efficacy of machines. She resolves this dilemma by deciding to merge with the system to better serve her clients.

Merger unsettles our notion of the cult of productivity and highlights the paradox of striving for both human perfection and labor infrastructures based on perfected automated systems, ultimately arriving at the idea that work will change dramatically in the future and in such a way that requires new definitions and understandings. **MB**

MAKE THE BREAST PUMP NOT SUCK / MAKE FAMILY LEAVE NOT SUCK HACKATHON 2014/2018
CATHERINE D'IGNAZIO, ALEXIS HOPE, COURTNEY LORD, ELIZABETH BAYNE, AND TEAM

Over the past five years, a multidisciplinary group of designers, technologists, and community partners have convened twice at MIT's Media Lab to reimagine the future of one of the most universal human experiences: breastfeeding. In 2014 over a hundred people gathered at the Make the Breast Pump Not Suck (MtBPNS) Hackathon to rethink the breast pump, a technology that has rarely been a focus of product and industrial design. Four years later, a second gathering widened the scope to include policies surrounding paid family leave, public breastfeeding, and postpartum health care. Three hundred collaborators from diverse backgrounds met to brainstorm better tools, programs, policies, and systems to more equitably support breastfeeding and pumping. By the end of the weekend, over thirty hackathon teams had produced concrete design prototypes for new services

and technologies, while attendees of the co-organized policy summit had envisioned new pathways to advocate for and build political will around paid family and medical leave.

Many social barriers to breastfeeding exist in the United States, a country without universal health care or a standardized family leave policy. In addition, US maternal mortality rates are on the rise.[1] These obstacles and outcomes disproportionately affect minority groups. MtBPNS solicits and embraces marginalized individuals and their perspectives, foregrounding the lived experiences of breastfeeding people and their bodies. By adapting (and co-opting) the hackathon format, MtBPNS has formed a participatory ecosystem that includes sites of co-creation, large cross-disciplinary networks, and a growing community of innovators where intersectional feminist approaches to design can

flourish and break taboos related to bodies and bodily fluids.

MtBPNS's organizers made a short documentary and a poster to communicate the ideas and outcomes of the gatherings, and they now consult with the designers of public spaces—including the *Designs for Different Futures* exhibition team—to create areas where parents and carers can feel welcome and supported. An interventionist project aiming to change the culture of technology design from the inside, MtBPNS creates space to imagine and build a plurality of possible utopias and preferable futures. MMF

1 Nina Martin and Renee Montagne, "U.S. Has the Worst Rate of Maternal Deaths in the Developed World," NPR, May 12, 2017, www.npr.org.

QUORI 2019 MARK YIM, SIMON KIM, AND MARIANA IBAÑEZ, UNIVERSITY OF PENNSYLVANIA SCHOOL OF ENGINEERING AND APPLIED SCIENCE, PHILADELPHIA

Quori is a robot developed with funding from the US National Science Foundation to help human-robot interaction researchers experiment with a socially interactive robot platform. The design team is led by the engineer Mark Yim and the architects Simon Kim and Mariana Ibañez at the University of Pennsylvania. They have carefully designed Quori's upper body to evoke gender-neutral human features. The robot is enlivened by a rear-projection head that displays cartoonish eyes, as well as two gesturing arms and a bowing spine. It moves on an omnidirectional base. As the designers note, gestures are a key part of natural communication in social interaction, so Quori mimics human movement; for example, its arms move on a ball joint modeled on the human shoulder. At just under a meter and a half (five feet) tall,

the robot is life-size without being physically intimidating to the people—who may range in age, height, and cognitive capacity, among other variables—with whom it is designed to interact.

Quori has a suite of sensors through which it can detect and interact with people, and they can be programmed to perform a hierarchy of behaviors that range from quite simple (moving in set directions and avoiding people) to very complex (conversational interactions, which are still in the testing and prototyping stage). The programming is designed by the computer scientist Ross Mead at Semio, an AI behavioral science company. These innovative behavioral possibilities emerge from interdisciplinary collaborations between computing, robotics, and social-science research, traditionally

carried out in relatively cloistered academic settings. Quori's designers are particularly interested in making the robot accessible. Each component is designed to be modular, expandable, customizable, and affordable so that more researchers can develop and test algorithms and conduct user studies in the real world, experimenting with the robot's limits and informing future development of its applications in real-life interpersonal scenarios. This community-driven research approach is at the very heart of Kim and Yim's project. The online surveys, hosted workshops, and conference presentations they conducted with researchers in the field of human-robot interaction as they developed their own prototype was the quorum that gave Quori its name. MMF

212

CITIEs

A Place We Build Together

FRANCIS KÉRÉ in conversation with MICHELLE MILLAR FISHER on neighborhoods, communities, and belonging

This conversation took place by phone on September 20, 2018, with Francis Kéré in Berlin and Michelle Millar Fisher in Philadelphia.

MMF
Can you describe yourself and your practice in two sentences? Can that be done, now that it's gotten so large?

FK
Well, I am an architect born in Burkina Faso, based in Europe, and trying to use whatever is available in terms of materials, but also in terms of location, to shape the world through architecture.

MMF
You are truly a cross-cultural figure. You've spent time working in many areas of the world. You trained as an architect in Germany, where you now live and teach. You've also realized projects in the United States and China, and in countries in Europe and Africa. Where do you think that we—as citizens, policy makers, and emerging architects or designers—should look for models for future cities? Who is getting it right and who is getting it wrong today?

FK
I don't think anyone is doing it right. Facing the growing population of cities, we have to consider a city as an organism, a holistic organism that needs to adapt. We have to take into account the strong, rapid growth and change of cities, and only then will we tackle our problems. Policy makers, architects, and even citizens—we have to consider seriously how we create cities to bring people together. An organism needs every component.

MMF
I wonder whether, in the near and perhaps very far future,

it's going to be citizens rather than policy makers or trained architects who will be able to get us to think of the city as an organism and work more collaboratively. I know that you crowdfunded one of your first projects, the school you built in Gando [in Burkina Faso]. What do you think is the role of citizens in the future of architecture and cities? Do they have an important role to play?

FK
Of course. We have to adapt to see our neighborhood as a community for ourselves. We have to be ready to engage ourselves, for [the good of] our cities. We cannot wait for something from outside to come and show us how to do things. We have to feel responsible for our cities. This is a beginning. Being a citizen means I am part of a community in a given place. We need to talk again about neighborhoods, about infrastructure, about sharing. In French it's *partager* [to share out]. That's how to make our cities places to work, spend our time, raise our children. Everyone has to feel responsible: you, me, the architect, the policy maker. We have to work together if we want this place—if *you* want this place— to be ready for the future, a place we build together where we like to stay.

MMF
Can you talk about some of the ways that you have engendered or helped that type of sensibility take root in the communities where you've worked—how you bring people into a project so that they feel ownership and a shared responsibility?

FK
When we were raising the addition to the school, children brought building materials. They brought a rock or a stone every day so they could participate in the building of the project themselves. Or they brought a bucket of

The Gando Primary School Extension, Burkina Faso, West Africa, designed by Francis Kéré, completed in 2008. Photograph by Iwan Baan

water to contribute to the construction process. It is important to connect; giving something makes people feel it is their project. It helps them be a part of the vision, but it also gives [the project] a soul and identity. How do you do this in a city, where you have to buy materials and structures are often complicated? I think that we must engage cities in dialogue. We must get neighborhoods to be part of the process, in terms of communicating the project, in terms of asking them to contribute—intellectually, socially—and in terms of community. I think it's better to take people with you, take them along with the idea, rather than force them to accept it. We need to take more time to explicate our projects. To explicate is to implicate.

MMF
Can the village be a model for building future cities? Can microscales be knitted together so that many small communities form one macro? Can rural areas offer the city of the future a template?

FK
Yes. Exactly. If you really analyze a city and break it down, what are the components? There are many, many neighborhoods linked to each other. I'm not just talking about infrastructure that we share, or even parks, things that we commonly use together. It's about the micro, the little parts of the city, the neighborhoods. I know my neighbor, and we talk together; we share the neighborhood; we share the same road. So we and our kids know each other. And if we knit together many, many spaces like this, we make a city. I'm not talking about the physical parts of the city; I'm talking about the people living there, raising their kids there, working there, sharing time. So if architecture and architects engage at this level, you will see that the city is not far from our village—from the way the community lives in the village: knowing each other, supporting each other, sharing what they have together. This is what a good, working neighborhood should do in the city. We have to build our comfort, in terms of security, in terms of trusting each other, in terms of quality of the community. Everybody should have someone that they can appreciate, you understand?

MMF
You make me think of different cities, from Lagos to Edinburgh to New York, that are experienced neighborhood by neighborhood. I think of the *arrondissements* in Paris, for example, where it's very, very localized. Is the future of the city perhaps something that's much more medieval than modern—in that it's not the modern gridded city, which is expansive or that can be surveilled and seen all at once, but something that's very much about the premodern, something that is rooted in the very local?

FK
The very local is so important. What is local? Local is now. If you know it, it gives you safety. It gives you security. It gives you comfort.

MMF
You've spoken before about migration as, in part, a building crisis. You have said that architects have a responsibility to help make sure they're not building in ways that keep people on the periphery—you've talked about the *banlieues* in Paris, for example—but instead to think about how existing structures can be creatively reused to shape community and embed people in a diverse way in a city.

FK
Absolutely. I think that is the key. We cannot just, as in the past, go to a green field and create a new place for new arrivals so that these people remain at a distance. If you integrate them, bring them into an existing community, they will try to behave like the existing community. If you want to learn to swim, you cannot sit at a desk and just learn to swim; you have to jump into the water. So you need to really integrate people. If you don't invest in this way, and you keep them distant, then don't be surprised to find isolation. It is a requirement for human beings to become part of a vital community. People will always migrate. People will migrate because of catastrophes, because of the destruction of their environment, because of their living space, even to find better opportunities. The United States is a result of migration.

The library of the Gando Primary School, Burkina Faso, West Africa, designed by Francis Kéré, completed 2001

Students in a classroom in the Gando Primary School. Photograph by Erik-Jan Ouwerkerk

MMF

Yes.

FK

I know that it's always a challenge when new arrivals come to a town. But every experience is a challenge, you know? It has been this way always. But healthy and wealthy communities have to have the potential to integrate new arrivals. That is the way we renew ourselves.

MMF

Earlier, you said that taking time was very important—that being able to take time to speak with people, to listen to their concerns, to make them feel invested in a project is necessary in thinking about future architecture. Is that something that you feel you can always do, especially when things move so quickly today? Is it something that you really insist upon having?

FK

I can say that it depends on the project. But, honestly, any project involves a community and affects a community, because constructing it will affect the neighborhood. Any building is like putting a new puzzle into an existing puzzle. So we have to find the time to explain the project and to communicate. I know that is not always easy, but it's important. I'll tell you an example: In Germany there is a train station in Stuttgart. A very successful colleague of mine won the huge competition to build the train station. And then there were some trees that had to be removed [in order to build], and there was a huge, huge demonstration, a real battle, between the building company's security and the community living there. There were the interests of the city, which wanted the train station to happen and the idea of the architect to be realized. And then there was the community, which

didn't want to see them kill trees that were hundreds of years old. It was really, really a huge battle. So, if you don't communicate, it becomes a mess. If you want things like this *not* to happen, you have to explicate right from the start. You have to give people ownership.

MMF

So, my last question is very short. Do you have hope for the future as an architect?

FK

Oh, yes. I'm very hopeful. Challenges make our profession even stronger. We just have to see the advantage in striving, in changing, in challenging, and to make the best of it. That is what will take us into the future. ⌗

City as Postcard /
City as Polis /
City as Poem

ALEXANDRA MIDAL in conversation with
MICHELLE MILLAR FISHER on the forms,
politics, and fantasies of the city

This conversation took place by phone on February 12, 2019, with Alexandra Midal in her apartment in Paris and Michelle Millar Fisher at the Philadelphia Museum of Art.

MMF
What to you is a city?

AM
What to me is a city? This question is appealing because I'm realizing I have never really asked it aloud. For me, the first example I have in mind is a book by Hannah Arendt [*The Human Condition*, 1958] that I read years ago. In a footnote, she describes the difference between the public and the private sphere. She explained that for the Greeks, the private was opposed to the public space. So far, this makes total sense. But in this formulation, the public sphere, in fact, was the place of freedom, and the private sphere was where people were unequal; they could be enslaved, and depending on their masters, they could be killed. There was absolutely no guarantee whatsoever to be able to express yourself in the private sphere. The place where you could be equal and express yourself was the city.

I think it's really interesting to look at the city on Arendt's terms, in a political way. Instead of thinking of the city as a physical structure, or as a network of streets—as different as Los Angeles with no city center or New York with its central nexus of Manhattan, to take two examples from the United States—I would highlight the political idea, the concept of the city as the place of equality, of freedom, where people can express themselves, and where there is a certain sharing of values. And privacy is the family, or the household. That's where violence is embedded, at least for Arendt and for the Greeks. And you realize right away that the Greek *polis* is the ideal of the gathering of the

community, a fantasy of a place where complete equality could happen.

MMF
I know you studied under [the Princeton professor of architectural history] Beatriz Colomina, and so your answer made me think about the ways in which she formulates public and private spheres, too.[1] And in terms of the Greek fantasy—just as everybody has their own social location from which they experience life, everybody then brings that experience to the city. So your New York is not my New York is not someone else's New York, and the same with Philadelphia, anywhere. Time changes cities, too. We say, "Oh, you should have seen New York in the seventies, those were the days."

AM
Yeah. But you know I live in a nineteenth-century city. It's so passé already. I spent a year in Rome, I commute to Switzerland every week, and I live in Paris. I'm living in a postcard. I am aware of that.

MMF
It might seem like a self-evident question, but why should we care about the future of cities?

AM
It's not at all a self-evident question. I find science-fiction novels that deal with the future of cities extremely important in this regard. Science fiction has often used the future as a tool to imagine the city of tomorrow. Many urban planners are looking to sci-fi, and especially sci-fi from the sixties and the seventies. The city is a trope in sci-fi. Two fantastic examples from earlier are [Isaac] Asimov's novels *Caves of Steel* [first published in 1953], which depicts the city of tomorrow, a city on Earth, and *The Naked Sun* [1957], which depicts a city

Alexandra Midal's film *Possessed*, 2018, on view in her solo exhibition *Drive-In* at CAPC, Musée d'Art Contemporain, Bordeaux, France

on the fictional planet of Solaria, a city in space. Asimov placed these two cities in opposition regarding the idea of privacy. On Earth, people live in extremely packed environments. They have no access to light. And many inhabitants share the same space. But on Solaria, in space, they have a very different experience because they have plenty of room. The idea of privacy does not exist anymore because everyone has so much space. For instance, when they make something similar to a Skype call—they call it *viewing* in the book—they can be naked. But they cannot stand the idea that someone would share their house. They have immense houses to avoid personal contact. But to be naked in front of a screen is absolutely not a problem. That's exactly the opposite of what's happening on Earth in *Caves of Steel*, where people are very close to each other. They share very, very tiny apartments, but the idea of nudity is absolutely unheard of.

This question of body, of sexuality, is even more important for *The World Inside* [1971] by Robert Silverberg. Silverberg's idea is that the population of the planet is seventy-five billion people and they all live in huge skyscrapers, thousand-floor skyscrapers. So this is another idea of how people live together, and of course sexuality becomes an issue. Each city is made of forty floors and people wander from one floor to another of these huge cities. According to your status, you can move a certain number of floors up or down. And men wander during the night and enter any home and have sex with anyone. These kinds of ideas transform the concept of the city, not only, as I said before, as a physicality, but as a way of examining how sexuality and politics are intertwined in a city context.

MMF
I'm listening to you talk about the city as both a trope and a text. So often, yes, sci-fi has been a catalyst for thinking about futures in general, and our moment in the present, and our architecture and design, our cityscapes. And as you were talking about clothes and nakedness it took me back to someone like [the nineteenth-century German historian Gottfried] Semper's thoughts [in *The Four Elements of Architecture* from 1851] about the architecture of what we wear and the notion of the first proto-architecture being really reliant on fabric.

Thinking back to another kind of genesis, one of your own—I wondered if you had a memory of the first imaginary place or city that you ever encountered, textile or otherwise?

AM
I have two for you. One is, when I was a kid in the seventies, I would play a game with my older brother that we called Year Two Thousand, in which we made a city. We stole sheets from my mum, put them into water with ink, and created surfaces and then structures. The idea was never to play with dolls or with soldiers, but to create cities. The city was really the core of my childhood's *Spielraum* [playroom, or free play].

Then the second thing is also very personal. I can't remember how many years I've been obsessed with the same recurring dream, a dream about cities. I'm in a city and I'm discovering new buildings, new houses, even places. I'm sure that any psychoanalyst could understand the meaning behind this, but I don't want to overanalyze, I just love the idea that my dream conveys a fantastic sense of architecture. They are fantastic

dreams of wandering, walking, discovering places. The city of my subconscious is very vivid during the night. And it's always whispering to me new paths, new ways to explore.

It is not a nineteenth-century idea of wandering in the city, but dreams that reflect continuity, expansion, and freedom, where the street becomes the room and the room becomes the street. [Walter] Benjamin constructs a model for the explosion of the interior outward, from the home into the streets, in his examination of Parisian arcades.[2] Similarly, in relation to my dreams, I speculate that there is no real gap between mind, psyche, and the city. Have you noticed that you can fly in your dreams? You can fly above the city. And that's linked also, of course, to this modernist idea that you could see the cities from above thanks to the plane, which is something that Le Corbusier invoked several times.[3]

MMF
So that's a perfect segue into the next question, thinking about the city as a ground zero for architects' fantasies, or thinking about it as a design device, as a playground for exploration. What is an example of a future city proposed by an architect that's a particular favorite or particularly interesting to you?

AM
I would go with a dystopian city—

MMF
Yeah, let's do that, let's do a *counter*example.

AM
—like [Ron Herron's] *Walking City* from the 1960s.[4] Herron's city is like a living organism that is delegating, contracting, multiplying, walking; it's as if the city is a natural component in a landscape of despair.

MMF
We've talked about imagination and Arendt's notion that the city is a place of freedom, at least from its Greek beginning, versus the contextless, roaming city that Herron's project evokes. Today, we might think about people who lack access to shelter and are forced into a nomadism, rather than that being an imaginary or playful state. So there's this duality between the possibilities of freedom offered by the city and the harsh realities of urban living for many today, a situation that's not generative or human or in any way fair. I guess that leads me to think about writers like David Harvey, for example, and *The Right to the City*,[5] and to wonder who has access to the city in our current capitalist nexus?

AM
We can also change this perspective. Currently, in the nave at the Museum of Contemporary Art in Bordeaux, there is an exhibition of my most recent series of films that analyze the failings and the dark side of the Industrial Revolution. One of the films, *Possessed*, explores the remains of a ghost town, examining the chiasmus between objects that were perceived as subjects and slaves that were perceived as things during the gold rush in California. In a ghost town, you can feel the sediment and the layers of all the pasts. But the authorities in charge of it fantasize it and try to keep it the way they

think it should have been. Of course it's not the past exactly, and yet you might see ghosts everywhere. It is so potent. American ghost towns are not the only examples; we could also conjure up Chernobyl as a more recent ghost town.

MMF
Cities, whether they are the cities of tomorrow or today, have the potential to shape the lives and ecosystems of the people who live within them for better or worse. So often their design is a project of imagination only; a lot of it does not get enacted. We can think about the term *utopia*, for example, literally meaning "no place." Are cities always destined to be aspirational no-places?

AM
The idea of the city of tomorrow is intriguing and very important for many designers—take Henry Dreyfuss's Democracity or Norman Bel Geddes's Futurama for the 1939 New York World's Fair, for example.[6] Their cities of tomorrow are places of imagination and fantasy that some would call utopic. Interestingly, Thomas More [in his 1516 book] gives his fictional Utopia spatial specificities. He even provides the exact number of kilometers of one side of the island of Utopia. So Utopia cannot be considered as a place that does not exist; it has a topography. It has a certain—

MMF
Dimension.

AM
—dimension, exactly. The island, of course, is not located somewhere, but More gives some insight into how it looks. And that's really interesting, because it's maybe not a nowhere, it's for sure a somewhere. All of us have

in our minds an idea of the way the world looks. It's not just that a country can look different from one map to another—each of us perceives the world differently. There is even the possibility that several worlds coexist, as in quantum mechanics, for example, or Philip K. Dick's novels, such as *The Eye in the Sky* [1957]. You can see a multiplicity of ideas, and parallel ideas, worlds of tomorrow, worlds of today, and even future cities. We might say that, in a way, the city of the future is not a no-place, it is a mental place. It's a place of imagination but also a place of overlapping—

MMF
Memory.

AM
—yes, exactly, memories. I don't know if it happens to you, but when I walk—for example when I would walk home while I was at Princeton, I would always think, "This was where, at one time, Einstein walked." And maybe I was walking in his very steps. I think that's maybe an aspirational idea of the no-place, linked to people who are above you, under you, all of them all together, across time.

MMF
Yes, it's a multilayered, multilevel sediment that is deeply human, acknowledging conflicts, acknowledging histories, seeing the many competing, contested, complex, but ultimately more careful ways you can understand where you stand, literally.

AM
I like the idea that we are part of a genealogy, a history, within our cities. That there's history before and after us, and that the city might be a specific location with memories of people, of lives, of people who loved each

other and who died, and that it can go on and on forever. I think if we single out the city as a constellation of lives and people, then we go far beyond the idea of rationalism or aesthetics to reach the poetry that lies in what a city is. ⌗

1   In *Privacy and Publicity*, Beatriz Colomina stipulates that mass media has moved the site of architectural production from the private to the public space, thereby transforming privacy. See Colomina, *Privacy and Publicity: Modern Architecture as Mass Media* (Cambridge, MA: MIT Press, 1994).

2   See Walter Benjamin, *The Arcades Project*, trans. Howard Eiland and Kevin McLaughlin (Cambridge, MA: Belknap Press of Harvard University Press, 1999), 406.

3   Le Corbusier's *Aircraft* discusses flight's ability to give us a bird's-eye view of cities: "The eye now sees in substance what the mind formerly could only subjectively conceive." Le Corbusier, *Aircraft: The New Vision* (London: The Studio, 1935), 96.

4   For more on Ron Herron's *Walking City*, see Zoë Ryan's essay in this volume, p. 43.

5   First used in the 1960s by the philosopher and sociologist Henri Lefebvre, "right to the city" was defined by David Harvey as "far more than the individual liberty to access urban resources: it is a right to change ourselves by changing the city … [that] inevitably depends upon the exercise of a collective power to reshape the processes of urbanization." Harvey, "The Right to the City," *New Left Review* 53 (September–October 2008), newleftreview.org.

6   For more on Norman Bel Geddes's vision of the city of tomorrow, see the discussion of his Futurama pavilion at the 1939 New York World's Fair in Andrew Blauvelt's essay in this volume, p. 91.

The Matrimandir (The Mother's Shrine), designed by the architect Roger Anger, in Auroville, India. According to its charter, Auroville, founded in 1968, "belongs to humanity as a whole. … Auroville wants to be the bridge between the past and the future." Photograph by Santosh Namby Chandran

MAD HORSE CITY   2017
OLALEKAN JEYIFOUS AND WALE LAWAL

Mad Horse City, shown at Art X Lagos, West Africa's principal art fair, in February 2019, and earlier at Munich's Architekturmuseum der Technischen Universität as part of the 2018 exhibition *African Mobilities: This Is Not a Refugee Camp*, is a futuristic vision of Lagos, Nigeria, in 2115. Created by the Nigerian-born, Brooklyn-based artist and designer Olalekan Jeyifous and the Lagos-based writer and editor Wale Lawal, the original installation included a video projection, a graphic novella, and a virtual-reality experience.

Jeyifous studied architecture at Cornell University in Ithaca, New York, and then worked at the communications agency DBOX as senior designer for four years before leaving to pursue his own "creative compulsions."[1] His current artistic practice challenges popular perceptions of architecture, the city, and its inhabitants through conceptual visual narratives. Inspired by science fiction, Jeyifous designs future worlds with buildings that begin as 3D computer models and then are combined with photographs into photomontage

assemblages. *Mad Horse City*—the name Lawal uses to describe the energy of Lagos—is the most recent iteration of Jeyifous's 2015 (and ongoing) speculative project *Improvised Shanty-Megastructures*, which creates discourse around the marginalizing of society in a Lagos of 2050, with massive shanty super-structures made of improvised materials towering over and sprawling in the coveted city center, overtaking the landscape and displacing poorer communities. As Jeyifous explained when the project was shown in Shenzhen, China, the structures "combine attributes of improvised settlements with the scale of imposing hi-end commercial developments."[2]

Mad Horse City projects even farther into the future. For the collaboration, Lawal was tasked with writing three stories that emphasized "organic interactions between the city's inhabitants and its spaces," which Jeyifous then animated with digital images.[3] *Offline*, the first of the three dystopian narratives, tells the story of a woman who pays to disconnect from the internet in an illegal botanical garden.

The second narrative, *Òminíra*, is set in a community that resembles Lagos's Makoko neighborhood, located on the lagoon, but where fishing has been privatized. *Dreamscape*, the final vignette, describes how dreams can be purchased as consolation at the time of one's prescribed death. Challenging the forms and expressions that cities take on for their inhabitants, *Mad Horse City* explores issues of freedom and mobility, and agency and surveillance, in one of the fastest-growing cities in the world. **KBH**

1   See Jeyifous's statement at Archinect: People, archinect.com/vigilism.

2   Jeyifous, "Urban Assemblages: Imminence and Immanence," Bi-City Biennale of Urbanism/Architecture, November 20, 2015, en.szhkbiennale.org.

3   Ashley Okwuosa. "This Is What Lagos Could Look Like in 2115," Okayafrica., May 29, 2018, www.okayafrica.com.

DOMA.PLAY   2018–19   MAKSYM ROKMANIKO AND FRANCESCO SEBREGONDI, DOMA (DOMA.CITY), KIEV, UKRAINE, IN COLLABORATION WITH FRANCIS TSENG

Owning a home has become a luxury that fewer and fewer people can afford. Around the world, a younger generation of city dwellers has resigned itself to a lifetime of paying rent. As concepts such as home ownership and real-estate value become unstable, we must rethink the future of housing, urban living, and belonging.

Coming out of the New Normal program at the Strelka Institute in Moscow, DOMA proposes a blockchain-based, shared-ownership platform for equitable housing. Manifesting as an online video game in the exhibition (*doma.play*), the interactive project redistributes the value of urban properties, breaking them down into tokens that can be purchased and traded. Ownership is exploded into multiple parts as

members make diminishing payments, supporting the community while providing themselves with increasing equity. This process demonstrates how such a system can enable access to the housing market and extend its benefits to people who are currently priced out of it as they progressively become homeowners. Players can also move from one property to another in the network without losing their equity, which offers both the flexibility of a twenty-first-century lifestyle and much-needed economic stability.

The distributed, nonprofit, cooperative nature of the project empowers its community to become players in the housing market and in their chosen cities. The initiative bridges the great divide between renting and owning a

home, proposing a model of urban regeneration that makes room for every kind of urban dweller and allows them to shape their urban neighborhoods. The originators of the project note that the housing crisis has become "a structural feature of the twenty-first century city" but that it is "not a crisis of production" but of distribution.[1] With its flexible and inherently redistributive premise, *doma.play* signals a shift toward affordable, inclusive, and sustainable cities. **MB**

1   &Beyond, ed., *Archifutures: A Field Guide to Surviving the Future of Architecture*, vol. 5, *Apocalypse* (Barcelona: DPR, 2019), 203, 206.

DRIVER LESS VISION 2017 URTZI GRAU AND GUILLERMO FERNÁNDEZ-ABASCAL, UNIVERSITY OF TECHNOLOGY SYDNEY, AND DANIEL PERLIN AND MAX LAUTER, MAKE_GOOD, NEW YORK

Driver Less Vision examines the tension between AI and humans when negotiating how cities of the future will be shaped for autonomous vehicles. Through an immersive installation, *Driver Less Vision* allows viewers to experience the city from the perspective of a self-driving vehicle. A film projected on an overhead dome narrates the wanderings of a self-driving car that finds itself in a postapocalyptic urban Seoul. Told from the car's point of view, the story is inspired by Sara Teasdale's poem "There Will Come Soft Rains" (1918) and Ray Bradbury's short story of the same name (1950).[1]

While self-driving cars are usually discussed from a technological standpoint, the real challenge is not the ability to design a self-driving car but our capacity to design cities and legal frameworks that are able to assimilate them. *Driver Less Vision* questions the disruptive effects on the built environment that the deployment of technologies for

autonomous mobility would cause. Our current traffic patterns are indicated through visual stimuli designed for human sight, such as repetitive signage. Driverless cars require an omnidirectional gaze to negotiate these existing visual codes, and they construct images using a combination of radar, cameras, ultrasonic sensors, and LiDAR (light detection and ranging) scanners[2]—a method that highlights the differences between human and AI perception. Starting from the view of the road ahead captured by the LiDAR scanner, the images in the film were overlaid with other 360-degree images. This technique precisely emphasizes dissonant perceptions, illustrating how the machine's intelligence corresponds less and less to a human understanding of the world.

There is a history of urban transformation related to changing vehicular technologies.[3] *Driver Less Vision* highlights this history and asks where discussions on the future effects

of self-driving cars should take place: "Can we bring together parliaments and labs under the same roof?"[4] The installation becomes not only an immersive visual exercise but also a space to gather and negotiate the sociotechnical controversies occasioned by autonomous vehicles. **MB**

1   Both the poem and the story imagine a future devoid of human life as the result of war, with the short story told from the point of view of an automated house.

2   Guillermo Fernández-Abascal and Urtzi Grau, "Learning to Live Together," *E-Flux*, accessed March 14, 2019, www.e-flux.com.

3   See Fernández-Abascal and Grau, "Learning to Live Together."

4   Urtzi Grau, Guillermo Fernández-Abascal, Daniel Perlin, and Max Lauter, "V.1.1 Driver Less Vision Seoul," project proposal for *Designs for Different Futures*, fall 2018.

LUNAR SETTLEMENTS   2012
FOSTER + PARTNERS, LONDON

As space exploration continues to reach new milestones, human habitation on the moon approaches reality. To prepare for that moment, Foster + Partners, in collaboration with the European Space Agency, explored the possibilities of constructing extraterrestrial shelters, as well as machines that could 3D print them. The design envisions a domed dwelling that can accommodate up to four astronauts.

Inspired by the natural landscape of the moon, the shelters are designed to be built from regolith—soil and loose rocks found on the lunar surface—that is fused into building material using microwaves. The base of the shelter is unpacked from a modular tube that can be transported to the moon on a rocket. An inflatable dome extends over the base to provide a support for construction. This frame would then be covered in layers of regolith using a robot-operated binder-jetting printer to create a lightweight, foamlike formation,

an element of the design that was ground-breaking in its demonstration of 3D printing's potential to mimic biological systems commonly found in nature. This external shell offers a protective shield to safeguard the settlement from extreme temperatures and radiation. "As a practice, we are used to designing for extreme climates on earth and exploiting the environmental benefits of using local, sustainable materials," said Xavier De Kestelier, then head of the special modeling group at Foster + Partners, in a 2013 statement. "Our lunar habitation follows a similar logic. It has been a fascinating and unique design process, which has been driven by the possibilities inherent in the material."[1]

Building on the methods they developed for the Lunar Settlements, Foster + Partners won NASA's 2017 3D-Printed Habitat Challenge with similar designs for dwellings on Mars. NASA's competition also aimed to

advance the technology employed to build sustainable housing solutions on and off Earth. The Mars Habitat uses research from previous designs to outline how a collaborative array of preprogrammed, semiautonomous robots will start to build the settlement before the arrival of the astronauts who will complete the work.   JRB

1   "Foster + Partners Works with European Space Agency to 3D Print Structures on the Moon," Foster + Partners website, January 31, 2013, www.fosterandpartners.com.

HYPERLOOP 2013– HYPERLOOP TRANSPORTATION TECHNOLOGIES (HYPERLOOPTT), CULVER CITY, CA

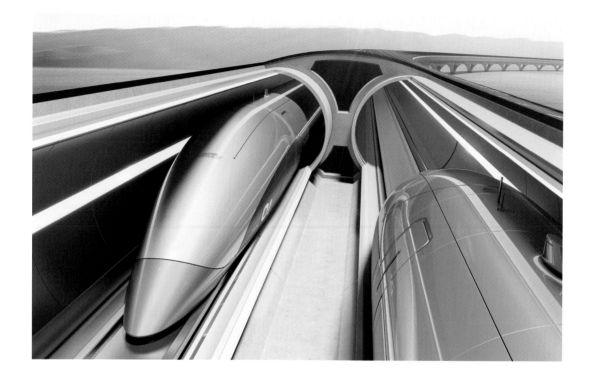

In August 2013, the South African–born technology designer and entrepreneur Elon Musk published a fifty-eight-page proposal on his SpaceX company blog for a new system of transportation that he called the Hyperloop.[1] Musk described it as a fifth mode of transport, after boats, cars, trains, and planes, intended for passenger and freight travel between cities less than nine hundred miles apart—as, for example, Los Angeles and San Francisco. The basic concept involves capsules or pods propelled at high speeds inside a nearly airless tube. Levitating magnetically on a layer of air, the pods are driven by electric motors and achieve extra acceleration through electro-magnets in the tube's walls, their high speed a result of the minimal friction and air resistance within the tube.

How the pod levitates and how it is propelled are two of the elements still being explored with different technologies, by start-up companies, students, and Musk himself, who open-sourced his original technical specifications. While Musk's Hyperloop concept for the future of freight and mass transportation was hailed by many as impossibly futuristic and thought to face overwhelming engineering, logistical, political, and economic challenges, none of its technical components lack historical precedent. The idea of passengers and goods traveling in airtight tubes was advanced before 1900,[2] pneumatic tunnels for rail travel operated in Great Britain in the mid-nineteenth century, and the American rocket pioneer Robert H. Goddard anticipated many of Musk's Hyperloop ideas with his vacuum-tube transportation system, which he proposed as early as 1909 but which was patented only in 1950, five years after his death.

The vision and promise of more efficiently moving people and cargo by means of the Hyperloop has inspired investors around the world. Hyperloop Transportation Technologies (or HyperloopTT), established in 2013, has since signed agreements to develop Hyperloop systems in Slovakia, Abu Dhabi, Indonesia, Korea, India, the Czech Republic, Ukraine, and China, and a feasibility study is underway to consider connecting Cleveland and Chicago, the first proposed use of such a system in the United States. **KBH**

1   Elon Musk, "Hyperloop Alpha," August 12, 2013, SpaceX, www.spacex.com/sites/spacex/files/hyperloop_alpha-20130812.pdf.

2   See Philip Strange, "Brunel's Atmospheric Railway," *Guardian*, April 4, 2014, www.theguardian.com.

Creating Our Sustainable Development Goals for Mars

DANIELLE WOOD on ethics, equality, and sustainability as an interplanetary species

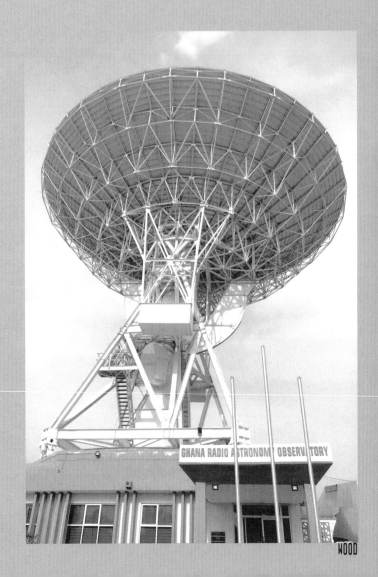

Radio Astronomy Observatory, Ghana, part of an international network of radio telescopes called the Square Kilometer Array, April 2018. Photograph by Danielle Wood

The Sustainable Development Goals, curated by the United Nations, provide an urgent vision for creating a safe and just society on Earth.[1] Technology from space offers one set of tools to help achieve the Sustainable Development Goals. We must also work to propose a vision for just communities beyond Earth as humans prepare to live as an interplanetary society.

> I'm a going to lay down my sword
>   and shield
> Down by the riverside
> Down by the riverside
> Down by the riverside
> Going to lay down my sword and shield
> Down by the riverside
> Ain't going to study war no more[2]

Technology that links people to outer space—satellites, rockets, telescopes, spaceships, and space stations—is available today in large part because of early investments in capabilities for military operations, national defense, and projects to bring prestige to individual countries. Later breakthroughs in space activity were motivated by scientific discovery and human curiosity for exploration as well. Since the early space era of the 1950s, the role of space technology has expanded dramatically. Today, space technology contributes to every facet of the infrastructure that enables our society, from transportation, communication, shipping, and energy to banking, construction, disaster response, and environmental management. Looking to the future, we will see a new relationship between humans and

space emerge as more commercial activity takes place in orbit around Earth, humans prepare to establish communities on the moon and on Mars, businesses investigate the money-making possibilities of harvesting materials from asteroids, and scientists uncover clues about the prospects of life beyond our home planet. As we prepare for this new cosmic future, in which space will continue to be woven into the fabric of human society, let us actively consider how we can reenvision space as a medium to advance justice for all rather than accrue benefits for a few.

The term *space* is used to describe the vast expanse of reality beyond Earth that humans at once fear and revere. When confronted with the enormous possibilities of space, humans should be reminded both of our insignificance and of our power to influence or destroy the systems around us. Today, climate change shows us the results of collective choices made by the human race that have caused long-term harm to people, animals, plants, and the planet. As we transition to living as an interplanetary species, we have the opportunity to learn from our experiences about the importance of caring for life and nature on Earth. We also have the opportunity to consciously choose to design a better culture to enjoy the next chapter of human learning on Earth and beyond. We must seize this opportunity before it slips by! Change will not happen automatically. Many people are inspired by the possibility of uncovering new mysteries by exploring space. We have the potential to achieve great change if we

1   See United Nations, "About the Sustainable Development Goals," www.un.org/sustainabledevelopment/sustainable-development-goals/.

2   "Down by the Riverside" dates to before the Civil War but was first published in 1918; for the version given here, see John Wesley Work, *American Negro Songs: 230 Folk Songs and Spirituals, Religious and Secular* (Mineola, NY: Dover Publications, 1998), 202–3.

CREATING OUR SUSTAINABLE DEVELOPMENT GOALS FOR MARS

3   See United Nations, "Transforming Our
        World: The 2030 Agenda for Sustainable
        Development" and "The 17 Sustainable
        Development Goals (SDGs) to Transform
        Our World," sustainabledevelopment.un
        .org/post2015/transformingourworld;
        www.un.org/development/desa/disabilities
        /envision2030.html.

4   "Follow the Drinking Gourd" dates back to
        the era of the Underground Railroad; it was
        first published in 1928. As with many folk
        songs and spirituals passed down orally, the
        lyrics vary; the version given here is that
        known to the author.

The Twin Peaks, modest-size hills to the southwest of the Mars Pathfinder landing site, September 1998. Photograph processed by Dr. Timothy Parker, NASA Jet Propulsion Laboratory

NASA Astronaut Karen Nyberg, the Expedition 36 flight engineer, works with the InSPACE-3 experiment in the Microgravity Science Glovebox in the Destiny laboratory of the International Space Station, August 1, 2013

harness this inspiration from space exploration and channel it to change our thinking about how humans live on Earth and in new locations. Let us decide to lay down our swords and shields and remake the space technology that was born of military urgency. Let us imagine a future of human communities in space that learn from history how to "study war no more."

Today's citizens of Earth face serious challenges that must be overcome in order to ensure that people of all backgrounds can live safely and flourish. One helpful summary of these challenges is the 2030 Agenda for Sustainable Development and the seventeen Sustainable Development Goals (SDGs), as developed through international dialogue curated by the United Nations.[3] The SDGs provide a holistic description of the progress that the global community must make to ensure equitable access to health care, economic opportunity, and political enfranchisement. The seventeen SDGs were developed through a process that included input from many types of people, including the young, the old, those with disabilities, and representatives from every UN member nation. SDGs 1 through 8 focus on basic human needs; 9 through 12 address global economic systems; and 13 through 17 highlight the need for globally minded leadership to reduce the harm from climate change and preserve biodiversity in the ocean and on land.

Underlying the seventeen Sustainable Development Goals are 169 aggressive, quantifiable Targets. The global community will need to use all the tools available to us to meet these Targets by 2030. There are six examples of space technology that have already been demonstrated to contribute toward achieving the SDGs by providing information, services, or useful technical capabilities. These six technologies are satellite Earth observation, satellite communication, satellite positioning, microgravity research, technology transfer, and fundamental scientific research. The following section examines several of these and their impact on sustainable development.

Follow the drinking gourd
Follow the drinking gourd

For the old man is waiting for to carry you to freedom
Follow the drinking gourd[4]

In the 1800s, runaways escaping slavery in the American South used the stars for navigation, such as by following the constellation the Big Dipper, known as the Drinking Gourd, in the northern sky. Humans have a long history of navigating using clues from the sky, such as the positions of the sun, moon, and stars. Today, satellite navigation systems help us identify our location and plan a route to our destination. Many cars, cell phones, and other electronic devices are equipped with radios that receive signals from satellite navigation systems. When a device receives positioning information from three or more navigation satellites, it is possible to calculate its location very accurately. Several national or international governments operate Global Navigation Satellite Systems, including the United States, the European Union, Russia, and China. These satellite systems provide both positioning and timing services that enable transportation, energy infrastructure, and even financial transactions, the latter often timed based on the highly precise clocks operated on navigation satellites.

In addition to navigation services, satellites offer other practical benefits to Earth's inhabitants via satellite observation services. Earth observation satellites provide information about changes on Earth's surface by taking pictures and measuring key features of the atmosphere, land, water, and ice. Earth observation satellites are designed with customized sensors that collect different types of energy in order to measure what is happening on the planet. For example, one set of satellites takes measurements using infrared and microwave energy that help us build computer models to predict the weather for the next few days. Another set of satellites uses radar and microwave energy to measure rainfall and snowfall. Yet another set uses infrared measurements to identify large sources of heat that may indicate fires. These and other environmental measurements are visualized and combined with information about human activity to help development leaders make decisions in support of SDG 14 (Life below Water) and SDG 15 (Life on Land).

Other societal benefits of space technology come from applying the knowledge and innovative ideas that result from activities such as human space flight, microgravity research, and fundamental scientific investigation in areas like astrophysics (study of the universe) and heliophysics (study of Earth's sun). Each of these activities can support the goals of sustainable development if the knowledge and skills derived from them are consciously transferred in support of development efforts. Consider the example of microgravity research. Since the year 2000, an international agreement between Russia, Canada, Japan, the United States, and several European countries has allowed people to live continuously on the International Space Station (ISS), which currently serves as the primary human destination in space. There are facilities on the ISS to support research experiments on the influence of the microgravity environment on the human body, plants, animals, and materials. Because the ISS is in orbit around Earth, the station itself and everything inside it are in a constant state of free fall, so the effects of gravity are not felt. For humans and animals, this causes physiological changes because bones and muscles are not working against gravity and the fluids in our bodies behave differently. Physical processes such as combustion, protein crystallization, and thermodynamics change. Researchers from all around the world send experiments to the ISS to investigate these changes, which inspires design innovations on Earth. For example, astronauts use exercise in space to keep their bones and muscles healthy; their experience is similar to that of people on Earth who face long periods of bed rest and must exercise to recover or maintain their strength. In addition, findings from studies on combustion, protein crystallization, and manufacturing are enabling potential new approaches on Earth for designing fire safety systems, medicines, and fiber-optic wire.

The Space Enabled research group within the MIT Media Lab is pursuing a mission "to advance justice in Earth's complex systems using designs enabled by space."[5] One way that Space Enabled is pursuing this mission is by redesigning the six space technologies to make them more useful in addressing the seventeen Sustainable Development Goals. Space Enabled builds relationships with development leaders who represent multilateral organizations, national governments, regional governments, native communities, universities, and entrepreneurs. When collaborating with these development leaders, Space Enabled utilizes a six-stage cycle to support their initiatives. This includes (1) applying DESIGN THINKING from engineering, business, and creative fields to understand the problems and needs of the collaborators; (2) practicing ART in response to the knowledge shared by the collaborators; (3) studying the context by drawing on methods from SOCIAL SCIENCES such as anthropology, history, and economics; (4) building COMPLEX SYSTEMS MODELS by using computer software to simulate interactions connecting human actions, environmental forces, and technology; (5) designing new SPACE SYSTEMS such as satellites that collect measurements about the environment; and (6) applying methods from DATA SCIENCE to create decision support systems that help pursue an SDG. In 2019, Space Enabled is working to apply this six-stage cycle in several projects, including working with government agencies and entrepreneurs in Benin and Ghana to monitor forests and invasive plants, working with city government leaders in Brazil to understand mangrove forest health, learning about goals for pursuing new space activities in Bermuda, and sharing ideas with national agencies on how they can apply satellite Earth observation data more effectively.

But *can* space-enabled designs advance justice and development? A graduate class I teach at the MIT Media Lab asks precisely this question. Before considering the potential role of space technology, the class reflects on the meaning of justice and development by reading the work of leading social scientists in these fields. For example, students are assigned the book *The Rise of "The Rest"* (2001) by the late Alice H. Amsden, who studied the experiences of countries in East Asia, the Middle East, and Latin America as they pursued industrialization in the decades after World War II. Another key text is *Stamped from the Beginning: The Definitive History of Racist Ideas in America* (2016), by Ibram X. Kendi of American University in Washington, DC. This book is used to explore one example of global-scale, systemic injustice in the form of racism. In his deeply researched text, Kendi shows how leaders who shaped the consciousness of each generation from the 1600s through today have used racist ideas to justify policies that led to systems of discrimination and injustice constructed along racial lines. Many of these racist ideas have been consistently held, with only slight variations, for more than four hundred years. The impact of the racist policies they inspired is deeply entrenched in the social, political, economic, and governing systems of the United States today. Kendi's book demonstrates how injustice resulting from social hierarchy is not primarily a problem of individual behavior but of systemic barriers built upon discrimination on the basis of race, gender, sexual orientation, national origin, immigration status, education level, income, or other human characteristics.

Why does a class seeking to apply space technology in support of the Sustainable Development Goals need to reflect on the writings of Amsden and Kendi? Before students work with development leaders and design systems using the six space technologies to support the SDGs, they need to ask themselves several challenging questions. The first of these is, What forces led to today's unequal society? The key texts assigned to the class help provide an answer through their emphasis on the roles of colonialism and racism, as powerful elites worked to maintain their hegemony at the expense of others. The second question the students are asked to consider is, How have long-standing forces such as colonialism, racism, gender inequality, and class structure formed our current societies—and how do these forces continue to create barriers to reaching the SDGs? Third, students are asked to examine the question, How have today's six space technologies grown out of the prevailing world, in which a few countries or groups dominate politically and economically?

5   Space Enabled overview, MIT Media Lab, www.media.mit.edu/groups/space-enabled/overview/.

The satellite Sputnik 1, the first human-made object to orbit Earth, launched on October 4, 1957

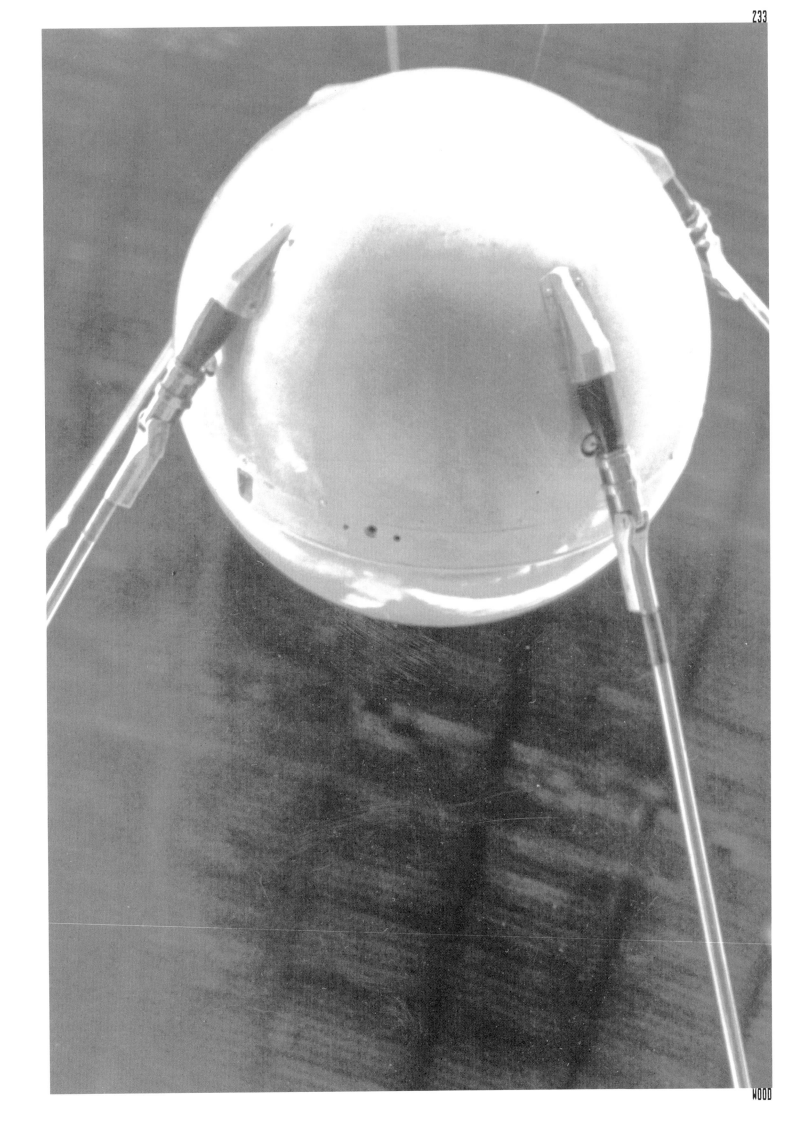

Space technology is the product of an unequal society, so students are invited to consciously imagine redesigning this technology to create an equal society. This redesign can be achieved by humbly listening to leaders who are working to achieve development and reduce the impact of social hierarchy in their own communities.

I'm gonna sit at the welcome table
I'm gonna sit at the welcome table
One of these days
I'm gonna sit at the welcome table
I'm gonna sit at the welcome table
One of these days[6]

Looking beyond the Sustainable Development Goals and the year 2030, now is also the time for humans to consider the interplanetary future before us. We can take a vital lesson from our history of living together on Earth: when we prioritize national or elite benefits over global well-being, we are acting in ways that harm current and future generations. The twentieth and twenty-first centuries have been defined by the initiatives of governments, nonprofits, and researchers seeking solutions to global challenges created by self-serving corporations and countries. These challenges include climate change, industrial pollution of our water and air, and systemic poverty for billions of people. As the lure of space beckons a new generation of humans to create societies

on the moon, in Earth's orbit, on asteroids, or on Mars, will we continue with our traditional patterns? Will the first companies and countries to land on these new space outposts set the local rules and create jurisdictions based on capitalist optimization or neocolonial expansion? This outcome is likely because the first people or robots to live and work on the moon or on Mars may represent a small set of companies or countries from Earth.

The Sustainable Development Goals can be applied now to consider how we want to build a future interplanetary society of humans beyond Earth. We don't need to wait until we have created problems that are almost impossible to solve, as we have done on Earth. Instead, we can put in place new guidelines, policies, and cultural norms that will establish our future space societies on foundations that support economic equality, gender equality, sustainable consumption, and sustainable communities. These ideas do not assume that future space societies will be utopias. Rather, they challenge us to grasp a narrow window of opportunity to consciously apply fresh thinking to future human societies in space, rather than allowing current human systems to become the default foundations.

What will it take for humans to consciously design future interplanetary societies based on the principles of the Sustainable Development Goals? It will require starting a movement

6 "I'm Gonna Sit at the Welcome Table" was written by Hollis Watkins during the civil rights era; for a partial quotation from that period, with slightly different lyrics, see Margaret Walker, *Jubilee* (Boston: Houghton Mifflin, 1966), 70.

that fosters reflection, dialogue, and action. It will take reflecting on the lessons of political economists such as Amsden and historians like Kendi, who teach us about the heritage of unjust human colonialism. It will require creating a global dialogue that invites people of many backgrounds to provide input, building on the lessons learned in creating the 2030 Agenda for Sustainable Development. It will necessitate deploying multidisciplinary design approaches, such as the six-stage cycle developed by the Space Enabled research group to learn through design thinking, art, social science, complex systems, engineering, and data science. It will require listening to those who have experienced the harms of violent colonial expansion, especially members of indigenous communities. Finally, it will entail inviting today's youth to envision designing interplanetary communities for their own future and for their children's future. What this will look like in practice is universities offering courses to invite students to study and debate how humans should live beyond Earth; government representatives continuing to dialogue on international space policy through established channels such as the United Nations Committee on the Peaceful Uses of Outer Space; and existing organizations such as the Space Generation Advisory Council continuing to invite students and young people all over the world to share their visions for human activity in space. It will look like governments and nonprofit organizations collaborating to host hackathons and to crowdsource the best ideas for the design of future interplanetary human societies. It will involve entrepreneurial companies forming in every region to propose novel business models that foster responsible and equitable human interactions in space.

Now is the time to start this movement and invite the next generation to help create a just, interplanetary human society built on the lessons of the Sustainable Development Goals. Let's ask the question, What type of society do we want to call home—on Earth, on the moon, on Mars, and beyond? 茻

The International Space Station, May 23, 2010.
Photograph by NASA/Crew of STS-132

236

The Fragile Frontier

COLIN FANNING on design in the
second space age

Writing soon after the climactic event of the twentieth-century space age—the United States' 1969 moon landing—the historian and critic Lewis Mumford argued in *The Pentagon of Power* that the concept of the "frontier" has been a driving force in Western thought and experience since at least the fifteenth century.[1] Fueled by desires to break free from geographic or historical boundaries, this perpetual search for a "New World utopia" is, in Mumford's critique, built on a technological and militaristic mindset—epitomized by the powerful symbol of the *Saturn V* rocket—that seeks to organize and control both the natural world and "the irreducible richness of human experience, and the inexhaustible promise of human potentialities [which are] not to be contained within any single system."[2] This same tension runs through our contemporary discourse on space travel, where narratives of progress and ingenuity sit alongside more difficult questions of access, equity, and ethics.

The rapid pace of scientific discovery and technological development in recent years has fostered a sense of exciting possibility for humanity's future in space. But this "second space age," as some have called it, also includes revenant strains of nationalism, territoriality, and militarization that uncomfortably parallel the Cold War frictions of the first space race.[3] Amid mounting evidence that a climate catastrophe awaits us on Earth, a note of urgency has crept into dialogues on space travel, shifting the tenor from the joys of discovery toward the imperative of survival.[4] As governments and private enterprises look to the stars as a site of scientific inquiry, an untapped trove of resources to exploit, or a home for some segment of humanity, the complexities of future-making design underscore the contested meanings of this new space age.

At the most personal scale, clothing and other wearables present challenging parameters for designers, who must grapple with the extreme inhospitality of space.

In his history of the Apollo space suit, Nicholas de Monchaux details how the suit made iconic by moon-landing photographs was initially rejected for its soft, baggy appearance. NASA engineers, however, ultimately had to abandon their favored rigid designs, whose sleek, machinelike forms presented numerous problems with fabrication, pressurization, and range of movement.[5] Many of today's forward-looking designs, such as Reebok's Space Boot SB-01 (see p. 247), build on the Apollo suit's enduring logic of soft, adaptive materials layered for resilience and redundancy.

Historically, space suits have been precisely fitted for specific wearers, a feasible process given the small number of people traveling into space. But if that number grows significantly, bespoke production could prove inefficient. One proposal for a next-generation space suit (the Z-2), from a NASA team led by Lindsay Aitchison, combines soft limbs with a rigid 3D-printed torso that is adjustable at the crucial waist and shoulder joints, offering the possibility of streamlined production and use.[6] Yet, as Aitchinson's intensive human-factors research reveals, the human organism is stubbornly resistant to standardization; treating spacefaring bodies as a regularized system of biometric data risks leaving behind those who don't conform to narrowly defined physical ideals, as disability advocates point out.[7] Space itself also changes the body, sometimes revealing the limitations of equipment decisions made on the ground. In March 2019, NASA had to cancel what would have been the first all-woman space walk when the astronaut Anne McClain discovered mid-mission that the suit torso initially prepared for her use on the International Space Station no longer fit correctly and she would need to use the same one as fellow astronaut Christina Koch.[8]

Given humans' inherent fragility, proxies like autonomous robots and AI represent an appealing design

Prototype for NASA's Z-2 space suit, 2014.
Photograph by NASA/Bill Stafford

direction for space agencies. Already, our most intrepid off-world explorers have been the robotic probes, landers, and rovers that have explored the lunar and martian surfaces and the deeper reaches of the solar system. Intended as our first ambassadors to the rest of the universe, the Pioneer and Voyager probes launched between 1972 and 1977 carried portraits of humanity on their plaques and golden phonograph records. The age of social media has fostered even more direct forms of personification, with space agencies tweeting in the chirpy first-person voices of their probes and landers.[9] A wave of new proposals for human proxies, such as NASA's Valkyrie robot prototype, further humanize the field of technological assistants with forms that suggest the charismatic droids of *Star Wars* rather than more sinister science-fiction AIs. Backing the design decisions behind these anthropomorphized machines is a broadly shared theory that astronauts can work more effectively with computer intelligences that borrow human characteristics, especially when so much of the high-stakes work of spaceflight is delegated to invisible software systems.

With such projects pointing toward a future of highly integrated human-machine interactions, the importance of data infrastructure for both space exploration and terrestrial life becomes clear. The invisible flows of information that structure our existence—communication, navigation, entertainment, weather tracking, geopolitical surveillance—rely on an aging satellite network that requires considerable funding and continual maintenance. Small-scale, modular designs like CubeSats (see p. 246) are beginning to put satellite technology within reach of a broader constituency, offering access to space research to those outside wealthy governments or elite research institutions. But for all the vastness of outer space, useful Earth orbits are a finite resource, and increasing orbital congestion represents a major problem for future efforts to reach the final frontier.[10]

No matter how advanced our orbital instruments or robotic explorers may become, the human desire to experience the universe for ourselves seems innate. Crewed spaceflight and habitation are long-term goals for many of the new generation of private space companies. Design is a powerful rhetorical tool for these young aerospace enterprises to promote their visions of space travel,

leveraging what de Monchaux calls the "visual seduction" of space.[11] Slick branding, computer-generated visualizations, and the spectacles of livestreamed launches and landings all lend a convincing presentness to the imagined futures these companies sell. Commercial craft—such as Virgin Galactic's sleek SpaceShipTwo, Bigelow Aerospace's Alpha Station, or a growing number of private space-station proposals—tend toward a greater aesthetic awareness as they build on the design precedents set by public space agencies. In some cases, these companies have commandeered the very infrastructure built for governmental efforts, claiming a continuity with past spaceflight legends while constructing their own futures. SpaceX, for instance, signed a twenty-year agreement with NASA to use the Kennedy Space Center's historic Launch Complex 39A (the launch site for the first moon landing) and renovated it with an elegant new gangway to match its corporate image.[12]

Though permanent off-Earth habitation is still only a prospective notion, the allure of otherworldly outposts has attracted major names in architecture and design, whether proposing methods for fabricating structures in situ, as in Foster + Partners' Lunar Settlements (see p. 227), or designing earthbound test beds like the planned Mars Science City outside Dubai by the Danish firm BIG-Bjarke Ingels Group. A more frankly speculative project, Sean Thomas Allen's *Platinum City* (see p. 249) points out one possible extreme of humanity's quest for resource extraction. Despite their futuristic cladding, moon bases, Mars colonies, and asteroid cities are in many ways the apotheosis of twentieth-century modernist dreams: technocratic cities dropped into an empty landscape, planned from a bird's-eye view as a unified aesthetic system, with no historical urban or social fabric to complicate the master vision.

Such proposals face immense challenges. Prolonged low- or zero-gravity living takes a physiological toll on the human body; hermetic environments make disease and injury major concerns; and long-term inhabitants of worlds like Mars—with thin atmospheres and weak magnetic fields—would live under the constant threat of radiation, dust storms, or human error and technical failures.[13] Even so, it's striking that most designs approach the problem of otherworldly dwelling through a technophilic lens. Rarely do they address the thorny

Rendering of BIG-Bjarke Ingels Group's Mars Science City, Dubai, 2017. The design uses vernacular architecture of the Middle East as a potential model for future Mars colonies.

Prototype for NASA's Valkyrie (R5) robot, 2013. Designed and built by the Johnson Space Center Engineering Directorate to compete in the 2013 DARPA (Defense Advanced Research Projects Agency) Robotics Challenge Trials

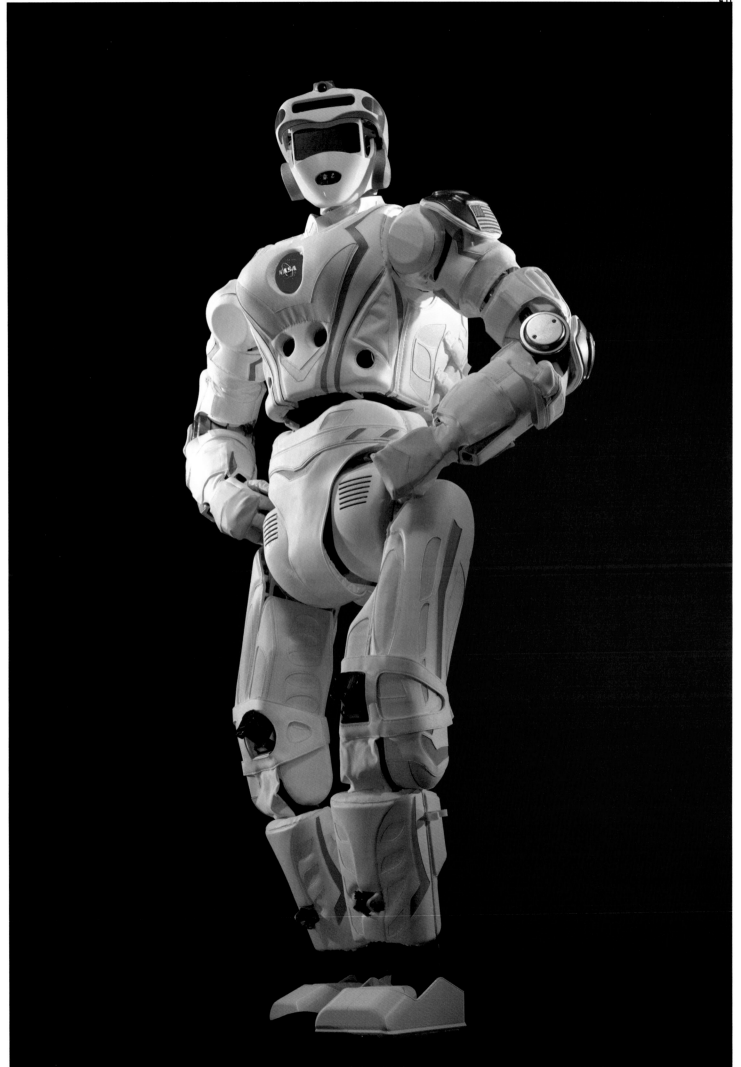

social, psychological, or philosophical questions like those Virginia Heffernan raises: "What would it feel like to live without Earth's sweet old topographical features? Without civilization inscribed into our stones and soil? What would happen to our bodies and brains, and would we even remember how much our essence as humans is an epiphenomenon of our beautiful, warm, blue and dying habitat?"[14] The artist Lisa Hartje Moura's project *Alien Nation* (see p. 245) is one reflective departure from typical colonization narratives, proposing space as a refuge for the dispossessed and nationless rather than a playground for the wealthy and adventurous—but the long-term viability of off-world living remains an open question.

Of course, the possibility of contact with alien life remains one of the great motivations for space exploration. We can't predict what forms such life might take, but design speculations can widen our imaginations beyond stage makeup and pointy ears; the 2016 film *Arrival*, for instance, explores potential modes of communication beyond our familiar verbal and textual vocabularies (see p. 248). Less frequently considered is how human exploration could affect alien life and its environment—particularly for life-forms that turn out to be microscopic or otherwise uncommunicative. Although space agencies follow "planetary protection" protocols to decontaminate equipment sent off-world, the process is imperfect and only a marginal part of space activity.[15] More broadly, the ethical implications of the imperialist attitudes involved in "colonizing" space are still largely unaddressed outside academic discourse.[16]

Ultimately, many of our narratives about space are really about exceptionalism—as individuals, as nations, as a species—and the triumph of human knowledge over hostile environments. The open-ended, non-narrative format of the 2017 video game *Everything* (see p. 244) attempts to impart a different lesson of interconnectedness. As "a magical playpen of being, rather than doing," the game embraces the beauty of our cosmic insignificance.[17] By contrast, the current paradigm of space design is largely built around *doing*, rooted in narratives of humans exercising their agency in and on outer space, rather than *being*, or considering how an interstellar existence might fundamentally change the conditions of humanness. If technical solutions continue

to overshadow the more demanding moral and cultural questions of space exploration, we may come to recognize, as Meghan O'Gieblyn argues, that "the billionaire space race is part of a larger war on symbolic spaces, one in which every utopia is literalized, its iconography appropriated and affixed a price tag."[18]

Design, at its most critical and human-centric, may be able to help expand discussions of space travel beyond the solely technocratic or utilitarian. But in the contexts of private and governmental enterprise, we face the increasing unlikelihood of the kind of communitarian, post-scarcity future envisioned in the more utopian strains of futurism. The arc of possibility seems to bend toward a fragmentary landscape of commercial competition, luxury tourism, resource extraction, and armed conflict—populated by Mumford's figure of the "encapsulated" human, whose autonomy is subservient to institutional masters.[19] How might we instead design a future in which extraterrestrial humans can live, grow, and create beyond narrow functionalist ends or nationalistic posturing? Is it possible to make room for the full diversity of humanity among the stars? Without design approaches that privilege these more fragile facets alongside the dominant ethos of technological advancement, our greatest off-world export ultimately may be the same earthly inequities our imaginings have tried to escape. ⌗

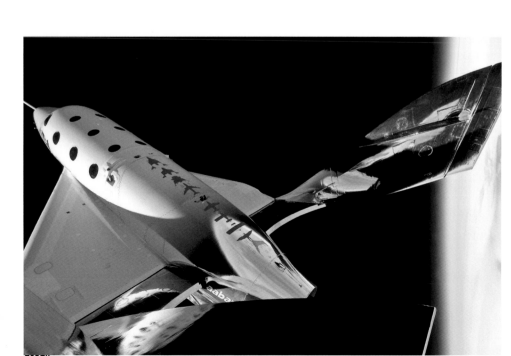

Burt Rutan and Virgin Galactic's VSS *Unity* (SpaceShipTwo), 2016, pictured during its third powered flight on July 26, 2018

Rendering of Bigelow Aerospace's Alpha Station in low Earth orbit. Bigelow has licensed NASA's canceled TransHab project, an inflatable system developed in the 1990s by Lockheed Martin and architect Constance Adams.

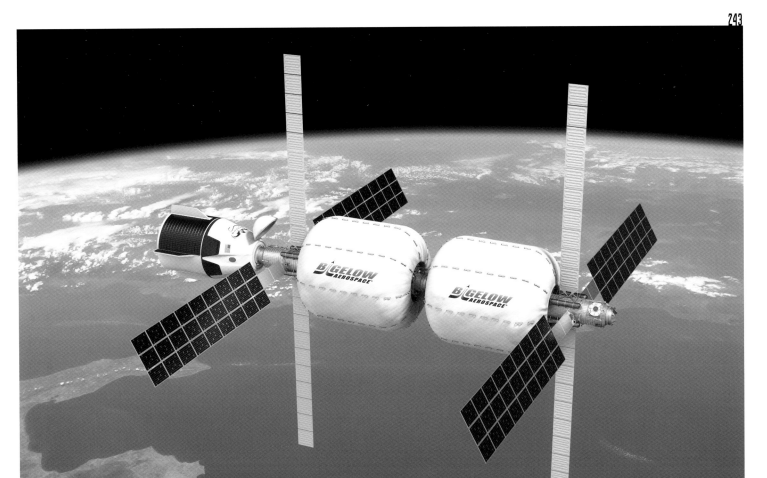

1   Lewis Mumford, *The Myth of the Machine*, vol. 2, *The Pentagon of Power* (New York: Harcourt Brace Jovanovich, 1970), 4.

2   Mumford, *Myth of the Machine*, 2:63.

3   See, for example, Todd Harrison and Nahmyo Thomas, "NASA in the Second Space Age: Exploration, Partnering, and Security," *Strategic Studies Quarterly* 10, no. 4 (Winter 2016): 2; and Garrett M. Graff, "The New Arms Race Threatening to Explode in Space," *Wired*, June 26, 2018, www.wired.com.

4   Elon Musk of SpaceX and Jeff Bezos of Blue Origin share this view; see Elon Musk, "Making Humans a Multi-Planetary Species," *New Space* 5, no. 2 (June 2017): 46.

5   Nicholas de Monchaux, *Spacesuit: Fashioning Apollo* (Cambridge, MA: MIT Press, 2011), 3.

6   Lindsay Aitchison, "Designing for Outer Space: Design, UX & HMI Development for Next Generation Space Suits" (PowerPoint presentation, National Aeronautics and Space Administration, May 17, 2017, ntrs .nasa.gov/archive/nasa/casi.ntrs.nasa.gov /20170004351.pdf).

7   See, for example, Ace Ratcliff, "Staircases in Space: Why Are Places in Science Fiction Not Wheelchair-Accessible?," *io9*, July 31, 2018, io9.gizmodo.com.

8   Matthew Cantor, "NASA Cancels All-Female Spacewalk, Citing Lack of Spacesuit in Right Size," *Guardian*, March 26, 2019, www .theguardian.com.

9   NASA's *Curiosity* rover (@MarsCuriosity on Twitter) is the paradigmatic example; others include the European Space Agency's ExoMars orbiter (@ESA_TGO) and the French-German-Japanese collaboration Mobile Asteroid Surface Scout (@MASCOT2018).

10 European Space Agency, *Annual Space Environment Report*, May 18, 2018, 2, www.sdo.esoc.esa.int/environment_report /Space_Environment_Report_latest.pdf.

11 De Monchaux, *Spacesuit*, 8.

12 See Anna Heiney, "Historic Launch Pad Is Liftoff Site for First SpaceX Crew Dragon," *NASA Blogs*, February 27, 2019, blogs .nasa.gov; and Robert Z. Pearlman, "SpaceX Adds New Astronaut Walkway to Historic NASA Launch Pad," Space.com, August 27, 2018, www.space.com.

13 Since the 1960s, a robust scientific litera-ture has explored the impacts of space travel on human health; see Leonard David, "Artificial Gravity and Space Travel," *BioScience* 42, no. 3 (March 1992): 155–59; and Leonard A. Mermel, "Infection Prevention and Control during Prolonged Human Space Travel," *Clinical Infectious Diseases* 56, no. 1 (January 2013): 123–30.

14 Virginia Heffernan, "An Infinite Space Utopia Can't Replicate Earth's Humanity," *Wired*, December 17, 2018, www.wired.com.

15 Committee on Space Research, *COSPAR Planetary Protection Policy*, October 20, 2002, amended March 24, 2011, cosparhq .cnes.fr/sites/default/files/pppolicy.pdf.

16 See Caroline Haskins, "The Racist Language of Space Exploration," The Outline, August 14, 2018, theoutline.com; and Kelly C. Smith and Keith Abney, eds., "Human Colonization of Other Worlds," special issue, *Futures* 110 (June 2019).

17 Colin Campbell, review of *Everything*, *Polygon*, March 21, 2017, www.polygon.com.

18 Meghan O'Gieblyn, "Objects of Despair: Mars," *Paris Review Daily*, March 27, 2019, www.theparisreview.org.

19 Mumford, *Myth of the Machine*, pls. 14–15.

Named one of the best video games of 2017 by the *New Yorker* and *Wired* and laureled at major design- and game-industry convocations (including the prestigious Ars Electronica Festival), *Everything* invites players into vibrant, algorithmically generated landscapes brimming with flora and fauna. Players begin as one of three thousand possible creatures and forms (humans are not included) and, as the game progresses, inhabit increasingly smaller entities. After reaching the subatomic level, players can then move more freely, assuming larger bodies, including moons, planets, and solar systems. If left idle, the game starts to generate its own play (creating what cultural critic Kat Brewster describes as "an ambient aquarium of universes"),[1] mimicking the way nature's dramas play out independent of human intervention.

The animations are deliberately simplified—for example, animals move by flipping forward instead of walking. While initially a function of the small design team and a consequent desire to keep production to the essentials, this logistical choice results in a whimsical aesthetic that players and critics have responded to positively. Snippets from hypnotic lectures given in the 1960s and 1970s by the

British American Episcopal priest–turned-philosopher Alan Watts accompany the game. Watts, who made Eastern philosophy accessible for a Western audience in the mid-twentieth century, speaks on topics ranging from reality to mortality, creating a trippy, transcendental soundtrack to the kaleidoscopic visuals. In her review of the game, Brewster describes one possible ludic path a player might take, and the cosmic existentialism that accompanies it: "A daisy creeps across a rocky landscape. It becomes a blade of grass, which, in turn, becomes a caterpillar, which then turns itself into a very miniature zebra."[2]

The game's success may be due in part to the human fascination with ordering the universe. Each object and form players inhabit is added to an encyclopedia that is completed once they have assumed all available forms. The game's designer, David OReilly, intended for players to see the world around them in new ways by considering what it might be like to partially inhabit other points of view.[3] This is the tension at the heart of the game—the friction between humans and the other life-forms and ecosystems around them, the Donna Harawayesque pursuit of empathy for nonhuman existence alongside the taxonomic

impulse to collect and interpret such experiences.[4] The possibility of encountering the world in all its wonders, scales, and shapes engenders an emotional response, an awareness of the nexus of life—and of its futures, the fates of which are simultaneously much larger than us and, just like the game controls, in the palms of our hands. **MMF**

1   Kat Brewster, "A Joyfully Expansive Dream of a Game," review of *Everything*, *Guardian*, March 24, 2017, www.theguardian.com.

2   Brewster, review of *Everything*.

3   Mitch Wallace, "Something about *Everything*: A Conversation with David OReilly," *Forbes*, March 27, 2017, www.forbes.com.

4   Haraway's scholarship has, in part, argued for a decentering of Western scientific approaches that construct knowledge of the world around us through a heavily anthropocentric, patriarchal, and capitalist lens. See, for example, Haraway, *Primate Visions: Gender, Race and Nature in the World of Modern Science* (New York: Routledge, 1990).

ALIEN NATION: PARADE 0   2017
LISA HARTJE MOURA, HEAD-GENÈVE

Alien Nation: Parade 0 is an installation comprising a five-minute film, two books, and futuristic, space-inspired costumes, scenography, and flags by the Portuguese designer Lisa Hartje Moura. Created as her 2017 thesis project in the master's program in space and communication at the Haute École d'Art et de Design (HEAD) in Geneva, Switzerland, the speculative project plays on the double meaning of the word *alien*. At once connoting something extraterrestrial and otherworldly (often as a feature of science fiction), the noun also describes foreigners or immigrants, those who find themselves on the wrong side of geopolitical boundaries without access to citizenship.

An open-ended and ongoing project, *Alien Nation* proposes a new sociopolitical model wherein belonging and territory are self-defined by individuals and collectives rather than dictated by nation-states and inflexible traditions. These identities are then manifested publicly in parades.[1] The accompanying books mine the past and speculate on the future of aliens working to emancipate themselves from present structures on Earth. The first is *A Pictorial Anthology of Alien Memorabilia*, which exhibits Moura's in-depth research into the portrayal of aliens of both kinds in cinema, television, and historical documents. The second book, *A Lyrical Anthology of Future History*, uses poetry to define the future aspirations of the new Alien Nation.

The white coveralls and bent-plexiglass masks covered in black elastic and neoprene geometric shapes obscure the bodies and faces within, hindering easy categorization or stereotyping. These ensembles, alongside the flags proclaiming new territory and the film depicting the journey ahead, all serve to play out the scenes imagined in the second tome. As Moura describes the project, "As alienated humans, we have been told we do not belong. But as sci-fi characters we have been told we came from Outer Space. So, as Aliens, we politely excuse ourselves from Earth and begin our journey to return to the home science fiction gave us, by establishing the Moon as our first step to the final frontier."[2] MMF

---

1   This is another play on words, alluding to the British English *identity parade* (a police lineup, in American English) as well as the traditional meaning of *parade* as a ceremonial procession.

2   "Alien Nation: Parade 0," Lisa Hartje Moura's website, lisahmoura.com.

EXOCUBE 1:1 SCALE MODEL   2019
POLYSAT, CALIFORNIA POLYTECHNIC STATE
UNIVERSITY, SAN LUIS OBISPO

Since Sputnik I—the first artificial satellite—was launched in 1957, over 2,500 satellites have been sent into space. Most of these satellites have been massive in scale, comparable in size to city buses, and with antennae spanning as wide as football fields. Today, as part of a growing trend toward miniaturization, a new fleet of smaller satellites roam the solar system. Aptly named *CubeSats*, these satellites are between ten and sixty centimeters cubed (closer in size to a loaf of bread than to an automobile) and weigh up to eight kilograms (eighteen pounds).[1] Though smaller in size, CubeSats don't have diminished value. Their modest scale makes their production more affordable; in turn, they are accessible to a wider group of users.

Launched in 2015, the ExoCube is a space-weather satellite that measures the density of hydrogen, oxygen, helium, and nitrogen in the upper atmosphere to verify climate models. The ExoCube was designed by PolySat, a student-run, multidisciplinary research lab at California Polytechnic State University with team members from a wide range of fields, including engineering, physics, business, journalism, and graphic design. Programs like PolySat allow research institutions to perform experiments and participate in space exploration using CubeSats.

Until recently, the size and cost of commercial satellites limited their use in scanning regions for the effects of environmental change. With fewer constraints on their use, CubeSats like ExoCube can more regularly and comprehensively observe Earth's environment by evaluating natural resources, assessing crops, and mapping cities. Their accessibility enables wider use and therefore closer monitoring of climate-change events, such as floods, earthquakes, volcanic eruptions, and fires. For example, during a string of catastrophic wildfires in late 2018, the California governor's office deployed CubeSats to map the spread of the fires and assess the damage to urban areas. And there are other projects on the horizon: the Ghanaian government recently launched its first satellite (a CubeSat), and there are hopes that its next CubeSat will include high-resolution cameras to track illegal mining and deforestation.[2]   **JRB**

1   See "CubeSats Overview," NASA website, last updated February 14, 2018, www.nasa.gov.

2   Jake Bright, "Africa Has Entered the Space Race, with Ghana's First Satellite Now Orbiting Earth," TechCrunch, August 6, 2018, techcrunch.com.

SPACE BOOT SB-01 2017 MATT MONTROSS AND MATT BOUDREAU, REEBOK, BOSTON

Space Boot SB-01, introduced in 2017, is Reebok's first footwear for use off-Earth. Designed in collaboration with the David Clark Company, it will be worn with the new space suit Clark has produced for Boeing's crewed mission to the International Space Station (ISS) with its Crew Space Transportation (CST)-100 Starliner spacecraft, planned for August 2019. Creating a new boot for astronauts shuttling to and from and working aboard the ISS took about two and a half years. Clark's brief to Reebok was to develop a hybrid of a running shoe/sandal/wrestling boot/aviator boot that would be lighter, more flexible, and more comfortable than the leather boots with firm soles traditionally worn by astronauts. Reebok integrated its

new Floatride foam—introduced earlier in 2017 in the Floatride Run line of running shoes—into the space boot.

The precisely engineered Floatride foam has a consistent cell structure that provides lightweight, soft, responsive support, according to Matt Montross of Reebok Innovation, with weight being a primary factor—"just a single pound having big financial implications."[1] Astronauts tested the calf-high boots by climbing in and out of the spacecraft capsule and running, jumping, and wearing them over long periods of time with their pressurized space suits. Unlike traditional space boots, the SB-01 was designed to be part of the space suit: it is built with a flame-resistant shell and an interior mesh that

stretches and then locks when the suit inflates. The space suit and the boot are also, for the first time, united by a shared aesthetic. In the initial version of SB-01, the boot shaded from white on the foot to blue at the top to integrate with the new space suit astronauts will wear on the mission. In this latest iteration, the entire boot is blue, save for the zipper, logo, and lower sole, with its stylish chevron pattern. **KBH**

1   Danil Boparai, "Reebok Unveils Lightweight Floatride Space Boot for Astronauts," Dezeen, July 20, 2017, www.dezeen.com.

LOGOGRAMS FROM THE FILM *ARRIVAL* 2016
MARTINE BERTRAND AND PATRICE VERMETTE

In the 2016 film *Arrival,* twelve forebodingly large spacecraft suddenly appear over disparate locations on Earth, provoking fear, wild speculation as to motive, and government and military responses by the world's superpowers.[1] The seven-legged alien "heptapod" lifeforms then begin to produce large, circular, logographic symbols, which appear in milky black rings, like ink extruded by giant squid into the ether. The film's protagonists—Louise Banks, a professor of linguistics, and Ian Donnelly, a physicist—are drafted by the United States government to translate the symbols and communicate with the extraterrestrials.

The *Arrival* producers consulted with a McGill University linguistics specialist as well as code analysts during the film's development. But it was the film's production designer, Patrice Vermette, and his wife, the artist Martine Bertrand, who developed the vocabulary of one hundred symbols that became the film's visual signature. Bertrand sketched the symbols one night, working with paper and ink, and Vermette translated the results into the segmented graphic language that Banks and Donnelly race to interpret as tensions mount on Earth. The logograms can communicate a single word or a more complex phrase, depending on their structure, with the thickness of the swirls communicating both tone and urgency.

One of the phrases translates ambiguously. Banks suggests it means "offer tool," but other world authorities interpret it as an antagonism, and conflict occurs. Banks finally ascertains that the heptapods have come in peace to offer a gift to humankind: their circular language allows them—and the humans who learn to speak it—to see into the past and the future, a complete subversion of linear time. Banks manages to diffuse the global standoff against the twelve spacecraft by using this tool, which allows her to foresee an event at the United Nations during which the Chinese prime minister thanks her for persuading him not to attack the spaceships, a feat she had managed by reciting his deceased wife's dying words to him.

The film goes beyond a traditional aliens-meet-humans scenario, its complex graphic language a powerful symbol for the ways in which we might—or might not—shape our collective morality, as well as our own personal choices, more carefully if we could foresee the consequences of our actions. The tense, deeply moving, profound exchanges between the human and alien interlocutors challenge the anthropocentric idea that linguistic communication is a uniquely human trait, and explore the question of how confrontations with life other than our own might unfold. **MMF**

1   *Arrival*, by Paramount Pictures, was directed by Denis Villeneuve and produced by Shawn Levy, Dan Levine, Aaron Ryder, and David Linde. The screenplay by Eric Heisserer was based on the novella "Story of Your Life," by Ted Chiang (1998).

PLATINUM CITY 2014 SEAN THOMAS ALLEN

Platinum City—with its towering spires, deep chasms, and filigree of spiraling ramps and conduits, all built into an asteroid in orbit around Earth's moon—originated as the architect Sean Thomas Allen's thesis project at the University of Greenwich School of Architecture in London. Awarded an innovation prize by the Jacques Rougerie Foundation and recognized in other design competitions (including *Eleven* magazine's Moontopia competition and the 2018 Laka Prize), the project highlights the power of the image in speculative design. A series of detailed black-and-white renderings and technical drawings depict a future industrial city built around the mining and refining of platinum, a metal with numerous applications, from chemical and electronics manufacturing to medicine.[1]

Modeled after densely layered computer circuitry and the hollow structure of bird bones, Allen's fractal-like design posits the post-human city as a kind of "organic computer" where biological and mechanical processes work in concert to form an adaptive, ever-changing urban fabric without the need for human labor or intervention. Its logic depends partly on current discussions in space technology—for example, habitats 3D printed by robots from regolith (see p. 227)—but also suggests more frankly speculative kinds of engineering. Carbon-nanotube cables reach from the city into space to transfer mined platinum, and the city itself is powered by both solar energy and a population of genetically modified electric eels living in subterranean reservoirs.[2]

Allen credits Jules Verne's 1865 novel *From the Earth to the Moon* as a conceptual guidepost, but he also builds on recent political and economic debates on space. The United States' 2015 Space Act authorized the extraction of extraterrestrial resources by private companies—a move some have criticized as a rejection of international collaboration and public good, leaving space to be plundered through commercial or military occupation.

Platinum City explores the potential ramifications of such policies, combined with our increasing reliance on automatic technologies.

The narrative of *Platinum City*, though cast in the confident rhetoric of an architectural manifesto or developer's brochure, also contains subtle shades of critique. The notions of extreme machine intelligence, self-sufficient architecture, and urban sentience uproot our assumptions about the importance of human agency in space exploration. *Platinum City* raises an unsettling question: In environments self-constructed by algorithms and machines, will humans be necessary—or even welcome? **CF**

1   "Jacques Rougerie—Grand Prix: Innovation in Space Architecture 2018," Sean Thomas Allen's website, www.seans.me. See also "Platinum City," the Laka Architektura website, lakareacts.com.

2   "Jacques Rougerie—Grand Prix."

MUNDANE AFROFUTURIST MANIFESTO   2013
MARTINE SYMS

Martine Syms, a designer, artist, and filmmaker who describes herself as a "conceptual entrepreneur," examines expressions of black identities in mass media and popular culture. Her Mundane Afrofuturist Manifesto is a thought-for-thought rewrite of the 2004 Mundane Manifesto, written by a workshop of "pissed off" science-fiction authors who posited a type of science fiction that eschews impossible ideas in favor of radically plausible ones.[1]

Manifestos are common tools in creative fields, signaling resistance to dominant narratives through exaggerated critique. The Mundane Manifesto advises science-fiction authors to avoid the hackneyed tropes and scientifically unlikely aspects of typical galaxy-hopping sci-fi—faster-than-light travel, humanoid aliens, instant communication over vast distances, and quantum hijinks, for example—and instead create a sense of wonder in the here and now. By embracing constraints and avoiding prevalent narrative shortcuts, the resulting stories are formally tighter; by situating their science fiction in a plausible future,

the authors model a responsibility toward the present in a genre that often promotes a false sense of cosmic and earthly abundance. As the manifesto states, "The most likely future is one in which we only have ourselves and this planet."

Where the 2004 manifesto skewered the ecologically irresponsible escapism of science fiction and its techno-optimistic paradigm, Syms's text deftly marries the language of futurity with the heavy history of racial oppression to tackle the seeming impossibility of the existence and survival of black futures within a system of white oppression. Framing the conversation of progress through the lens of culture, identity, power, and injustice, Syms's provocative and lyrical manifesto rejects futures that rely on "normative, white validation" and instead advocates for mundane Afrofuturism as "the ultimate laboratory for worldbuilding outside of imperialist, capitalist, white patriarchy." Syms's mundane future recognizes that the lies of egalitarian cyberspace, cosmic origin stories, and breezy postracial

futures epitomized by utopian-leaning science fiction like *Star Trek* run counter to the majority of black experiences and expectations.

Syms also turns her critical gaze toward the aesthetic clichés of Afrofuturism itself, arguing instead for more quotidian wonders of blackness. Echoing the original manifesto's desire to embed our creative visions of the future in the realities of the present, she highlights "the sense that the rituals and inconsistencies of daily life are compelling, dynamic, and utterly strange." In doing so, she reminds us that visions of the future, while often inspiring and transformational, might reify constructs that impede the productive reimagining of even more vital alternative presents. **EB**

1   The Mundane Manifesto was written and signed by Geoff Ryman and the attendees of the 2004 Clarion West Writers Workshop. The full text of the manifesto is available at *SFGenics: Notes on Science, Fiction, and Science Fiction* (blog), July 4, 2013, sfgenics.wordpress.com.

MUNDANE AFROFUTURIST MANIFESTO

The undersigned, being alternately pissed off and bored, need a means of speculation and asserting a different set of values with which to re-imagine the future. In looking for a new framework for black diasporic artistic production, we are temporarily united in the following actions.

THE MUNDANE AFROFUTURISTS RECOGNIZE THAT:

We did not originate in the cosmos.

The connection between Middle Passage and space travel is tenuous at best.

Out of five hundred thirty-four space travelers, fourteen have been black. An all-black crew is unlikely.

Magic interstellar travel and/or the wondrous communication grid can lead to an illusion of outer space and cyberspace as egalitarian.

This dream of utopia can encourage us to forget that outer space will not save us from injustice and that cyberspace was prefigured upon a "master/slave" relationship.

While we are often Othered, we are not aliens.

Though our ancestors were mutilated, we are not mutants.

Post-black is a misnomer.

Post-colonialism is too.

The most likely future is one in which we only have ourselves and this planet.

THE MUNDANE AFROFUTURISTS REJOICE IN:

Piling up unexamined and hackneyed tropes, and setting them alight.

Gazing upon their bonfire of the Stupidities, which includes, but is not exclusively limited to:

- Jive-talking aliens;
- Jive-talking mutants;
- Magical negroes;
- Enormous self-control in light of great suffering;
- Great suffering as our natural state of existence;
- Inexplicable skill in the martial arts;
- Reference to Wu Tang;
- Reference to Sun Ra;
- Reference to Parliament Funkadelic and/or George Clinton;
- Reference to Janelle Monáe;
- Obvious, heavy-handed allusions to double-consciousness;
- Desexualized protagonists;
- White slavery;
- Egyptian mythology and iconography;
- The inner city;
- Metallic colors;
- Sassiness;
- Platform shoes;
- Continue at will...

WE ALSO RECOGNIZE:

The harmless fun that these and all the other Stupidities have brought to millions of people.

The harmless fun that burning the Stupidities will bring to millions of people.

The imaginative challenge that awaits any Mundane Afrofuturist author who accepts that this is it: Earth is all we have. What will we do with it?

The chastening but hopefully enlivening effect of imagining a world without fantasy bolt-holes: no portals to the Egyptian kingdoms, no deep dives to Drexciya, no flying Africans to whisk us off to the Promised Land.

The possibilities of a new focus on black humanity: our science, technology, culture, politics, religions, individuality, needs, dreams, hopes, and failings.

The surge of bedazzlement and wonder that awaits us as we contemplate our own cosmology of blackness and our possible futures.

The relief of recognizing our authority. We will root our narratives in a critique of normative, white validation. Since "fact" and "science" have been used throughout history to serve white supremacy, we will focus on an emotionally true, vernacular reality.

The understanding that our "twoness" is inherently contemporary, even futuristic. DuBois asks how it feels to be a problem. Ol' Dirty Bastard says "If I got a problem, a problem's got a problem 'til it's gone."

An awakening sense of the awesome power of the black imagination: to protect, to create, to destroy, to propel ourselves towards what poet Elizabeth Alexander describes as "a metaphysical space beyond the black public everyday toward power and wild imagination."

The opportunity to make sense of the nonsense that regularly—and sometimes violently—accents black life.

The electric feeling that Mundane Afrofuturism is the ultimate laboratory for worldbuilding outside of imperialist, capitalist, white patriarchy.

The sense that the rituals and inconsistencies of daily life are compelling, dynamic, and utterly strange.

Mundane Afrofuturism opens a number of themes and flavors to intertextuality, double entendre, politics, incongruity, polyphony, and collective first-person—techniques that we have used for years to make meaning.

THE MUNDANE AFROFUTURISTS PROMISE:

To produce a collection of Mundane Afrofuturist literature that follows these rules:

1. No interstellar travel—travel is limited to within the solar system and is difficult, time consuming, and expensive.
2. No inexplicable end to racism—dismantling white supremacy would be complex, violent, and have global impact.
3. No aliens unless the connection is distant, difficult, tenuous, and expensive—and they have no interstellar travel either.
4. No internment camps for blacks, aliens, or black aliens.
5. No Martians, Venusians, etc.
6. No forgetting about political, racial, social, economic, and geographic struggles.
7. No alternative universes.
8. No revisionist history.
9. No magic or supernatural elements.
10. No Toms, Coons, Mulattoes, or Bucks.
11. No time travel or teleportation.
12. No Mammies, Jezebels, or Sapphires.
13. Not to let Mundane Afrofuturism cramp their style, as if it could.
14. To burn this manifesto as soon as it gets boring.

—Martine Syms & whomever will join me in the future of black imagination.

REFERENCE

Glossary

affective computing:
An interdisciplinary field including engineering, computer science, psychology, cognitive science, and big data that seeks to enable robots to interact with humans through interpreting and simulating their emotions. Sometimes called *artificial emotional intelligence*.

Afrofuturism:
Coined by the cultural critic Mark Dery. A field of literature, music, art, and philosophy that uses science fiction or futuristic narratives to examine themes of the African diaspora and black histories and cultures.

AI:
An abbreviation for *artificial intelligence*. The simulation of human intelligence by machines, especially computer systems, through the ability to perform tasks that require processes such as speech recognition, decision-making, learning, and reasoning. *See also* Turing test.

Anthropocene:
The proposed geological time period defined by the influence of human activities on Earth, materially, culturally, and environmentally.

app:
An abbreviation for *application*. A software program, most often used on a mobile device, that performs a specific function.

associative design:
A design practice that combines a straightforward approach to materials with an inventive approach to fabrication processes and a focus on artistic speculation rather than design for production.

autonomous robots:
Robots that can navigate and perform tasks without direct human control (e.g., Roomba).

binder jetting:
A 3D-printing process in which a liquid binding agent is deposited on a bed of powder or granular particles (made of metals, sand, ceramics, foodstuffs, etc.) following a computer program, bonding them together to form an object, part, or food, layer by layer, without using heat.

biodesign:
The use of or collaboration with living organisms in design (e.g., plants as architectural structures, materials made with bacteria).

biofabrication:
The fabrication of materials and products using living organisms, cells, and tissues. In biomedical research, biofabrication includes tissue engineering, regenerative medicine, bioprinting, and bioassembly. Medical products created through biofabrication include titanium jawbones, cochlear implants, and thought-controlled bionic limb replacements.

biohacker:
One who performs biological experimentation (often on their own body) outside traditional laboratory or medical settings with the intent of improving the quality or capabilities of living organisms.

bioink:
A suspension of living human cells that can be printed on a surface along with a matrix material to create a scaffold for regeneration and repair of tissue. *See also* bioprinting.

biomimicry:
(1) In design, the copying of functional forms from nature in human-made products (e.g., burr hooks translated into fasteners for Velcro; friction-reducing properties of sharkskin incorporated into high-tech swimsuits). (2) In science, the imitation of biological materials or their functions (e.g., replicating organ structures and their environments within the body on a microchip for research purposes).

biopolymer:
Large molecules produced by living organisms that are composed of many repeating small structural units (e.g., RNA, DNA, cellulose, rubber).

bioprinting:
The (currently experimental) 3D printing of tissue and organs using bioink. *See also* biofabrication.

bioprospecting:
Systematic research and harvesting of plants and animals to find new sources (e.g., chemical compounds, genes) that could be used for medicines and other beneficial products, often for commercial gain.

bioreceptive:
The ability of an environment to host microorganisms and nurture biocolonization.

biotechnology:
The use of biological processes, organisms, or systems—either in their natural form or in an altered version through breeding or genetic engineering—for the benefit of humankind. *See also* clean meat; synthetic bacteria; synthetic biology.

bitmap printing:
The printing of a computer image file, such as one constructed of pixels (e.g., a JPEG or TIFF).

blockchain:
A digital public ledger system that records online transactions openly and verifiably; can be used by virtual currencies, or *cryptocurrencies* (e.g., Bitcoin).

carbon nanotube:
A cylindrical carbon molecule with walls made of graphene (one-atom-thick sheets of carbon) that is strong, lightweight, elastic, and has excellent thermal and electrical conductivity properties; could be used in the medical field (e.g., in drug-delivery systems, biomedical devices, or implants that take advantage of its hollow cavities).

cellular agriculture:
Growing agricultural products (e.g., coffee, vanilla) using a microorganism, such as yeast or bacteria, or making cell-based alternatives to animal products (e.g., clean meat).

circular economy:
A system aimed at minimizing waste and maximizing resources through processes such as durable design, repair, reuse, and recycling.

clean meat:
Animal protein grown in a laboratory from in vitro cultivation of animal cells using processes from tissue engineering, biotechnology, and molecular biology. *See also* cellular agriculture; disembodied agriculture.

CRISPR:
An abbreviation for *clustered regularly interspaced short palindromic repeats*. A molecular-level gene-editing technology that can alter or modify gene function; used in fields like medicine, biology, agriculture, and animal husbandry to improve climate tolerance or disease resistance, correct genetic mutations, or make organ transplantation between species possible. Can be used as shorthand for *CRISPR-associated 9*. *See also* genetically modified organisms (GMOs).

CRISPR-associated 9:
CRISPR technology that makes use of the protein Cas9, an enzyme that can recognize specific DNA strands and edit or cut them. Also called *CRISPR-Cas9*.

critical design:
Coined by the designer Anthony Dunne. A thought process that challenges preconceptions and expectations of the present to provoke new ways of thinking about an object and its environment. *See also* speculative design.

cryonics:
The practice of human preservation through freezing the recently deceased at very low temperatures (cryopreservation) with the hope of revival and continued life at a future time.

cultured meat:
See clean meat.

CV:
An abbreviation for *computer vision*. An interdisciplinary computer field for automatically processing and analyzing digital images. Used, for example, in autonomous vehicle programming, medical-image analysis, facial recognition, and biometrics.

defuture:
Coined by the designer, theorist, and philosopher Tony Fry. To negate sustainability due to the inability to move beyond current ideas and methods, theoretically resulting in the lack of a future for the world.

diachronic:
Concerned with phenomena, such as language or culture, as they occur and change over time.

digital fabrication:
Computer-controlled design combined with digital manufacturing processes.

digital governance:
A framework for establishing standards, strategies, policies, accountability, and oversight for an organization's digital presence (e.g., websites, internet, mobile sites, and other digital services).

disembodied agriculture:
Animal protein or plant-based food produced outside of sources found in nature (e.g., clean meat; plants grown in a controlled, engineered environment).

DIY:
An abbreviation for *do-it-yourself*. A practice or culture of self-sufficiency; the ethic of not requiring experts to complete a task. In art and design, the circumvention of professional studios and commercial manufacturing to create and distribute high-quality work.

drone:
An uncrewed aerial vehicle (UAV) that can be remote controlled or that can fly autonomously through the use of embedded computer programs.

electromagnets:
Magnets in which the magnetic field is created by interactions between electrically charged particles in a wire coiled around a magnetic core, such as iron. Uses of electromagnets include MRI scanners, hyperloops, and doorbells.

entomophagy:
The consumption of insects (generally by humans).

exaflop:
In computing, one billion billion calculations per second.

exoskeleton:
(1) An artificial external supporting frame for the human body that robotically simulates and/or enhances body movement (e.g., enables a wearer to carry heavy loads without fatigue or injury; enables a disabled person to walk or move by simulating body movement). (2) The external supportive covering of an invertebrate animal.

fractal:
A complex, never-ending pattern that repeats itself at different scales. Fractals occur in nature (e.g., branching rivers, trees, blood vessels) and in mathematics (geometric figures produced by calculating a simple equation over and over).

fused deposition modeling:
A 3D-printing process in which a plastic filament is heated to its melting point and extruded in layers that fuse together to build an object.

futurist:
A person who imagines possibilities for the future.

genetic algorithm:
A method of finding the best or most fitting solution for a given computational problem through methods that imitate natural biological or reproductive processes.

genetic engineering:
The artificial manipulation of genes using biotechnology to modify an organism or group of organisms. *See also* biotechnology; posthuman; synthetic biology.

genetically modified organisms (GMOs):
Organisms whose genetic material has been modified in some way using gene-editing techniques, such as CRISPR.

Gulf Futurism:
Coined by the musician Fatima Al Qadiri and the artist and writer Sophia Al-Maria. An aesthetic that builds on the oil-rich Gulf region's youth subculture (with its focus on modern technology and entertainment), desert landscape populated by soaring skyscrapers, adoration of kitsch (e.g., the ubiquity of the shopping mall), and ultraconservative religious and social codes to critique a dystopian future-turned-reality.

hack:
(1) In computing, to circumvent barriers to make a point, share information, or subvert code. (2) In design, to take apart or deconstruct an old or iconic design and rethink and reconfigure it into a new design or design statement.

hackathon:
Short for *hack marathon*. A group meeting or workshop of designers, engineers, programmers, and others, sometimes including subject experts and laypeople, to develop a new, functioning program or product.

haptic:
Feedback based on the sense of touch; interaction between a person and a tactile user interface (e.g., phone vibration, braille, pressure exerted by a prosthetic device).

imaginary (*noun*):
The system of meanings or collective picture that governs a given society or social structure in its time period.

Kansei engineering:
A method of translating human impressions and feelings into product design and performance in order to improve them.

LiDAR:
Short for *light detection and ranging*. A remote sensing method that uses pulsed laser light to measure distance and gather information for the production of precise 3D representations (e.g., maps, models, or other visuals). Used by self-driving cars.

liquid-metal printing:
A 3D-printing process in which liquified metal droplets are propelled by an electromagnetic field to precisely produce a solid object very quickly, accurately, cheaply, and with almost no waste.

logogram:
A symbol, sign, or character that represents a word or phrase (e.g., Chinese characters, Egyptian hieroglyphics).

material ecology:
Coined by the architect and designer Neri Oxman. A cross-disciplinary approach at the intersection of biology, materials science, engineering, and digitally controlled 3D manufacturing that focuses on the relationship between products, buildings, systems, and their environment to produce environmentally friendly structures, clothing, and products (e.g., photosynthetic building facades, plastics made from corn, clothing with wearable microbiomes).

microfluidic technology:
A multidisciplinary field that can design systems that process very small volumes of fluids with precise control and manipulation on a submillimeter scale. Uses a nanoscale chip etched with microchannels for fluid or gas to pass through.

nanopore technology:
The process of passing an ionic current through a nanoscale hole (pore) and measuring changes in the current as biological molecules pass through or near the nanopore; the change in current can be used to identify the molecule. Used for simple and fast genome sequencing.

neural network:
In information technology, an artificial learning technology made of densely interconnected computer processors working simultaneously in parallel to mimic the operation of the neurons and nerves in the brain. Can be used for complex data analyses and applications and for solving complex pattern recognition problems (e.g., handwriting recognition, facial recognition, speech-to-text transcription).

objectile:
A concept that replaces the idea of a solid object with a mathematical function that allows the object to be thought of as having endless variations of shape.

panopticon:
(1) A circular institutional building (e.g., prison, hospital, library) that allows for continuous observation of its occupants by a single person or CCTV. (2) By extension to the digital realm, the systems that make possible surveillance or monitoring of information transfer through computers and networks.

post-human:
A condition in which humans and intelligent technology have become more intertwined as the focus of humanness shifts from the human as a body to a set of information patterns. Also used to refer to the modification of human capabilities through advances in science and technology, such as genetic engineering.

programmable materials:
Materials (e.g., carbon fiber, printed wood grain, textile composites, rubbers, plastics) that are programmed to form or assemble themselves into objects when exposed to an external stimulus (such as water, heat, light, or air pressure), but are cost effective and can be shipped or packaged flat and unformed.

regolith:
The layer of loose, mixed deposit—dust, soil, pebbles, etc.—that covers solid rock on the surfaces of planetary bodies such as Earth, Mars, the moon, and some asteroids.

render farm:
A group of networked computers devoted to producing image sequences for computer-animated films.

roboethics:
The area of study concerning what rules should be made to ensure the ethical creation, use, and behavior of robots.

robotic fabrication:
Manufacture by a preprogrammed device. Can be used in factories for repetitive manufacturing tasks or for tasks that require extreme precision.

rotoscoping:
A visual-effects technique in which an object is traced over and copied from one scene so that it can be reproduced on another background. Used to create live-action scenes in movies.

selective sintering:
A 3D-printing process that fuses granular or powdered material into a desired shape, layer by layer, using heat but without melting the granules.

self-assembly:
A process in which individual parts, components, cells, etc. move autonomously in an uncontrolled environment toward a final assembled state in which each part has a precise position.

Singularity, the:
The hypothetical future transformation of society that occurs when superintelligent machines with the ability to self-direct surpass human intelligence, ultimately resulting in the extinction of humans.

smart textiles:
(1) Fabrics that have fiber optics, digital components, tiny electronic devices, etc. incorporated in their fabrication. (2) Human-made materials enabled to sense and react to functions of the human body or changes in the environment and alter their properties according to those stimuli.

social innovation:
New practices developed to better meet social needs or to challenge current issues or solutions in order to promote social progress.

social robotics:
A field of robot development focused on designing robots that interact socially with humans or with other robots.

speculative architecture:
Architecture expressed in unbuilt projects that take into account technologies, systems, and networks that shape cities, buildings, and public spaces and how they function; a combination of design, fiction, community, and space.

speculative design:
Introduced by designers Anthony Dunne and Fiona Raby as an expansion of critical design. An open-ended, often deliberately provocative approach to design using or proposing novel or developing technologies outside traditional solution-based design that challenges or critiques current design thinking and enables thinking about futures or alternative presents.

STEM:
An abbreviation for *science, technology, engineering, and mathematics.*

Sustainable Development Goals (SDGs):
A collection of seventeen broad-based and interdependent goals set by the United Nations General Assembly in 2015 to be attained by 2030. They include, but are not limited to, eliminating poverty and hunger, establishing good health and education, achieving gender equality, providing clean water and energy, and creating sustainable cities.

synthetic biology:
An interdisciplinary branch of biology and engineering that combines fields like biotechnology, molecular biology, biophysics, computer science, and genetic engineering to design or construct new artificial parts of natural biological pathways or to redesign existing ones.

synthetic meat:
See clean meat.

technosphere:
The modifications that humans have made to the environment, including buildings, streets, landfills, and technological systems.

techno-utopianism:
An ideology based on the premise that technological advances should bring about a perfect society whose values and culture are modeled on those technologies.

teledildonics:
Technology for remote sex over a data link. Also called *cyberdildonics.*

terraforming:
The hypothetical transformation of another planet, moon, or other celestial body to resemble Earth and its atmosphere so that it can support human and other life-forms found on Earth.

3D printing:
The laying down of superimposed layers of material following a computer design file to create a three-dimensional object (e.g., furniture, food, bioceramic scaffolding material). Also called *additive manufacturing. See also* binder jetting; fused deposition modeling; liquid-metal printing; selective sintering.

transformative normality:
A new normal in a local community created by acceptance and adaptation to changes in or to it.

Turing test:
In AI, a test developed by Alan Turing to determine whether a computer showed intelligent behavior indistinguishable from that of a human.

virtual reality:
An artificial environment usually created on a computer to emulate a real environment. Used for video games, pilot training, and visualizations.

virtual witnessing technology:
A medium (e.g., newspapers, television, film) by which individuals can indirectly experience an event or natural phenomena without being present.

voxel:
The three-dimensional, or physical, equivalent of a pixel and the tiniest distinguishable element of a 3D object. Also called a *volumetric pixel*.

wabi-sabi:
A Japanese aesthetic that accepts the imperfect and the transitory and honors the marks of use.

wearable(s):
A category of small smart electronic devices with sensor technologies that can be embedded in clothing, worn as accessories, or implanted in the body to track biofeedback, location, and other information about their surroundings.

Contributor Biographies

JULIANA ROWEN BARTON is a Philadelphia-based architecture and design historian and curator completing her doctorate at the University of Pennsylvania, Philadelphia. As the Andrew W. Mellon Graduate Fellow in modern and contemporary design at the Philadelphia Museum of Art in 2017–18, she co-organized *Design in Revolution: A 1960s Odyssey* (2018). Previously she worked on exhibitions at the Center for Architecture and the Museum of Modern Art in New York.

ANDREW BLAUVELT is director of the Cranbrook Art Museum in Bloomfield Hills, Michigan, and curator-at-large for the Museum of Arts and Design, New York. Prior to these positions he was senior curator of design, research, and publishing at the Walker Art Center in Minneapolis, where he served for seventeen years, organizing numerous exhibitions and leading various print and online publishing and audience-engagement initiatives.

MAITE BORJABAD LÓPEZ-PASTOR is the Neville Bryan Assistant Curator of Architecture and Design at the Art Institute of Chicago. She is an architect and curator educated at the Universidad Politécnica de Madrid and Columbia University, New York. She was the author and curator of *Scenographies of Power: From the State of Exception to the Spaces of Exception* (2017). Her work revolves around diverse forms of critical spatial practices, operating across architecture, art, and performance.

V. MICHAEL BOVE JR. holds an SB in electrical engineering, an SM in visual studies, and a PhD in media technology, all from MIT, where he is currently head of the Media Lab's Object-Based Media group and teaches the course "Objectification: How to Write (and Talk, and Think) about Objects" with Nora

Jackson. He holds numerous patents for inventions relating to video recording, hardcopy, interactive television, medical imaging, holographic displays, and sporting equipment.

EMMET BYRNE is the design director and associate curator of design at the Walker Art Center in Minneapolis. He provides creative leadership and strategic direction for the Walker in all areas of visual communication, branding, and publishing, while overseeing the award-winning in-house design studio. He was one of the founders of the Task Newsletter in 2009 and is the creator of the Walker's Intangibles platform.

CHRISTINA COGDELL is chair of the Design department at the University of California, Davis. Her research broadly encompasses the history and theory of design and science and is most recently focused on biodesign. Cogdell's books—*Toward a Living Architecture? Complexism and Biology in Generative Design* (2019) and *Eugenic Design: Streamlining America in the 1930s* (2004/2010)—illuminate her practice of working across the disciplines of design, art, architecture, science, and technology.

GABRIELLA COLEMAN holds the Wolfe Chair in Scientific and Technological Literacy at McGill University, Montreal. Trained as a cultural anthropologist, she researches, writes, and teaches on computer hackers and digital activism. Her publications include *Hacker, Hoaxer, Whistleblower, Spy: The Many Faces of Anonymous* (2014) and *Coding Freedom: The Ethics and Aesthetics of Hacking* (2012).

COLIN FANNING is a PhD candidate at the Bard Graduate Center, New York, and a consulting curator for the Philadelphia Museum of Art, where he organized the exhibition

Dieter Rams: Principled Design (2018). His research interests cover a broad range of architecture and design history, with specialties in the material culture of childhood, postwar craft and counterculture, and design education in the United States.

MICHELLE MILLAR FISHER is the Ronald C. and Anita L. Wornick Curator of Contemporary Decorative Arts at the Museum of Fine Arts, Boston. Her work investigates intersections of power, people, and design. She is currently collaborating on the book and exhibition *Designing Motherhood*. Previously she worked at the Philadelphia Museum of Art and the Museum of Modern Art, New York, where she co-organized *Items: Is Fashion Modern?* (2017) and *Design and Violence* (2015).

FORMAFANTASMA (Andrea Trimarchi and Simone Farresin) is an Italian design duo based in Amsterdam. Formafantasma's body of work is characterized by experimental material investigations that explore the relationship between tradition and local culture, critical approaches to sustainability, and the significance of objects as cultural conduits. Their work has been published and acquired by the Art Institute of Chicago; the Museum of Modern Art, New York; and the Museum of Applied Arts, Vienna.

MARINA GORBIS is executive director of the Institute for the Future (IFTF), the world's longest-running futures-thinking organization, based in Silicon Valley. Her current research focuses on transformations in the world of work and new forms of value creation. Most recently she launched the Workable Futures Initiative at IFTF with the aim of developing a deeper understanding of new work patterns and prototyping a generation of "Positive Platforms" for work.

AIMI HAMRAIE is assistant professor of medicine, health, and society and American studies and director of the Mapping Access Project at Vanderbilt University, Nashville. Hamraie's interdisciplinary scholarship bridges critical disability, race, and feminist studies; architectural history; and science and technology studies. Their publications include *Building Access: Universal Design and the Politics of Disability* (2017).

KATHRYN B. HIESINGER is the J. Mahlon Buck, Jr. Family Senior Curator of European Decorative Arts after 1700 at the Philadelphia Museum of Art. Her work focuses on decorative arts and design from the mid-nineteenth century to the present, and includes the exhibitions and publications *Zaha Hadid: Form in Motion* (2011), *Out of the Ordinary: The Architecture and Design of Robert Venturi, Denise Scott Brown and Associates* (2001), *Japanese Design: A Survey since 1950* (1994), and *Design since 1945* (1983).

SREĆKO HORVAT is a philosopher, author, and political activist. He is the author of *Poetry from the Future* (2019), *Subversion!* (2017), *The Radicality of Love* (2015), and *What Does Europe Want?* (with Slavoj Žižek, 2014). He has written for the *Guardian*, *New York Times*, *Al Jazeera*, and many other leading news media. He is the cofounder of DiEM25 (Democracy in Europe Movement).

NORA JACKSON teaches memoir and writing at MIT. In 2017 she and Michael Bove created a course on writing about design—"Objectification: How to Write (and Talk, and Think) about Objects"—at the MIT Media Lab. She works as a translator of academic publications in Dutch, French, and English. Her research interests include nineteenth- and twentieth-century literature, linguistics, aesthetics, and design.

FRANCIS KÉRÉ established the Kéré Foundation while studying at the Technical University of Berlin and founded Kéré Architecture in Berlin in 2005. He received the Aga Khan Award for Architecture in 2004 and the Arnold W. Brunner Memorial Prize in 2017. Kéré has undertaken projects across four continents and continues to work in his home country, Burkina Faso. He currently teaches architectural design and participation at the Technical University of Munich.

DAVID KIRBY is chair of the Interdisciplinary Studies in Liberal Arts department at California Polytechnic State University, San Luis Obispo. Previously he was professor of science communication studies at the University of Manchester, England, and program director of the MSc in science communication. He has a PhD in evolutionary genetics from the University of Maryland, College Park, and is the author of *Lab Coats in Hollywood: Science, Scientists, and Cinema* (2011).

HELEN KIRKUM graduated from the Royal College of Art, London, in July 2016 with a master's in footwear. That same year she exhibited her graduate collection at International Talent Support, where she won both the accessories award and the Vogue Talents award. Kirkum designs sneakers from parts of used and discarded shoes, conceptualizing connections with commerce and materiality through textures, graphics, and silhouettes.

BRUNO LATOUR, trained first as a philosopher and then as an anthropologist, is now emeritus professor of the médialab and the program in political arts (SPEAP) of Sciences Po Paris. Since January 2018 he has been a fellow at the Center for Art and Media and professor at the University of Arts and Design, both in Karlsruhe. His most recent publication is *Facing Gaia: Eight Lectures on the New Climate Regime* (2018).

MARISOL LEBRÓN is assistant professor of Mexican American and Latina/o studies at the University of Texas at Austin. An interdisciplinary scholar, her research and teaching focus on social inequality, policing, violence, and protest. Her recent book, *Policing Life and Death: Race, Violence, and Resistance in Puerto Rico* (2019), traces the growth of punitive governance in contemporary Puerto Rico.

EZIO MANZINI has worked in the field of sustainable design for over two decades. His interest in social innovation as a driver of sustainable changes led him to form DESIS, an international network of design schools active in this field. Manzini is distinguished professor at ELISAVA, Barcelona; honorary professor at the Politecnico di Milano; and guest professor at Tongji University, Shanghai, and Jiangnan University, Wuxi. His most recent publication is *Politics of the Everyday* (2019).

JILLIAN MERCADO is an American fashion model and disabilities activist. She studied merchandising management at New York's Fashion Institute of Technology and worked for the photographer Patrick McMullan and Tumblr before becoming creative director of We the Urban magazine. She has appeared in print and television campaigns for Diesel denim, Olay, Bumble 100, Calvin Klein fragrance, Nordstrom, Target, and Tommy Hilfiger. In September 2018 she was featured on *Teen Vogue*'s first digital cover.

ALEXANDRA MIDAL is an independent curator and professor in the master's in Design, Space and Communication program at HEAD-Genève. She has curated exhibitions at the Musée d'Art Moderne, Luxembourg; Musée d'Art Moderne de la Ville de Paris; Wolfsonian Museum, Miami; and Artists Space, New York. Midal is the director of the annual Festival of Invisible Films at HEAD, presenting experimental films by designers. Her most recent publication is *Design by Accident: For a New History of Design* (2019).

NERI OXMAN is a professor at MIT and director of the Media Lab's Mediated Matter group, an experimental design practice creating enabling technologies for commissioned works. Her research lies at the intersection of computational design, digital fabrication, materials science, and synthetic biology, applied across scales and disciplines. Oxman pioneered the field of material ecology and is the recipient of the San Francisco Museum of Modern Art's Contemporary Vision Award (2019) and the Cooper Hewitt National Design Award (2018).

CHRIS RAPLEY is professor of climate science at University College London (UCL). He is the chair of the UCL Policy Commission on Communicating Climate Science and chair of the London Climate Change Partnership, which seeks to ensure that London is the most climate-resilient city. Previously he was director of the Science Museum, London; the British Antarctic Survey; and the International Geosphere-Biosphere Programme.

ZOË RYAN is the John H. Bryan Chair and Curator of Architecture and Design at the Art Institute of Chicago. She is the editor of *As Seen: Exhibitions That Made Architecture and Design History* (2017) and curator of *In a Cloud, in a Wall, in a Chair: Six Modernists in Mexico at Midcentury* (2019) and the 2014 Istanbul Design Biennial, *The Future Is Not What It Used to Be*. Her projects explore the impact of architecture and design on society.

MAUDE DE SCHAUENSEE was associate editor of the Hasanlu Publications Series at the University of Pennsylvania Museum of Archaeology and Anthropology, Philadelphia, and keeper of that museum's Near Eastern section. She has authored articles on horse gear, textiles, and furniture from Hasanlu, Iran, and a book on two lyres from the Royal Cemetery of Ur. She is the senior author of a forthcoming monograph on the horse gear from Hasanlu.

ORKAN TELHAN is an interdisciplinary artist, designer, and researcher whose investigations engage critical issues in social, cultural, and environmental responsibility. An associate professor of emerging design practices at the University of Pennsylvania School of Design, Philadelphia, his individual and collaborative work has been exhibited at the Istanbul Biennial; Milan Design Week; Armory Show Special Projects, New York; Ars Electronica, Linz; Museum of Contemporary Art Detroit; and New Museum of Contemporary Art, New York.

EYAL WEIZMAN is the founding director of Forensic Architecture and professor of spatial and visual cultures at Goldsmiths, University of London, where he founded the Centre for Research Architecture. He was a Global Scholar at Princeton University and a professor at the Academy of Fine Arts in Vienna. He serves on the boards of the International Criminal Court and the Centre for Investigative Journalism and is a founding member of the architectural collective DAAR in Beit Sahour/Palestine.

DANIELLE WOOD is the Benesse Corporation Career Development Assistant Professor of Research in Education within the Program in Media Arts and Sciences and the Aeronautics and Astronautics department at MIT and founder of MIT Media Lab's Space Enabled research group, whose mission is to advance justice in Earth's complex systems using designs enabled by space. Wood is a scholar of societal development with a background in satellite design, systems engineering, and technology policy for the United States and emerging nations.

LINYEE YUAN is a design journalist as well as the editor and founder of *MOLD* magazine (thisismold.com), which examines issues relating to food from a futuristic design perspective, including investigating how design and innovation can create food solutions for an expanding global population. Yuan has written about design and art for Core77, *Riposte*, Food52, Design Observer, Cool Hunting, *Elle Decor*, and *Wilder Quarterly*.

EMMA YANN ZHANG is a PhD student in computer science at the Imagineering Institute, Malaysia, and a committee member of the annual conference on Love and Sex with Robots. Her research interests are multisensory communication, haptic technologies, and pervasive and wearable computing. In 2015 she developed, with Adrian David Cheok, the Kissenger, the world's first mobile device for transmitting realistic kiss sensations over the internet.

Further Readings

Alloway, Lawrence, Reyner Banham, and David Lewis. *This Is Tomorrow*. Exh. cat. London: Whitechapel Art Gallery, 1956.

Ambasz, Emilio, ed. *Italy: The New Domestic Landscape; Achievements and Problems of Italian Design*. Exh. cat. New York: Museum of Modern Art in collaboration with Centro Di, Florence, 1972.

Appadurai, Arjun. *The Future as Cultural Fact: Essays on the Global Condition*. London: Verso, 2013.

Arendt, Hannah. *The Human Condition: A Study of the Central Dilemmas Facing Modern Man*. Chicago: University of Chicago Press, 1958.

Asimov, Isaac. *I, Robot*. Garden City, NY: Doubleday, 1950.

Asma, Stephen T. *The Evolution of Imagination*. Chicago: University of Chicago Press, 2017.

Bleecker, Julian. "Design Fiction: A Short Essay on Design, Science, Fact, and Fiction." Near Future Laboratory, March 2009. drbfw5wfjlxon.cloudfront.net/writing /DesignFiction_WebEdition.pdf.

Brand, Stewart. *The Clock of the Long Now: Time and Responsibility*. New York: Basic Books, 1999.

Bridle, James. *New Dark Age: Technology, Knowledge and the End of the Future*. London: Verso, 2018.

Bundy, Robert, ed. *Images of the Future: The Twenty-First Century and Beyond*. Buffalo, NY: Prometheus Books, 1976.

Butler, Octavia E. *Parable of the Talents: A Novel*. New York: Seven Stories, 2016.

Carrington, André M. *Speculative Blackness: The Future of Race in Science Fiction*. Minneapolis: University of Minnesota Press, 2016.

Clarke, Adele, and Donna Haraway, eds. *Making Kin Not Population: Reconceiving Generations*. Chicago: Prickly Paradigm, 2018.

Coleman, Gabriella. *Hacker, Hoaxer, Whistleblower, Spy: The Many Faces of Anonymous*. London: Verso, 2014.

De Monchaux, Nicholas. *Spacesuit: Fashioning Apollo*. Cambridge, MA: MIT Press, 2011.

Dery, Mark, ed. *Flame Wars: The Discourse of Cyberculture*. Durham, NC: Duke University Press, 1994.

Dunne, Anthony. *Hertzian Tales: Electronic Products, Aesthetic Experience, and Critical Design*. Rev. ed. Cambridge, MA: MIT Press, 2008.

Dunne, Anthony, and Fiona Raby. *Speculative Everything: Design, Fiction, and Social Dreaming*. Cambridge, MA: MIT Press, 2013.

Emre, Merve, ed. *Once and Future Feminist*. Forum 7. Cambridge, MA: Boston Review, 2018.

Erisman, Porter. *Alibaba's World: How a Remarkable Chinese Company Is Changing the Face of Global Business*. New York: Palgrave Macmillan, 2015.

Frase, Peter. *Four Futures: Life after Capitalism*. London: Verso, 2016.

Fry, Tony. *A New Design Philosophy: An Introduction to Defuturing*. Sydney: UNSW Press, 1999.

Godfrey-Smith, Peter. *Other Minds: The Octopus and the Evolution of Intelligent Life*. London: William Collins, 2018.

Haraway, Donna. "A Cyborg Manifesto." In *The Cultural Studies Reader*, 2nd ed., edited by Simon During, 271–91. London: Routledge, 1999.

Hayden, Dolores. *Seven American Utopias: The Architecture of Communitarian Socialism, 1790–1975*. Cambridge, MA: MIT Press, 1976.

Helvert, Marjanne van, ed. *The Responsible Object: A History of Design Ideology for the Future*. Amsterdam: Valiz, 2016.

Hernandez, Robb, and Tyler Stallings, eds. *Mundos Alternos: Art and Science Fiction in the Americas*. Exh. cat. Riverside, CA: UCR ARTSblock, 2017.

Hertz, Betti-Sue, and Ceci Moss, eds. *Dissident Futures*. Exh. cat. San Francisco: Yerba Buena Center for the Arts, 2013.

Hyde, Rory, et al. *The Future Starts Here*. Exh. cat. London: V&A Publishing, 2018.

Indiana, Rita. *La Mucama de Omiculné* [Tentacle]. Cáceres, Spain: Editorial Periférica, 2015.

Jongerius, Hella, Louise Schouwenberg, and Angelika Nollert, eds. *Beyond the New: On the Agency of Things*. Exh. cat., Die Neue Sammlung/The Design Museum, Munich. London: Koenig Books, 2017.

Kafer, Alison. *Feminist, Queer, Crip*. Bloomington: Indiana University Press, 2013.

Kaku, Michio. *Physics of the Future: The Inventions That Will Transform Our Lives*. London: Penguin, 2012.

Keith, Vanessa, and Studio TEKA Design. *2100: A Dystopian Utopia/The City After Climate Change*. New York: Terreform, 2017.

Kilgore, De Witt Douglas. *Astrofuturism: Science, Race, and Visions of Utopia in Space*. Philadelphia: University of Pennsylvania Press, 2003.

Kirby, David A. *Lab Coats in Hollywood: Science, Scientists, and Cinema*. Cambridge, MA: MIT Press, 2011.

Kolbert, Elizabeth. *The Sixth Extinction: An Unnatural History*. New York: Henry Holt, 2015.

Kries, Mateo, Christoph Thun-Hohenstein, and Amelie Klein, eds. *Hello, Robot: Design between Human and Machine*. Exh. cat. English edition. Weil am Rhein, Germany: Vitra Design Museum, 2017.

Kubler, George. *The Shape of Time: Remarks on the History of Things*. New Haven: Yale University Press, 1962.

Lang, Peter, and William Menking. *Superstudio: Life without Objects*. Exh. cat., Design Museum, London. Milan: Skira, 2003.

Lanier, Jaron. *Who Owns the Future?* New York: Simon and Schuster, 2013.

Lin, Patrick, Ryan Jenkins, and Keith Abney, eds. *Robot Ethics 2.0: From Autonomous Cars to Artificial Intelligence*. New York: Oxford University Press, 2017.

Lowrey, Annie. *Give People Money: How a Universal Basic Income Would End Poverty, Revolutionize Work, and Remake the World*. New York: Crown, 2018.

McGuirk, Justin, and Gonzalo Herrero Delicado, eds. *Fear and Love: Reactions to a Complex World*. Exh. cat., Design Museum, London. London: Phaidon, 2016.

Montfort, Nick. *The Future*. Cambridge, MA: MIT Press, 2017.

Mumford, Lewis. *The Myth of the Machine*. Vol. 2, *The Pentagon of Power*. New York: Harcourt Brace Jovanovich, 1970.

Myers, William. *Bio Design: Nature, Science, Creativity*. London: Thames and Hudson, 2012.

O'Connell, Mark. *To Be a Machine: Adventures among Cyborgs, Utopians, Hackers, and the Futurists Solving the Modest Problem of Death*. New York: Doubleday, 2017.

Papanek, Victor. *Design for the Real World: Human Ecology and Social Change*. 2nd ed. London: Thames and Hudson, 1985.

Phillips, Rasheedah, ed. *Black Quantum Physics: Theory & Pratice*. Philadelphia: Afrofuturist Affair/House of Future Sciences Books, 2015.

Polak, Fred L. *The Image of the Future: Enlightening the Past, Orientating the Present, Forecasting the Future*. 2 vols. Leyden: A. W. Sythoff, 1961.

Prigogine, Ilya. *The End of Certainty: Time, Chaos, and the New Laws of Nature*. New York: Free Press, 1997.

Rao, Vyjayanthi Venuturupalli, Prem Krishnamurthy, and Carin Kuoni, eds. *Speculation, Now: Essays and Artwork*. Durham, NC: Duke University Press, 2015.

Rose, David. *Enchanted Objects: Innovation, Design, and the Future of Technology*. New York: Scribner, 2014.

Ryman, Geoff, et al. "The Mundane Manifesto." 2004. Available at *SFGenics: Notes on Science, Fiction, and Science Fiction* (blog), July 4, 2013. sfgenics.wordpress.com.

Semper, Gottfried. *The Four Elements of Architecture and Other Writings*. Translated by Harry Francis Mallgrave and Wolfgang Herrmann. Cambridge: Cambridge University Press, 2010. First published in 1851.

Silverberg, Robert. *The World Inside*. Garden City, NY: Doubleday, 1971.

Srnicek, Nick, and Alex Williams. *Inventing the Future: Postcapitalism and a World without Work*. Brooklyn, NY: Verso, 2015.

Sterling, Bruce, ed. *Twelve Tomorrows*. MIT Technology Review SF Annual 2014. Cambridge, MA: MIT Technology Review, 2014.

Stocker, Gerfried, Christine Schöpf, and Hannes Leopoldseder, eds. *AI: Artificial Intelligence; Das andere Ich*. Exh. cat., Ars Electronica Festival, Linz, Austria. Berlin: Hatje Cantz, 2017.

Taylor, Keeanga-Yamahtta, ed. *How We Get Free: Black Feminism and the Combahee River Collective*. Chicago: Haymarket Books, 2017.

Toffler, Alvin. *Future Shock*. New York: Random House, 1970.

UN General Assembly. "Report of the United Nations Conference on Environment and Development: Annex I, Rio Declaration on Environment and Development." A/CONF.151/26 (Vol. I). August 12, 1992, Rio de Janeiro. www.un.org/documents/ga/conf151/aconf15126-1annex1.htm.

Ward, Alex. *Curious Minds: New Approaches in Design*. Exh. cat. Jerusalem: Israel Museum, 2012.

Womack, Ytasha L. *Afrofuturism: The World of Black Sci-Fi and Fantasy Culture*. Chicago: Lawrence Hill Books, 2013.

World Commission on Environment and Development. *Our Common Future*. New York: Oxford University Press, 1987.

Yee, Joyce, Emma Jefferies, and Lauren Tan. *Design Transitions: Inspiring Stories, Global Viewpoints, How Design Is Changing*. Amsterdam: BIS Publishers, 2013.

Yelavich, Susan, and Barbara Adams, eds. *Design as Future-Making*. London: Bloomsbury Academic, 2014.

Yusoff, Kathryn. *A Billion Black Anthropocenes or None*. Forerunners: Ideas First 26. Minneapolis: University of Minnesota Press, 2018.

Žižek, Slavoj. *Interrogating the Real*. Edited by Rex Butler and Scott Stephens. Reprint, London: Bloomsbury Academic, 2017.

Žižek, Slavoj. *Welcome to the Desert of the Real! Five Essays on September 11 and Related Dates*. London: Verso, 2002.

Acknowledgments

In May 2019, during an exhibition preview event for *Designs for Different Futures*, an audience member asked, in response to the Scottish artist Katie Paterson's *Future Library* (p. 83), whether we thought there would be printed books in the future. We hope there will, if only to record on paper the tremendous debt of gratitude and thanks to those whose names follow, without whom the exhibition and this accompanying book would not have been possible. Every exhibition requires the efforts of many people with many different skills. But the complexity of conceptually and physically organizing a show across three partnering institutions—the Philadelphia Museum of Art, the Walker Art Center in Minneapolis, and the Art Institute of Chicago—and communicating effectively among five co-curators presented significant challenges. We are truly in awe of the colleagues who shaped this enterprise alongside us.

We have been incredibly fortunate to collaborate on this exhibition with some of the best designers, architects, artists, and cross-disciplinary practitioners working today. The past few years of research have been a true privilege because they have afforded us studio visits, conversations, and extended communications with such a breadth of creative minds. We extend our sincerest thanks to everyone who has allowed us to include their work in the exhibition and catalogue, and we are grateful for the considerable time they have given to this endeavor. Find each of them listed on the project pages throughout this volume. We owe particular thanks to those who provided us with extensions of their work or completely new commissions, which we are honored to present at our institutions. Most of the works in the exhibition were lent by their creators, and some are presented here for the first time. This took many leaps of faith on the part of these varied practitioners, as well as from our colleagues who helped realize their installations. We are proud and appreciative of all these efforts.

We owe special recognition to three collaborating guest curators—Andrew Blauvelt, Colin Fanning, and Orkan Telhan—who have helped guide and inform the exhibition at crucial junctures. In addition, the following people's support and counsel inspired the curatorial team as the exhibition took shape: Glenn Adamson, Paola Antonelli, Zara Arshad, Jurgen Bey, Jan Boelen, Carla Bonilla, V. Michael Bove Jr., Linda Brenner and Bill Christensen, Sean M. Carroll, Caroline Coates, Christina Cogdell, Anthony Dunne and Fiona Raby, Kira Eng-Wilmot, Claire L. Evans, Will Fenton, Ilona Gaynor, Simon Heijdens, Rory Hyde, David Kirby, Scott Klinker, Matylda Krzykowski, Andrea Lipps, Shannon Maldonado, Alexandra Midal, Helen Maria Nugent, Dietmar Offenhuber, Andy Ogden, Deniz Ova, Mariana Pestana, Erica Petrillo, Erika Pinner, Paul Rabinow, Alice Rawsthorn, Libby Sellers, Ala Tannir, Mark Tropea, Mark Wigley, and Jerry Wind.

Designs for Different Futures was pushed forward by the tireless work of certain colleagues who—with each piece of paperwork, Skype call, or spreadsheet—made our partnership possible. Without them, there would have been no exhibition: Yana Balson, associate director of exhibition planning, Cassandra DiCarlo, exhibition project manager, and Suzanne Wells, director of exhibition planning, in Philadelphia; Siri Engberg, senior curator and director of exhibitions management, and Erin McNeil, exhibitions administrator, at the Walker; and Megan Rader, director of exhibitions, and Lindsay Washburn, exhibition project manager, in Chicago.

In Philadelphia, several staff members deserve special mention for their work on the exhibition and catalogue. Collections assistant Rebecca Murphy deserves rousing thanks for her tireless and graceful organizational force and her calm, collected communications over so many of the granular arrangements, both internally and externally. Over several years, with intelligence and humor, Maude de Schauensee mastered a complex technical vocabulary and translated it for a general readership in the catalogue's indispensable glossary. We were also exceedingly happy to retain the wonderful Andrew W. Mellon Graduate Fellow Juliana Rowen Barton as an exhibition assistant.

We are extremely fortunate to host the work of incisive and brilliant thinkers, writers, activists, philosophers, and all-around polymaths in this publication, and we truly appreciate the time that the following contributors took to write or to be interviewed for the book: Juliana Rowen Barton, Andrew Blauvelt, V. Michael Bove Jr., Christina Cogdell, Gabriella Coleman, Colin Fanning, Formafantasma (Andrea Trimarchi and Simone Farresin), Marina Gorbis, Aimi Hamraie, Srećko Horvat, Nora Jackson, Francis Kéré, David Kirby, Helen Kirkum, Bruno Latour, Marisol LeBrón, Ezio Manzini, Jillian Mercado, Alexandra Midal, Neri Oxman, Chris Rapley, Orkan Telhan, Eyal Weizman, Danielle Wood, LinYee Yuan, and Emma Yann Zhang. We also thank all those who made images available to us for illustration purposes. A full list of credits can be found on page 271.

This book would not exist without Philadelphia's William T. Ranney Director of Publishing, Katie Reilly, who deftly managed the coordination of five curators at three institutions, as well as the work of nearly thirty contributors. The diversity of thought and ideas presented here were clarified and amplified by Philadelphia's masterful editors of this volume, Kathleen Krattenmaker and Katie Brennan. The design is the beautiful work of

Ryan Gerald Nelson, senior graphic designer at the Walker, in conversation with exhibition co-curator Emmet Byrne. Philadelphia's Rich Bonk oversaw production and ushered this book into being, aided by the expert coordination of publications colleague Sydney Holt. Our gratitude also goes to Chicago's Department of Publishing, led by Greg Nosan, for their collaboration and assistance.

Many teams of wonderful, smart, and talented people went above and beyond to help make the exhibition possible. In Philadelphia, our colleagues in the Division of Education have been working on this project for almost as long as we have, and we are so proud of the result of our collaboration, the Futures Therapy Lab. Emily Schreiner, the Zoë and Dean Pappas Curator of Education, Public Programs, suggested including this space, an integral part of the exhibition in Philadelphia, for visitors to catch their breath in a visually and conceptually abundant exhibition. It is also due to her trust and vision, and that of her departmental colleagues, that we have been able to devise twenty weeks of rich and experimental self-directed and programmed activities that appeal to students, teachers, families, and drop-in visitors of all ages and stages. We are incredibly grateful to Emily, to the ever-incisive Marla K. Shoemaker, the Kathleen C. Sherrerd Senior Curator of Education, and to their colleagues Elizabeth Yohlin Baill, Barbara Bassett, Caitlin Deutsch, Justine Kelley, Ah-Young Kim, Claire Kloss, Shivon Love, Suzannah Niepold, Chloe Pinero, Damon Reaves, Catherine Ricketts, Sarah Shaw, Greg Stuart, and Linnea West. Thanks also go to Jenni Drozdek, assistant director for interpretation, who was a dream to work with, and Dani Murano of Visitor Services, whose sensitivity reading was a most welcome and thoughtful review.

We would also like to thank Jack Schlechter, Helen Cahng, and Jorge Galvan in Exhibition Design, who went the extra mile every time and were true thought partners; Sharon Hildebrand, our master framer, and Tae Smith, independent consultant for costume and textile installation, whose craftwork is inspiring; Kara Willig in the Department of the Registrar, who was a delight to work alongside; Jeffrey Blair, in the Executive Offices, who kept us on the straight and narrow; Sally Malenka, Sara Reiter, Behrooz Salimnejad, and especially the indefatigable Kate Cuffari in Conservation; Stephen A. Keever and Brandon Straus in Audio-Visual Services, who know more about technology than we could ever hope to; Luis Bravo, Nick Massarelli, Greta Skagerlind, and Neil Spencer for the most wonderful graphic design across all platforms, Tammi Coxe for managing production, and Gretchen Dykstra, the most efficiently kind (and kindly efficient) person we know, as well as the elegant grammarians Nisa Qazi and Erika Remmy, all in the Editorial and Graphic Design Department; the brilliant Tracey Button and her supportive and thoughtful team in Marketing and Communications, Marcia Birbilis, Joy Deibert, Norman Keyes, Caitlin Mahony, Justin Rubich, and Claire Stidwell; Justyna Badach, Joseph Hu, and Timothy Tiebout in the Photography Studio, who saved us on several occasions; William Weinstein and Ariel Schwartz, who headed up the dream team of Sid Rodríguez, Jennifer Schlegel, and Blake Schreiner in Information and Interpretive Technologies; Richard Sieber and Mary Wassermann in Library and Archives, who provided a research refuge; our Development colleagues, led by Jonathan Peterson and David Blackman, without whom our work would not be possible, including Kate Brett, Nico Hartzell, Jackie Killian, Tina DiSciullo-Acker, and Kandra Bolden; Camille Focarino, who made our special events perfect in every way; Martha Masiello, Eric Griffin, Rebecca Kolodziejczak, and the outstanding team in Installations and Packing, as well as Jeanine Kline, James Keenan, and the brilliant construction crew, without whom nothing would have been built. We are also deeply grateful to the entire Department of European Decorative Arts and Sculpture for their support and encouragement over the course of this project: Dirk H. Breiding, Jack Hinton, Olivier Hurstel, Mary Anne Dutt Justice, Rebecca Murphy, and Emma Perloff, as well as Andrew W. Mellon Undergraduate Fellow Suji Kanneganti.

At the Walker, many individuals have made key contributions to the realization of this project. We are enormously grateful to curatorial fellow of visual arts William Hernández Luege, who brought many talents to bear on every aspect of the exhibition; Stephanie Nusser, who attended to myriad details surrounding the project; director and curator of education and public programs Nisa Mackie and her colleagues Jacqueline Stahlmann and Aysha Mazumdar Stanger for their vital insights and tireless dedication in pursuit of audience-focused interpretation and public programming; our stellar Design Studio, including design studio manager Alanna Nissen for keeping everything on track, senior editor Pamela Johnson and assistant editor Annie Jacobson for working closely with our education teams to "awaken the stories," senior imaging specialist Greg Beckel for transforming the look of the show, media producer Andy Underwood-Bultmann for his beautiful video work, photographer Bobby Rogers for creating stylish images for the catalogue and marketing campaign, and production artist Ian Babineau for assisting with the design process; the exhibition's registrar, Bryan Stusse, who kept the project's many moving parts in good order, guided by the steady hand of director of registration Joe King, and with the knowing help of Dave Bartley and Evan Reiter; the inimitable installation team, overseen by Ben Geffen and Doc Czypinski, and expertly led by Jonathan Karen, who—along with David Dick, Joel Schwarz, Kirk McCall, Peter Hannah, Peter Murphy, Jeffrey Sherman, and many others—worked with us to realize such an unusual and diverse exhibition in the Walker's galleries; chief of advancement Christopher Stevens, director of special projects fundraising Marla Stack, and development associate, special projects, Aaron Mack, whose tireless fundraising efforts enabled us to mount an exhibition of this breadth, and to Mary Polta, chief financial officer, who kept the many complex financial details in order; chief of marketing and strategic communications Alfredo Martel, associate director of public relations Rachel Joyce, associate director of strategic marketing Michelle Wood, and their colleagues Michelle Bastyr, Destanie Martin-Johnson, and Elizabeth Camp in Marketing and Public Relations for skillfully sharing the exhibition's multifaceted content with local, national, and international audiences; web editor Paul Schmelzer, who oversaw a series of related online materials that expanded on the themes of the exhibition and endeavored to ask vital and challenging questions on the exhibition's themes.

In Chicago, this exhibition would not have been possible without the indomitable cheer and unfailing capabilities of our colleagues in the Department of Architecture and Design, especially Lori Boyer, exhibitions and collections manager, who provided exceptional stewardship of the exhibition; Christopher Rosenberg, who carefully oversaw the installation; and Kim Chin, who offered unwavering support. Our heartfelt thanks also go to Alison Fisher, Elizabeth Mescher, and department intern Omar Dyette. We extend our gratitude to Leticia Pardo, Richard Jason Ferrer, and Jason Stec in Exhibition Design for the compelling installation design that emphasized the narrative and ideas of the exhibition; the Time-Based Media group, in particular the fundamental work developed by Kristin M. MacDonough and Solveig Nelson, as well as Tom Riley and William Foster from Media Production and Services, for their unconditional enthusiasm; the Department of Learning and Public Engagement, led by Jacqueline Terrassa, for making the exhibition accessible, relevant, and vibrant, and especially Emily Lew Fry for her commitment and collaboration throughout the whole process, and Fawn Ring and Melissa Tanner for their help developing and planning programming within the walls of the museum and beyond; the Department of Design, led by Jeffrey Wonderland, who collaborated with the Walker's graphic design team on the dynamic graphics for the exhibition, and editor Lisa Meyerowitz for her brilliant work on the labels and didactics to make the exhibition accessible, along with the support of the Publishing Department in the development of the exhibition; the Department of Experience Design, headed by Michael Neault, with Bronwyn Kuehler, Gina Giambalvo, and Andrew Meriwether; the Department of Collections and Loans, especially Cayetana Castillo and Sara Patrello, for skillfully coordinating the transportation of many loans and working with our peer institutions on this process; Leslie Carlson, Craig Cox, Michael Hodgetts, and the shipping and installation crew for their care in ensuring each work was received and installed safely; members of our Conservation staff, led by Francesca Casadio, namely Kristi Dahm, Christine Fabian, Isaac Facio, Emily Heye, Kathleen Kiefer, Allison Langley, Antoinette Owen, Sylvie Pénichon, and Suzanne Schnepp, for their expert and meticulous care and excitement for this exhibition; Joseph Vatinno and his colleagues in Museum Facilities, who worked carefully and skillfully on the show; Corey Burrage, Russell Collett, Lu Ventura, and their team in Museum Operations for overseeing security and visitor experience with unfailing professionalism; Troy Klyber and Maria Simon in the Department of Legal/General Counsel for their invaluable guidance; and the Department of External Affairs, led by Eve Coffee Jeffers, with Nora Gainer, Stephanie Henderson, Amanda

Hicks, Bridget Horgan, Kati Murphy, Kathryn Rahn, Lauren Schultz, and Robert Sexton, as well as the Department of Development, especially James Allan, Amy Allen, Erika Clauson, Joe Iverson, George Martin, Aleks Matic, Jennifer Moran, Erika Lowe, Jennifer Oatess, Jenny Sturrock Petkovic, Catherine Reckelhoff, and Sarah Van Loon for their diligent work. Over the course of planning and executing this exhibition and catalogue, we have found considerable benefit in the assistance and counsel of deputy director Sarah Guernsey, senior vice president Andrew Simnick, and the staff of the Office of the President and Director, including Claire Burdulis, Kate Tierney Powell, and Maureen Ryan.

Finally, and with sincere admiration and gratitude, we would like to thank our directors and senior curatorial affairs mentors for their trust in, commitment to, and support of this project and for their vision of design as an integral part of our institutional mandates: Timothy Rub, the George D. Widener Director and Chief Executive Officer, and Alice O. Beamesderfer, the Pappas-Sarbanes Deputy Director for Collections and Programs, at the Philadelphia Museum of Art; Mary Ceruti, executive director, and Olga Viso, former executive director, at the Walker Art Center; and James Rondeau, President and Eloise W. Martin Director, at the Art Institute of Chicago.

Our greatest love and thanks go to our closest family members, those who put our pasts, presents, and futures in sharpest perspective: Ulrich Hiesinger, Austin Fisher, Matthew Luken, Douglas Brull, Carmen López-Pastor, Santos Borjabad, Daniel Borjabad, Ramona Fernández del Moral, Federico López-Pastor, Ryan Palider, Wren Palider-Ryan, Max Ryan, Amy Ryan, Richard Heason, Clara Heason-Ryan, Edward Heason-Ryan, Peggy Palider, and Norman Palider.

Kathryn B. Hiesinger, Michelle Millar Fisher, Emmet Byrne, Maite Borjabad López-Pastor & Zoë Ryan

Index of Names and Projects

Reproduction and Lender Credits

Images have been supplied by the designers/ artists, photographers, and/or the following:

p. 21: © Condé Nast; p. 24: © suitX; p. 29: Photograph by Adam Richardson; p. 31: Courtesy of Arnav Kapur and Jimmy Day; pp. 33, 38, 61, 86, 135 (bottom), 136, 189, 208: Photographs by Timothy Tiebout/Philadelphia Museum of Art; p. 42: Courtesy of the Estate of R. Buckminster Fuller; pp. 43, 221: © 2019 Artists Rights Society (ARS), New York / DACS, London. Digital Image © The Museum of Modern Art/Licensed by SCALA / Art Resource, NY; p. 44: Courtesy of Superstudio/ Adolfo Natalini and Cristiano Toraldo di Francia; p. 57: Jonathan Prime/Netflix; p. 62: Courtesy of the designer/Laura Mainiemi; p. 80: Grace Chuang, courtesy of Harvard University Herbaria. © Ginkgo Bioworks, Inc., The Herbarium of the Arnold Arboretum of Harvard University; p. 81: Photograph by Yoram Reshef; p. 83: Photo © Bjørvika Utvikling by Kristin von Hirsch; p. 84: Photograph by Samuel Picas; p. 85: Photograph by Rachel Dray; p. 87: Photograph by Ryan Mario Yasin, 2018; p. 89: © John Corbett; p. 92: © Val Wilmer/CTSIMAGES; p. 94: Adapted from Stewart Brand, *The Clock of the Long Now: Time and Responsibility* (New York: Basic Books), 37, and Jan Jansson, *Atlantis majoris quinta pars, orbem maritimum seu omnium marium orbis terrarium … ,* c. 1650, distributed under a CC-PD-Mark license; p. 100: Digital Image © The Museum of Modern Art/Licensed by SCALA / Art Resource, NY; p. 105: iStock.com/Uros Poteko; p. 106: Courtesy of MoMA PS1 and Jenny Sabin Studio. Digital Image © The Museum of Modern Art/Licensed by SCALA / Art Resource, NY; p. 107: Distributed under a CC–BY-2.0 license; pp. 109, 111: © Trustees of the British Museum; p. 110: Pip Mothersill,

MIT Media Lab; p. 112: Photograph by Chris Tubbs; p. 114: Photograph by Pierre Yves Dinasquet; p. 115: Photograph by Petr Krejci; p. 116: Photo © Nagami Design S.L.; p. 117: Photo © Yannis Vlamous; p. 123: background photograph by Justinas Vilutis for *MOLD* magazine; p. 127: Roger Harris/Science Photo Library via Getty Images; p. 128: NNehring via Getty Images; p. 129: © Mitchell Joachim, Terreform ONE; pp. 131, 132: Photographs by Orkan Telhan; p. 133: Courtesy of Jo Wei, the curator of Quasi-Nature. Photograph by JJYPHOTO; p. 135 (top): Photograph by Tony Luong; p. 142: Adapted by Wikimedia Commons from H. E. Strickland and A. G. Melville, *The Dodo and Its Kindred* (London: Reeve, Benham, and Reeve, 1848); p. 148: Courtesy of the artist and The Third Line; p. 149: © Andrea Ferro Photography; p. 150: Photo © Tom Harris. Courtesy of the School of the Art Institute of Chicago and the University of Chicago; p. 153: 3D photograph by Brian Hart; pp. 166, 173: Forensic Architecture, 2018; p. 174: Forensic Architecture and Dr. Salvador Navarro-Martinez, 2017; p. 180: Courtesy of Google Ideas, DDoS data © 2019 and Arbor Networks, Inc., NetScout Systems, Inc.; p. 183: dezeen.com; pp. 184, 215: © Iwan Baan; p. 190: Photo © Dor Kedmi; p. 191: Photo © Manuel Herz Architects; p. 192: Courtesy of MGM Television; pp. 193, 198: NASA Earth Observatory; p. 203: Nearmap/ DigitalVision via Getty Images; p. 210: Photograph by Vanessa Simmons; pp. 216, 217: Courtesy of Kéré Architecture; p. 219: © F. Ribon & CAPC; p. 222: Distributed under a CC-BY-SA-2.5 license; pp. 230, 233, 241: NASA; p. 243: © Bigelow Aerospace LLC 2019; p. 244: © David OReilly 2019; p. 245: Photo © HEAD-Genève, Baptiste Coulon, 2017; p. 248: ARRIVAL © Paramount Pictures Corp. All Rights Reserved; p. 249: © 2014

Sean Allen. All rights reserved; p. 251: Text originally published in Rhizome, December 17, 2013, rhizome.org. Reprinted by permission of the author.

All projects in the catalogue were provided to the exhibition courtesy of the designers/ artists, with the exception of the following:

p. 25: Philadelphia Museum of Art: Purchased with funds contributed by Collab: The Group for Modern and Contemporary Design at the PMA, 2019-69-1; pp. 33, 61 (left), 135 (bottom left): Purchased for the exhibition; p. 38: Courtesy of the University of Pennsylvania; p. 58: Courtesy of the artists and Virtue Nordic, Copenhagen; p. 82: Exhibition display courtesy of the USDA Agricultural Research Service, National Laboratory for Genetic Resources Preservation; p. 83: Philadelphia Museum of Art: Gift of the Future Library Trust, 2018-100-2, 5, 6, 10 (Margaret Atwood, David Mitchell, Elif Shafak, Sjón, and Han Kang booklets) / Purchased with the European Decorative Arts Revolving Fund, 2018-101-1a–d (certificate); p. 116: Philadelphia Museum of Art: Purchased with funds contributed by Grant L. Greapentrog in memory of his parents, Walter and Mabel Greapentrog; p. 120: Philadelphia Museum of Art: Gift of Lisa S. Roberts, 2018-187-1; p. 148: Courtesy of the artist and The Third Line; p. 188: Courtesy of the artist and Fridman Gallery, New York; p. 189: Courtesy of the *Designs for Different Futures* curators; p. 192: Courtesy of Metro-Goldwyn-Mayer Studios; p. 207: Courtesy of the designers and Hochschule für Gestaltung, University of Applied Sciences, Schwäbisch Gmünd, Germany; p. 245: Private collection; p. 248: Courtesy of the designers and Paramount Pictures.

PUBLISHED ON THE OCCASION OF
THE EXHIBITION
Designs for Different Futures

Philadelphia Museum of Art
October 22, 2019–March 8, 2020

Walker Art Center, Minneapolis
September 12, 2020–January 3, 2021

The Art Institute of Chicago
February 6–May 16, 2021

Designs for Different Futures was organized by the Philadelphia Museum of Art, the Walker Art Center, Minneapolis, and the Art Institute of Chicago.

In Philadelphia, the exhibition was generously supported by the Annenberg Foundation Fund for Major Exhibitions, the Robert Montgomery Scott Endowment for Exhibitions, the Kathleen C. and John J. F. Sherrerd Fund for Exhibitions, the Lisa Roberts and David W. Seltzer Endowment Fund, the Women's Committee of the Philadelphia Museum of Art, the Laura and William C. Buck Endowment for Exhibitions, the Harriet and Ronald Lassin Fund for Special Exhibitions, the Jill and Sheldon Bonovitz Exhibition Fund, and an anonymous donor.

In Chicago, support was provided by the Exhibitions Trust and other generous sponsors.

Credits as of June 27, 2019

The organizing curators are Kathryn B. Hiesinger, the J. Mahlon Buck, Jr. Family Senior Curator of European Decorative Arts after 1700, Philadelphia Museum of Art; Michelle Millar Fisher, the Louis C. Madeira IV Assistant Curator of European Decorative Arts, Philadelphia Museum of Art; Emmet Byrne, design director and associate curator of design, Walker Art Center, Minneapolis; Maite Borjabad López-Pastor, Neville Bryan Assistant Curator of Architecture and Design, the Art Institute of Chicago; and Zoë Ryan, John H. Bryan Chair and Curator of Architecture and Design, the Art Institute of Chicago; with guest curators Andrew Blauvelt, director, Cranbrook Art Museum, Bloomfield Hills, Michigan, and curator-at-large, Museum of Arts and Design, New York; Colin Fanning, independent scholar, Bard Graduate Center, New York; and Orkan Telhan, associate professor of fine arts, emerging design practices, University of Pennsylvania School of Design, Philadelphia.

PRODUCED BY
The Publishing Department
Philadelphia Museum of Art
Katie Reilly, The William T. Ranney
Director of Publishing
2525 Pennsylvania Avenue
Philadelphia, PA 19130-2440 USA
philamuseum.org

PUBLISHED IN ASSOCIATION WITH
Yale University Press
302 Temple Street
P.O. Box 209040
New Haven, CT 06520-9040
yalebooks.com/art

EDITED BY
Katie Brennan and
Kathleen Krattenmaker

PRODUCTION BY
Richard Bonk

DESIGN BY
Ryan Gerald Nelson
and Emmet Byrne

INDEX BY
Enid Zafran

PRINTED AND BOUND IN ITALY BY
Musumeci S.p.A., Valle d'Aosta

SEPARATIONS BY
Musumeci S.p.A., Valle d'Aosta

Cover design by Ryan Gerald Nelson

Collages (pp. 19, 49, 67, 97, 123, 139, 153, 171, 201, 213, 237) designed by Ryan Gerald Nelson with photography by Bobby Rogers and models Devon Walker (p. 19), Samantha Surma-Heine (p. 49), Douglas Brull (p. 153), Nolan Mao (p. 171), Taoheed Bayo (p. 201), and Amal Abdinur (p. 237)

LIBRARY OF CONGRESS
CATALOGING-IN-PUBLICATION DATA

Names: Hiesinger, Kathryn B., 1943– editor. | Philadelphia Museum of Art, organizer, host institution. | Walker Art Center, organizer, host institution. | Art Institute of Chicago, organizer, host institution.
Title: Designs for different futures / Kathryn B. Hiesinger, Michelle Millar Fisher, Emmet Byrne, Maite Borjabad López-Pastor, and Zoë Ryan; with Andrew Blauvelt, Colin Fanning, and Orkan Telhan.
Description: Philadelphia, PA : Philadelphia Museum of Art ; New Haven : in association with Yale University Press, 2019. | "Published on the occasion of the exhibition *Designs for Different Futures*, Philadelphia Museum of Art, October 22, 2019–March 8, 2020; Walker Art Center, Minneapolis, September 12, 2020–January 3, 2021; The Art Institute of Chicago, February 6–May 16, 2021." | Summary: "*Designs for Different Futures* records the concrete ideas and abstract dreams of designers, artists, academics, and scientists engaged in exploring how design might reframe our futures—socially, ethically, and aesthetically. Centered on ninety-two innovative contemporary design objects, projects, and speculations, this handbook asks readers to contemplate our cultural attitudes toward technology, consumption, beauty, and the social and environmental challenges we face on both a local and global scale in futures near and far. Thought-provoking projects are explored through interpretive texts and interviews by the designers themselves and the core curatorial team. Interspersed with the project pages are newly commissioned texts by academics, scientists, designers, artists, curators, and futurists that explore wide-ranging issues, from historical visions of the future to the use of biological/living materials in products and production processes"— Provided by publisher.
Identifiers: LCCN 2019021193 | ISBN 9780876332900 (paperback)
Subjects: LCSH: Design—History—21st century—Exhibitions. | Civilization—Forecasting—Exhibitions. | BISAC: DESIGN / Industrial. | DESIGN / Product. | ART / Collections, Catalogs, Exhibitions / Group Shows.
Classification: LCC NK1397 .D483 2019 | DDC 745.409/05—dc23 LC record available at https://lccn.loc.gov/2019021193